Annals of Mathematics Stu

Number 203

Eisenstein Cohomology for GL_N and the Special Values of Rankin–Selberg L-Functions

Günter Harder
A. Raghuram

PRINCETON UNIVERSITY PRESS
PRINCETON AND OXFORD
2020

Published by Princeton University Press,
41 William Street, Princeton, New Jersey 08540

In the United Kingdom: Princeton University Press,
6 Oxford Street, Woodstock, Oxfordshire OX20 1TR

press.princeton.edu

ISBN: 978-0-691-19788-3
ISBN (pbk): 978-0-691-19789-0

British Library Cataloging-in-Publication Data is available

Editorial: Vickie Kearn, Susannah Shoemaker, and Lauren Bucca
Production Editorial: Nathan Carr
Text Design: Leslie Flis
Jacket/Cover Design: Leslie Fils
Production: Jacquie Poirier
Publicity: Matthew Taylor and Katie Lewis

The publisher would like to acknowledge the authors of this volume for providing the camera-ready copy from which this book was printed.

This book has been composed in LaTeX.

For *Lilo* and *Nita*

Faust: ...Da muß sich manches Rätsel lösen.
Mephistopheles: Doch manches Rätsel knüpft sich auch...

(Faust: Eine Tragödie - Kapitel 24, von Goethe)

Contents

Preface

The first author of this monograph studied Eisenstein cohomology for GL_2 over a number field F in [24] and proved rationality results for critical values for the L-functions attached to algebraic Hecke characters of F. With a view to generalizing, he then studied the situation of rank-one Eisenstein cohomology for $\mathrm{SL}_3/\mathbb{Q}$ to deduce information about ratios of successive critical values for the Hecke L-function $L(s, f)$ of a holomorphic modular form f. The results of this work appeared in [29]. On attempting a similar study for Eisenstein cohomology for $G = \mathrm{GL}_4$, especially in regard to the contribution from the maximal parabolic P with Levi quotient $M_P = \mathrm{GL}_3 \times \mathrm{GL}_1$, he observed that certain maps were always degenerating to the zero map. After he asked Birgit Speh for an explanation, she wrote him a letter to show that a particular standard intertwining operator (from an induced representation $\mathrm{Ind}_{P(\mathbb{R})}^{G(\mathbb{R})}(\sigma)$, where σ is a cohomological representation of $M_P(\mathbb{R})$, to its companion induced representation that is induced from the parabolic subgroup associate to P) happens to always give the zero map in relative Lie algebra cohomology. Around this time, the second author was visiting Germany, and both authors discussed Speh's letter to see what really was going on. Such discussions form the genesis of this project.

The main aim of this monograph is to study rank-one Eisenstein cohomology for the group GL_N/F, where F is a totally real field extension of $\mathbb{Q}$. This is then used to prove rationality results for ratios of successive critical values for Rankin–Selberg L-functions for $\mathrm{GL}_n \times \mathrm{GL}_{n'}$ over F with the parity condition that nn' is even. The key idea is to interpret Langlands's constant term theorem in terms of Eisenstein cohomology. The short announcement [33] contained the main results for $\mathrm{GL}_n \times \mathrm{GL}_{n'}$ over $\mathbb{Q}$ and in the case when n is even and n' is odd. In the meantime it became clear that the methods and results can be extended to the case when both n and n' are even and also to the situation when the base field is a totally real field. There is a great deal of current work in collaborations involving the authors and others (most notably with Chandrasheel Bhagwat and Muthu Krishnamurthy) about many other Langlands–Shahidi L-functions to which the methods of this monograph are applicable.

It is a special pleasure to thank Uwe Weselmann, who gave elegant proofs of two very technical ingredients: his proof of the combinatorial lemma appears in Sect. 7.2.4, and his calculation of the archimedean contribution to our main global results is all of Chap. 9. We also thank Chandrasheel Bhagwat whose discussions with the second author on the failure of the main results when nn' is odd, appear in Sect. 7.4.1.

It is a pleasure for both authors to acknowledge the support of several institutions and mathematical colleagues over the past decade during which time this project developed. It all started during a hike in the hills in the Black Forest near the Mathematisches Forschugsinstitut Oberwolfach (MFO) in 2008 when the authors met for a conference organized by Steve Kudla and Joachim Schwermer. It evolved further during meetings at the Erwin Schrödinger Institute, Vienna; their invitations are gratefully acknowledged. The authors are thankful to Don Blasius, Harald Grobner, Benedict Gross, Michael Harris, and Freydoon Shahidi for their interest and helpful discussions. The second author is grateful to the Max Planck Institut für Mathematik (MPIM), an Alexander von Humboldt Fellowship, a National Science Foundation grant (DMS-0856113), and travel grants from Oklahoma State University and IISER Pune that funded his visits to MPIM on innumerable occasions. Finally, both authors are grateful to MFO for hosting them as part of their *Research in Pairs* program in May, 2014, where a first draft of the manuscript was written. The second author acknowledges funding from the Charles Simonyi Endowment during his stay at the Institute for Advanced Study, Princeton, in spring and summer of 2018, during which time this monograph was prepared for publication in the *Annals of Mathematics Studies*.

The authors are grateful to their wives, Lilo and Nita, for their unconditional support during this arduous journey that started a decade ago and culminated with this book, which in turn opens the doors for so many questions that seem to guarantee long sleepless nights of mathematical wonder.

Günter Harder & A. Raghuram
August 28, 2019

The authors (GH on the right and AR on the left) at the Mathematisches Forschungsinstitut Oberwolfach, Germany, in May of 2014, where they spent a week as part of a *Research in Pairs* program.

Eisenstein Cohomology for GL_N and the Special Values of Rankin–Selberg L-Functions

Chapter One

Introduction

This monograph should be thought of as a disquisition of the general principle:

The cohomology of arithmetic groups and the L-functions $L(s, \pi, r)$ attached to irreducible "pieces" π have a strong symbiotic relationship with each other.

The symbiosis goes in both directions:

(A) Expressions in the special values $L(k, \pi, r)$ enter in the transcendental description of the cohomology. Since the cohomology is defined over $\mathbb{Q}$ we can deduce rationality (algebraicity) results for these expressions in special values.

(B) These special values in turn influence the structure of the cohomology as a Hecke module; prime numbers dividing these values occur in the denominators of Eisenstein classes.

Let G be a connected reductive group over $\mathbb{Q}$. A fundamental problem is to understand the geometry of a locally symmetric space $\mathcal{S}^G_{K_f}$ attached to G for some level structure K_f. In particular, one seeks to understand the cohomology groups $H^\bullet(\mathcal{S}^G_{K_f}, \widetilde{\mathcal{M}})$ of $\mathcal{S}^G_{K_f}$ with coefficients in a local system $\widetilde{\mathcal{M}}$ arising from a finite-dimensional representation of the algebraic group G. On the other hand, there are families of Langlands–Shahidi L-functions $L(s, \pi, r)$, attached to an automorphic representation π of a Levi subgroup M_P of G and an algebraic representation r of the Langlands dual group of M_P. The above dictum takes the shape: How do the L-functions $L(s, \pi, r)$ influence the structure of $H^\bullet(\mathcal{S}^G_{K_f}, \widetilde{\mathcal{M}})$? Let's illustrate this with the help of two well-known examples:

Let $\mathbb{A}$ denote the adele ring of $\mathbb{Q}$. A basic problem in modern number theory is to study the decomposition of the space of automorphic forms $\mathcal{A}(G(\mathbb{Q})\backslash G(\mathbb{A}))$ as a representation of $G(\mathbb{A})$. This space has two parts: the discrete spectrum and the continuous spectrum. Within the discrete spectrum, we have the cuspidal spectrum and the residual (discrete, but not cuspidal) spectrum. Since Langlands's groundbreaking work on Eisenstein series one knows that the holomorphy properties, such as the location of poles, of various L-functions influence the description of these spectra, and this in turn carries over to the cohomology groups $H^\bullet(\mathcal{S}^G_{K_f}, \widetilde{\mathcal{M}}_{\mathbb{C}})$ after changing the base field to $\mathbb{C}$ and then relating them to the relative Lie algebra cohomology of the space of automorphic forms. This point plays a decisive role in this monograph when we discuss the implications of the work of Mœglin–Waldspurger on the discrete spectrum for the structure of the cohomology; see Sect. 4.3.

A more subtle influence of L-functions on cohomology is afforded by the study of congruences. Primes that appear in the algebraic part of a special value of some L-function are usually congruence primes. For example, consider the Ramanujan Δ-function $\Delta(z) = \sum_{n=1}^{\infty} \tau(n) e^{2\pi i n z}$, the unique cusp form (up to scalars) of weight 12 for $\Gamma = \mathrm{SL}_2(\mathbb{Z})$. Ramanujan had empirically observed that $\tau(p) \equiv p^{11} + 1 \pmod{691}$ for all primes p. The prime 691 appears in the numerator of the special value $\zeta(-11) = 691/32760$ of the Riemann zeta function. These ingredients give us information about the cohomology of the quotient of the upper half-space $\mathfrak{H}$ by Γ. More generally, if $\widetilde{\mathcal{M}}_n$ is the sheaf on $\Gamma\backslash\mathfrak{H}$ corresponding to the irreducible representation of SL_2 of dimension $n+1$, then the denominator of an Eisenstein class in $H^1(\Gamma\backslash\mathfrak{H}, \widetilde{\mathcal{M}}_n)$ is intimately linked to the numerator of $\zeta(-1-n)$. For an elaboration of this very beautiful example, see [30, Chap. V].

In this monograph we discuss essentially only results of type (A), but these results allow us to formulate conjectures in the direction of (B) (see [28]). In fact, we study a different kind of influence, albeit somewhat related to the second example given earlier, which we now proceed to describe.

Let $P/\mathbb{Q}$ be a maximal parabolic subgroup of G with M_P its Levi quotient. Let $\underline{\sigma}_f$ be an irreducible Hecke summand appearing in the cohomology of M_P (with coefficients in some local system). Then the algebraically and parabolically induced module ${}^{\mathrm{a}}\mathrm{Ind}_P^G(\underline{\sigma}_f)$ appears in the cohomology of the Borel–Serre boundary of $\mathcal{S}^G_{K_f}$ for some K_f. If Q is the parabolic subgroup associated to P, then we may also consider the induced module ${}^{\mathrm{a}}\mathrm{Ind}_Q^G({}^w\underline{\sigma}_f)$, which appears in a different part of the boundary cohomology; here ${}^w\underline{\sigma}_f$ is a certain conjugate of $\underline{\sigma}_f$ by a Weyl group element w. The machinery of Eisenstein cohomology, which mitigates a relation between total cohomology and the cohomology of the boundary, lets us compare these two induced modules working entirely within the world of the cohomology of $\mathcal{S}^G_{K_f}$. On changing the base to $\mathbb{C}$, and invoking the relation to automorphic forms, these modules may be compared using the standard intertwining operator, which, by a famous calculation of Langlands, contains within it information about the product $\prod_{i=1}^{m} L(is, \underline{\sigma}, \tilde{r}_i)$, where on the dual group side we have ${}^L U_P = r_1 \oplus \cdots \oplus r_m$ as a representation for the adjoint action of ${}^L M_P$.

Let's simplify to our context: $G = \mathrm{GL}_N$ and P is the maximal parabolic subgroup with Levi subgroup $M_P = \mathrm{GL}_n \times \mathrm{GL}_{n'}$, with $n + n' = N$; then $\underline{\sigma} = \sigma \otimes \sigma'$ and ${}^w\underline{\sigma} = \sigma' \otimes \sigma$; also $m = 1$ and $L(s, \underline{\sigma}, \tilde{r}_1)$ is nothing but the Rankin–Selberg L-function $L(s, \sigma \times \sigma'^{\mathsf{v}})$, where σ'^{v} is the contragredient of σ'. It turns out that the sum

$$ {}^{\mathrm{a}}\mathrm{Ind}_P^G(\underline{\sigma}_f) \oplus {}^{\mathrm{a}}\mathrm{Ind}_Q^G(\underline{\sigma}_f^*) $$

(where $\underline{\sigma}_f^*$ is closely related to ${}^w\underline{\sigma}_f$) splits off as a Hecke summand from the cohomology of the boundary, and a certain ratio of L-values

$$\frac{L(s_0, \sigma \times \sigma'^{\mathsf{v}})}{L(s_0 + 1, \sigma \times \sigma'^{\mathsf{v}})}$$

is a key ingredient in the description of the image of global cohomology in the sum of these two induced modules. Furthermore, this description is possible if and only if s_0 and $s_0 + 1$ are both critical (in the sense of Deligne) for the L-function at hand. One expects the primes appearing in the algebraic part of this ratio to have a delicate bearing on the structure of integral cohomology leading to a study of congruences.

In other words, the special values of various Rankin–Selberg L-functions, which may be defined as Langlands–Shahidi L-functions using an ambient GL_N, influence the structure of the cohomology of GL_N. This monograph is an extended cerebration of this idea.

1.1 THE GENERAL CONTEXT

For the moment, let $G = \mathrm{Res}_{F/\mathbb{Q}}(\mathrm{GL}_n/F)$, where F is a totally real field extension of $\mathbb{Q}$. We fix a maximal torus T inside a Borel subgroup B of G. For an open-compact subgroup $K_f \subset G(\mathbb{A}_f)$, where $\mathbb{A}_f$ is the ring of finite adèles of $\mathbb{Q}$, let $\mathcal{S}^G_{K_f}$ be the adèlic locally symmetric space; see Sect. 2.1. Let E be a Galois extension of $\mathbb{Q}$ that contains a copy of F. For a dominant integral weight $\lambda \in X^*(T)$ we let $\mathcal{M}_{\lambda,E}$ be the algebraic irreducible representation of $G \times E$ and let $\widetilde{\mathcal{M}}_{\lambda,E}$ be the associated sheaf on $\mathcal{S}^G_{K_f}$; see Sect. 2.2. The fundamental objects of interest are the sheaf-theoretic cohomology groups

$$H^\bullet(\mathcal{S}^G_{K_f}, \widetilde{\mathcal{M}}_{\lambda,E}) = \bigoplus_q H^q(\mathcal{S}^G_{K_f}, \widetilde{\mathcal{M}}_{\lambda,E}).$$

The main tool that we use is the Borel–Serre compactification $\bar{\mathcal{S}}^G_{K_f}$ of $\mathcal{S}^G_{K_f}$. If $\partial\mathcal{S}^G_{K_f}$ is the boundary of $\bar{\mathcal{S}}^G_{K_f}$, then we have the following long exact sequence (Sect. 2.3.4):

$$\cdots \longrightarrow H^i_c(\mathcal{S}^G_{K_f}, \widetilde{\mathcal{M}}_{\lambda,E}) \xrightarrow{\mathfrak{i}^\bullet} H^i(\bar{\mathcal{S}}^G_{K_f}, \widetilde{\mathcal{M}}_{\lambda,E}) \xrightarrow{\mathfrak{r}^\bullet} H^i(\partial\mathcal{S}^G_{K_f}, \widetilde{\mathcal{M}}_{\lambda,E}) \longrightarrow$$
$$\longrightarrow H^{i+1}_c(\mathcal{S}^G_{K_f}, \widetilde{\mathcal{M}}_{\lambda,E}) \longrightarrow \cdots$$

The image of cohomology with compact supports inside the full cohomology is called *inner* or *interior* cohomology and is denoted $H^\bullet_! := \mathrm{Image}(\mathfrak{i}^\bullet) = \mathrm{Im}(H^\bullet_c \to H^\bullet)$. (See Sect. 2.3.4.) All these cohomology groups are finite dimensional vector spaces over E. They are Hecke-modules; i.e., there is an action of $\pi_0(G(\mathbb{R})) \times \mathcal{H}^G_{K_f}$. The inner cohomology is a semisimple module under the Hecke algebra and if $E/\mathbb{Q}$ is large enough then we get an isotypical decomposition:

$$H^\bullet_!(\mathcal{S}^G_{K_f}, \widetilde{\mathcal{M}}_{\lambda,E}) \;=\; \bigoplus_{\pi_f \in \mathrm{Coh}_!(G,K_f,\lambda)} H^\bullet_!(\mathcal{S}^G_{K_f}, \widetilde{\mathcal{M}}_{\lambda,E})(\pi_f),$$

where $\mathrm{Coh}_!(G, K_f, \lambda)$ is the finite set of isomorphism types of any absolutely irreducible $\mathcal{H}^G_{K_f}$ module that occurs with (a finite) nonzero multiplicity in this decomposition. We may also pass to the limit over all open-compact K_f and get an action of $\pi_0(G(\mathbb{R})) \times G(\mathbb{A}_f)$ on $H^\bullet(\mathcal{S}^G, \widetilde{\mathcal{M}}_{\lambda,E})$, and to retrieve the cohomology group for a particular level-structure K_f, we can take invariants: $H^\bullet(\mathcal{S}^G, \widetilde{\mathcal{M}}_{\lambda,E})^{K_f} = H^\bullet(\mathcal{S}^G_{K_f}, \widetilde{\mathcal{M}}_{\lambda,E})$; the definitions of the Hecke algebra and such Hecke actions are reviewed in Sect. 2.3.2.

We may pass to a transcendental level by taking an embedding $\iota : E \to \mathbb{C}$, and then use the theory of automorphic forms on GL_n to study $H^\bullet(\mathcal{S}^G_{K_f}, \widetilde{\mathcal{M}}_{{}^\iota\lambda,\mathbb{C}})$. It is well known that inner cohomology over $\mathbb{C}$ contains cuspidal cohomology and is contained in square-integrable cohomology; i.e., we have a chain of inclusions:

$$H^\bullet_{\mathrm{cusp}}(\mathcal{S}^G_{K_f}, \widetilde{\mathcal{M}}_{{}^\iota\lambda,\mathbb{C}}) \subset H^\bullet_!(\mathcal{S}^G_{K_f}, \widetilde{\mathcal{M}}_{{}^\iota\lambda,\mathbb{C}}) \subset$$
$$H^\bullet_{(2)}(\mathcal{S}^G_{K_f}, \widetilde{\mathcal{M}}_{{}^\iota\lambda,\mathbb{C}}) \subset H^\bullet(\mathcal{S}^G_{K_f}, \widetilde{\mathcal{M}}_{\lambda,E}) \otimes_{E,\iota} \mathbb{C};$$

see Sect. 3.3.4. The square-integrable cohomology, via Borel–Garland [8] and Borel–Casselman [7], is captured by the discrete spectrum of GL_n; see Sect. 3.2.3. Furthermore, cuspidal cohomology is understood using results about the possible infinite components of cohomological cuspidal representations; we use here the fact that cuspidal representations are globally generic (i.e., have a Whittaker model) and hence locally generic; the local components at infinity are reviewed in Sect. 3.1.

1.2 STRONGLY INNER COHOMOLOGY AND APPLICATIONS

Our first theorem in this monograph is an arithmetic characterization of cuspidal cohomology. We identify a subspace $H^\bullet_{!!}(\mathcal{S}^G_{K_f}, \widetilde{\mathcal{M}}_{\lambda,E})$ of inner cohomology, which we call strongly inner, that by definition is spanned by all those Hecke modules inside inner cohomology whose isotypic component in global cohomology is captured already by the isotypic component in inner cohomology; see Sect. 5.1. Strongly inner cohomology splits off in global cohomology via a Manin–Drinfeld principle, and we get a canonical decomposition

$$H^\bullet(\mathcal{S}^G_{K_f}, \widetilde{\mathcal{M}}_{\lambda,E}) \;=\; H^\bullet_{!!}(\mathcal{S}^G_{K_f}, \widetilde{\mathcal{M}}_{\lambda,E}) \oplus H^\bullet_{\mathrm{Eis}}(\mathcal{S}^G_{K_f}, \widetilde{\mathcal{M}}_{\lambda,E}),$$

which gives an arithmetic definition of Eisenstein cohomology inside global cohomology; see (5.3). If we go to a transcendental level via $\iota : E \to \mathbb{C}$ then we have

$$H^{\bullet}_{!!}(\mathcal{S}^G_{K_f}, \widetilde{\mathcal{M}}_{{}^\iota\lambda,\mathbb{C}}) = H^{\bullet}_{\text{cusp}}(\mathcal{S}^G_{K_f}, \widetilde{\mathcal{M}}_{{}^\iota\lambda,\mathbb{C}}).$$

See Thm. 5.2, where we summarize all the known characterizations of cuspidal cohomology.

The proofs of the assertions about strongly inner cohomology involve understanding the cohomology of the boundary $H^{\bullet}(\partial\mathcal{S}^G_{K_f}, \widetilde{\mathcal{M}}_{\lambda,E})$, and especially being able to pick out the Hecke modules that do not appear in boundary cohomology; this is the subject matter of all of Sect. 4. The boundary $\partial\mathcal{S}^G_{K_f}$ is stratified as $\cup_P \partial_P \mathcal{S}^G_{K_f}$, with a stratum corresponding to every $G(\mathbb{Q})$-conjugacy class of parabolic subgroups. There is a spectral sequence built from the cohomology of $\partial_P\mathcal{S}^G_{K_f}$ that converges to the cohomology of boundary; this is briefly reviewed in Sect. 4.1. Furthermore, for a single stratum $\partial_P\mathcal{S}^G_{K_f}$, its cohomology $H^{\bullet}(\partial_P\mathcal{S}^G_{K_f}, \widetilde{\mathcal{M}}_{\lambda,E})$ may be described in terms of the algebraic induction of the cohomology of the Levi quotient M_P of P with coefficient systems depending on λ and the set W^P of Kostant representatives for P in the Weyl group of G; see Prop. 4.3. The proofs about strongly inner cohomology also make use of multiplicity one and strong multiplicity one results of Jacquet–Shalika [38], [39] and Mœglin–Waldspurger [52].

1.2.1 The relative periods

As a first application of strongly inner cohomology, we describe how to attach certain periods that play an important role in the results on special values of L-functions. For this paragraph we take n to be <u>even</u> and let $\pi_f \in \text{Coh}_{!!}(G, \lambda)$, i.e., π_f is an irreducible Hecke module contributing to strongly inner cohomology. Taking E to be large enough, we know that for every character ε of $\pi_0(G(\mathbb{R}))$, and for the cohomology degree q being an extremal degree b_n^F or $\tilde{t}_n^F$ (see Prop. 3.4), the module $\pi_f \times \varepsilon$ appears in $H^q_{!!}(\mathcal{S}^G_{K_f}, \widetilde{\mathcal{M}}_{\lambda,E})$ with multiplicity one. We fix an arithmetic identification $T^{\varepsilon}_{\text{arith}}(\lambda, \pi_f)$ between the occurrences of $\pi_f \otimes \varepsilon$ and $\pi_f \otimes -\varepsilon$ in degree b_n^F; see (5.4). Then, we go to a transcendental level via $\iota : E \to \mathbb{C}$, and fix an identification $T^{\varepsilon}_{\text{trans}}({}^\iota\lambda, {}^\iota\pi_f)$ between the occurrences of ${}^\iota\pi_f \otimes \varepsilon$ and ${}^\iota\pi_f \otimes -\varepsilon$ in cuspidal cohomology in degree b_n^F by a map described purely in terms of the relative Lie algebra cohomology groups at infinity; see Sect. 5.2.2. We define a nonzero complex number, which we call a *relative period*, as the discrepancy between these two identifications:

$$\Omega^{\varepsilon}({}^\iota\lambda, {}^\iota\pi_f)\, T^{\varepsilon}_{\text{trans}}({}^\iota\lambda, {}^\iota\pi_f) \;=\; T^{\varepsilon}_{\text{arith}}(\lambda, \pi_f) \otimes_{E,\iota} \mathbf{1}.$$

Varying ι, the family of relative periods attached to π_f gives a well defined element of $(E\otimes\mathbb{C})^{\times}/E^{\times}$. Sometimes we suppress λ and write $\Omega^{\varepsilon}({}^\iota\pi_f)$ for $\Omega^{\varepsilon}({}^\iota\lambda, {}^\iota\pi_f)$.

1.2.2 Manin–Drinfeld principle for boundary cohomology

As a second application of strongly inner cohomology, we go back to boundary cohomology and prove a strong form of the Manin–Drinfeld principle, by showing that a certain E-subspace of $H^\bullet(\partial\mathcal{S}^G_{K_f}, \widetilde{\mathcal{M}}_{\lambda,E})$ splits off as an isotypical Hecke module; see Thm. 5.12.

To explain this result, take $N = n + n'$ and now let $G = \mathrm{Res}_{F/\mathbb{Q}}(\mathrm{GL}_N/F)$, and $P = \mathrm{Res}_{F/\mathbb{Q}}(P_0/F) = M_P U_P$ be the standard (i.e., contains the standard Borel subgroup of upper triangular matrices) maximal parabolic subgroup with Levi quotient

$$M_P = \mathrm{Res}_{F/\mathbb{Q}}(M_{P_0}) \;=\; G_n \times G_{n'} \;:=\; \mathrm{Res}_{F/\mathbb{Q}}(\mathrm{GL}_n/F) \times \mathrm{Res}_{F/\mathbb{Q}}(\mathrm{GL}_{n'}/F).$$

Let $Q = \mathrm{Res}_{F/\mathbb{Q}}(Q_0) = M_Q U_Q$ be the standard associate parabolic subgroup of P with Levi quotient $M_Q = G_{n'} \times G_n$. There is a distinguished element $w_P \in W$ that conjugates P into the opposite Q_- of Q and induces an identification $w_P^* : M_P \xrightarrow{\sim} M_Q$. Take pure weights μ and μ' for G_n and $G_{n'}$; only pure weights can support cuspidal or strongly inner cohomology; see § 3.1.2. Let $\sigma_f \in \mathrm{Coh}_{!!}(G_n, \mu)$ and $\sigma'_f \in \mathrm{Coh}_{!!}(G_{n'}, \mu')$.

We make a crucial combinatorial assumption on the weights μ and μ', that there is a Kostant representative $w \in W^P$ such that if we write $w = (w^\tau)_{\tau:F\to E}$ then the length $l(w^\tau)$ of each component is $\dim(U_{P_0})/2 = nn'/2$ and $w^{-1}\cdot(\mu+\mu')$ is dominant as a weight for G. An obvious necessary condition for the existence of such a w is that $\dim(U_{P_0}) = nn'$ is even. Without loss of generality we will take n even and n' of any parity. This condition on μ and μ' has other equivalent formulations that are captured by our *combinatorial lemma*; see Sect. 7.2.3. Sect. 7.2.4 has a proof of this lemma by Uwe Weselmann. A consequence is that the representation algebraically (un-normalized) parabolically induced from $\sigma_f \otimes \sigma'_f$ appears in boundary cohomology:

$${}^{\mathrm{a}}\mathrm{Ind}_{P(\mathbb{A}_f)}^{G(\mathbb{A}_f)}(\sigma_f \otimes \sigma'_f) \;\hookrightarrow\; H^{b_N^F}(\partial_P\mathcal{S}^G, \widetilde{\mathcal{M}}_{\lambda,E}),$$

where $b_N^F = b_n^F + b_{n'}^F + \dim(U_P)/2$ and $\lambda = w^{-1}\cdot(\mu + \mu')$.

Pick a level structure $K_f \subset G(\mathbb{A}_f)$, which satisfies $w_P K_f w_P^{-1} = K_f$ and so that the induced representation has K_f-fixed vectors to get a Hecke-stable k-dimensional subspace (for some k) in $H^{b_N^F}(\partial_P\mathcal{S}^G, \widetilde{\mathcal{M}}_{\lambda,E})^{K_f}$. We denote this k-dimensional space as $I_b^{\mathsf{S}}(\sigma_f, \sigma'_f, \varepsilon')_{P,w}^{K_f}$, where w is coming from the combinatorial lemma, and ε' is a signature discussed in (5.21). The element $w \in W^P$ has an associate $w' \in W^Q$ that is also balanced in the sense that $l(w'^\tau) = \dim(U_{Q_0})/2 = nn'/2$. Another consequence is that

$${}^{\mathrm{a}}\mathrm{Ind}_{Q(\mathbb{A}_f)}^{G(\mathbb{A}_f)}(\sigma'_f(n) \otimes \sigma_f(-n')) \;\hookrightarrow\; H^{b_N^F}(\partial_Q\mathcal{S}^G, \widetilde{\mathcal{M}}_{\lambda,E}),$$

and taking K_f-invariants we get another k-dimensional Hecke-stable subspace, denoted $I_b^{\mathsf{S}}(\sigma_f, \sigma'_f, \varepsilon')^{K_f}_{Q,w'}$, in $H^{b_N^F}(\partial_Q \mathcal{S}^G, \widetilde{\mathcal{M}}_{\lambda,E})^{K_f}$. The Manin–Drinfeld principle, stated in Thm. 5.12, says that the space

$$I_b^{\mathsf{S}}(\sigma_f, \sigma'_f, \varepsilon')^{K_f}_{P,w} \oplus I_b^{\mathsf{S}}(\sigma'_f(-n), \sigma_f(n'), \varepsilon')^{K_f}_{Q,w'},$$

which is of dimension $2\mathsf{k}$, splits off as an isotypical Hecke summand inside $H^{b_N^F}(\partial \mathcal{S}^G, \widetilde{\mathcal{M}}_{\lambda,E})^{K_f}$. Furthermore, the sum of the duals (or contragredients) of these modules splits off as an isotypical summand inside $H^{\tilde{t}_N^F - 1}(\partial \mathcal{S}^G, \widetilde{\mathcal{M}}_{\lambda^{\mathsf{v}},E})^{K_f}$, where the degree $\tilde{t}_N^F - 1$ is coming from the top degrees; see Sect. 5.3.7. For this introduction, we let

$$\begin{array}{c} H^{b_N^F}(\partial \mathcal{S}^G, \widetilde{\mathcal{M}}_{\lambda,E})^{K_f} \\ \Big\downarrow \mathfrak{R} \\ I_b^{\mathsf{S}}(\sigma_f, \sigma'_f, \varepsilon')^{K_f}_{P,w} \oplus I_b^{\mathsf{S}}(\sigma'_f(-n), \sigma_f(n'), \varepsilon')^{K_f}_{Q,w'} \end{array}$$

denote this Hecke projection in the bottom degree.

1.3 THE MAIN THEOREM ON EISENSTEIN COHOMOLOGY

We now come to our main result on rank-one Eisenstein cohomology which is stated as Thm. 6.2. It states that the image of full cohomology under the composition of the maps $\mathfrak{R} \circ \mathfrak{r}^*$ as in

$$\begin{array}{c} H^{b_N^F}(\mathcal{S}^G_{K_f}, \widetilde{\mathcal{M}}_{\lambda,E}) \\ \Big\downarrow \mathfrak{r}^* \\ H^{b_N^F}(\partial \mathcal{S}^G_{K_f}, \widetilde{\mathcal{M}}_{\lambda,E}) \\ \Big\downarrow \mathfrak{R} \\ I_b^{\mathsf{S}}(\sigma_f, \sigma'_f, \varepsilon')^{K_f}_{P,w} \oplus I_b^{\mathsf{S}}(\sigma'_f(-n), \sigma_f(n'), \varepsilon')^{K_f}_{Q,w'} \end{array}$$

is a k-dimensional E-subspace of the $2\mathsf{k}$-dimensional target space. There are two aspects to the proof:

1. One aspect is purely cohomological and says that the Eisenstein part of boundary cohomology is a maximal isotropic subspace under Poincaré duality (Prop. 6.1) bounding the dimension of the image by k; see Sect. 6.2.2.2.
2. The other aspect of the proof is transcendental and appeals to the theory of

automorphic forms; it also gives information about the internal structure of this image.

Let's elaborate on (2): Base change to $\mathbb{C}$ via an embedding $\iota : E \to \mathbb{C}$, and then use Langlands's constant term theorem, recalled in Thm. 6.3, which says that the constant term relative to Q of an Eisenstein series built from a section f of an induced representation from P is the same as the standard intertwining operator $T_{\rm st}$ applied to f. For the interpretation of Langlands's theorem in cohomology, the reader should compare the diagrams (6.17) and (6.18). This interpretation implies in particular that the required image contains all classes of the form $(\xi, T^*_{\rm st}\xi)$; see (6.14); i.e., the image is at least a k-dimensional subspace in a $2\mathsf{k}$-dimensional vector space.

Putting both the aspects together, we conclude that

$$\mathrm{Image}(\mathfrak{R} \circ \mathfrak{r}^*) \;=\; \left\{ (\xi, T^P_{\rm Eis}(\xi)) \;:\; \xi \in I^{\mathsf{S}}_b(\sigma_f, \sigma'_f, \varepsilon')^{K_f}_{P,w} \right\},$$

for an operator $T^P_{\rm Eis} : I^{\mathsf{S}}_b(\sigma_f, \sigma'_f, \varepsilon')^{K_f}_{P,w} \to I^{\mathsf{S}}_b(\sigma'_f(-n), \sigma_f(n'), \varepsilon')^{K_f}_{Q,w'}$ defined over E. It is of course possible that the Eisenstein series constructed from a section f picks up a pole at the point of evaluation that happens to be $s = -N/2$; but in this case we record in Prop. 7.10 that if we are *to the right of the unitary axis* then this Eisenstein series is in fact holomorphic. If, on the other hand, we are to the left of the unitary axis, then we start from Q and, in this case, the image would consist of classes of the form $(T^Q_{\rm Eis}(\psi), \psi)$ with $\psi \in I^{\mathsf{S}}_b(\sigma_f, \sigma'_f, \varepsilon')^{K_f}_{Q,w'}$. It is helpful to think about a situation when $\mathsf{k} = 1$; then the image is a line in a two-dimensional space, and we will see later that the *slope* of this line carries arithmetic information about ratios of L-values.

1.4 THE MAIN THEOREM ON ARITHMETIC OF RANKIN–SELBERG L-FUNCTIONS

Our main result on special values of L-functions (see Thm. 7.21) follows from considering the *slope* mentioned earlier; i.e., we analyze classes of the form $(\xi, T^P_{\rm Eis}(\xi))$, or of the form $(T^Q_{\rm Eis}(\psi), \psi)$. Passing to a transcendental level, we rewrite such a class in terms of the standard intertwining operator, which is given by an integral. Normalizing the local standard intertwining operator using appropriate L-values, we get an operator denoted $T_{\rm loc}$.

At finite places one checks that the operator is nonzero and holomorphic using results of Mœglin–Waldspurger [52]. Furthermore, at every finite place, one observes that the intertwining operator has nice rational properties.

At an archimedean place, using Speh's results (see, for example, [51, Theorem 10b]) on reducibility for induced representations for $\mathrm{GL}_N(\mathbb{R})$, one sees that the induced representations at hand are irreducible; next, using Shahidi's results [64] on local factors and the fact that $-N/2$ and $1-N/2$ are critical, we deduce that the standard intertwining operator is both holomorphic and nonzero, and therefore induces an isomorphism at the level of relative Lie algebra cohomology groups that are one-dimensional, and after making certain careful choices of generators, this scalar turns out to be a power of $(2\pi i)$ as shown in Chap. 8, and independently in Chap. 9. Then we descend to an arithmetic level via the relative periods and the arithmetic identification mentioned earlier. This exercise gives us a rationality result for such a ratio of L-values divided by the relative periods.

Before stating the result, let us mention that the Rankin–Selberg L-functions at hand may be thought of from different points of view:

1. As the motivic L-function attached to the tensor product of the pure motives $\mathbf{M}(\sigma_f)$ and $\mathbf{M}(\sigma'_f)^\vee$ that are conjecturally attached to σ_f and σ'_f.
2. As the cohomological L-function attached to σ_f and σ'_f, and this is entirely from the perspective of Hecke action on the cohomology of arithmetic groups; we will denote this as $L^{\rm coh}\left(\iota, \sigma \times \sigma'^{\vee}, s\right)$ for any embedding $\iota : E \to \mathbb{C}$.
3. As the automorphic Rankin–Selberg L-functions attached to a pair of cuspidal automorphic representations.

The interplay among these three points of view is reviewed in Sect. 7.1.

Suppose n is even and n' is odd then Thm. 7.21 says:

$$\frac{1}{\Omega^{\varepsilon'}({}^{\iota}\sigma_f)} \frac{L^{\rm coh}\left(\iota, \sigma \times \sigma'^{\vee}, \mathsf{m}_0\right)}{L^{\rm coh}\left(\iota, \sigma \times \sigma'^{\vee}, 1+\mathsf{m}_0\right)} \in \iota(E), \tag{1.1}$$

and moreover this ratio of L-values divided by the period is well behaved under the absolute Galois group of $\mathbb{Q}$. On the other hand, if both n and n' are even, then we have

$$\frac{L^{\rm coh}\left(\iota, \sigma \times \sigma'^{\vee}, \mathsf{m}_0\right)}{L^{\rm coh}\left(\iota, \sigma \times \sigma'^{\vee}, 1+\mathsf{m}_0\right)} \in \iota(E), \tag{1.2}$$

which is also well behaved under Galois automorphisms. The point of evaluation m_0 corresponds to the point $-N/2$ for the automorphic L-function.

The combinatorial lemma mentioned earlier has an important consequence: we may replace σ_f, say, by Tate twists $\sigma_f(k)$ for $k \in \mathbb{Z}$, and try to get other ratios of L-values; the lemma, however, imposes some restrictions on the set of such permissible k, and it turns out that *we get a rationality result for exactly all the successive pairs of critical values, no more and no less!*

1.5 SOME GENERAL REMARKS

We may refine this entire discussion and consider integral structures on the cohomology. Such a refinement of our reasoning implies that the periods $\Omega^{\varepsilon'}({}^\iota\sigma_f)$ are well defined modulo a group of units $\mathcal{O}_{E,S}^\times$ where we inverted primes out of a small well-controlled set S of primes. Then we can speak of the prime factorizations of the expressions in (1.1) and (1.2) and ask whether the primes appearing in such expressions are visible in the structure of the cohomology. A first instance of such an event is discussed in Harder [28], where this choice of the periods was simply an educated guess. Such and related issues are also addressed in Harder [27] and [31].

The preceding results on critical values are compatible with Deligne's conjecture [15] on the critical values of motivic L-functions. This compatibility is proved by Bhagwat and Raghuram [3] by proving an appropriate period relation for the ratio c^+/c^- of Deligne's periods for the tensor product motive $\mathbf{M}(\sigma_f) \otimes \mathbf{M}(\sigma'_f)$. These period relations were later generalized in Bhagwat [2], and subsequently put into a general framework in Deligne and Raghuram [16]. It is interesting to note that such period relations exist for all possible combinations of parities of n and n'. However, the methods of this article seem to break down when n and n' are both odd. In Sect. 7.4.1, Bhagwat and the second author show that the analogue of the combinatorial lemma does not hold: if we ask for a situation in which we have two successive critical values then it is shown in Prop. 7.26 that there does not exist an element of the Weyl group, let alone a Kostant representative, which would be needed to arrange for an induced module to appear in boundary cohomology. (Recall the comment in the Preface about Speh's letter, which is indeed in such a situation; $n = 3$ and $n' = 1$.)

In the papers by Kazhdan, Mazur, and Schmidt [44]; Kasten and Schmidt [43]; and Grenie [21], the authors use the theory of modular symbols to get information on the special values of L-functions for $\mathrm{GL}_n \times \mathrm{GL}_{n-1}/\mathbb{Q}$ (and $\mathrm{GL}_n \times \mathrm{GL}_n$ over a totally imaginary field in [21]). They even get rationality (algebraicity) results for the individual L-values, instead of ratios of L-values; see [43, Thm. A]. The question they raise in their introduction on the nonvanishing of their periods $P^{\pm}_{\infty,j}(j+\frac{1}{2})$ has been settled in a paper by Sun [68]. But it is not clear to us whether they have enough control on $P^{\pm}_{\infty,j}(j+\frac{1}{2})$ to derive our result from their theorem (only in the case of $\mathrm{GL}_n \times \mathrm{GL}_{n-1}$). In a sense, we have two entirely different approaches to get an understanding of the nature of special values of these L-functions at critical points, both starting from the study of the cohomology of arithmetic groups.

In the present monograph we carry out a step in a larger program that is outlined by Harder [30], where the general program of investigating Eisenstein cohomology is discussed. This includes a detailed study of the restriction map from global cohomology to the cohomology of the boundary: $H^i(\bar{\mathcal{S}}^G_{K_f}, \widetilde{\mathcal{M}}_{\lambda,E}) \xrightarrow{\mathfrak{r}^i}$

$H^i(\partial \mathcal{S}^G_{K_f}, \widetilde{\mathcal{M}}_{\lambda,E})$. One has to understand the target space and determine the image of $\mathfrak{r}^\bullet$. Our guiding dictum may be phrased as: theorems on the structure of these cohomology groups—as modules under the Hecke algebra—have number theoretic consequences. We also hope to get interesting number theoretic applications if we go deeper into the cohomology, or in other words, to higher filtration steps coming from the spectral sequence that converges to the cohomology of the boundary.

One may also study such number-theoretic applications of Eisenstein cohomology for other reductive groups G. The techniques and methods enunciated in this monograph have already begun to bear results in some other contexts: The reader is referred to Raghuram [58] for an analogous study of Rankin–Selberg L-functions for $\mathrm{GL}_n \times \mathrm{GL}_{n'}$ over a totally imaginary field; to Bhagwat and Raghuram [4] for the case of Eisenstein cohomology for $\mathrm{SO}(n+1, n+1)$ over a totally real base field and the special values of L-functions for $\mathrm{SO}(n,n) \times \mathrm{GL}_1$; and to Krishnamurthy and Raghuram [48] for the case of Eisenstein cohomology for unitary groups and arithmetic consequences for Asai L-functions attached to GL_n over a real quadratic field.

Chapter Two

The Cohomology of GL_n

2.1 THE ADÈLIC LOCALLY SYMMETRIC SPACE

2.1.1 The base field

Let F be a totally real number field of degree $\mathsf{r} = \mathsf{r}_F = [F : \mathbb{Q}]$. By a number field we mean a finite extension of $\mathbb{Q}$. Let $\mathcal{O}_F$ be the ring of integers of F. Let $\mathsf{S}_\infty = \mathrm{Hom}(F, \mathbb{R})$ denote the set of archimedean places. Let $\bar{\mathbb{Q}}$ be the field of algebraic numbers in $\mathbb{C}$. We identify the sets

$$\Sigma_F := \mathrm{Hom}(F, \bar{\mathbb{Q}}) = \mathrm{Hom}(F, \mathbb{C}) = \mathrm{Hom}(F, \mathbb{R}) = \mathsf{S}_\infty.$$

2.1.2 The groups

Let $G_0 = \mathrm{GL}_n/F$, and put $G = \mathrm{Res}_{F/\mathbb{Q}}(G_0) = \mathrm{Res}_{F/\mathbb{Q}}(\mathrm{GL}_n/F)$. An F-group will be denoted by G_0, H_0, etc., and the corresponding $\mathbb{Q}$-group, via Weil restriction of scalars, will be denoted by the same letter without the subscript. For any $\mathbb{Q}$-algebra A, we have

$$G(A) = G_0(A \otimes_{\mathbb{Q}} F) = \mathrm{GL}_n(A \otimes_{\mathbb{Q}} F).$$

Let B_0 stand for the standard Borel subgroup of G_0 consisting of all upper triangular matrices, T_0 the standard torus of all diagonal matrices in B_0, and U_0 the unipotent radical of B_0. The center of G_0 will be denoted by Z_0. These groups define the corresponding $\mathbb{Q}$-groups, and we have

$$G \supset B = TU \supset T \supset Z \supset S,$$

where S is the maximal $\mathbb{Q}$-split torus in Z. Let's identify $S \cong \mathbb{G}_m/\mathbb{Q}$, by sending $x \in \mathbb{G}_m(\ )$ to the diagonal matrix with all entries equal to x.

We have the norm character $N_{F/\mathbb{Q}} : Z \to \mathbb{G}_m$, and if we restrict to S then it becomes $x \mapsto x^{\mathsf{r}}$. The character group $X^*(\mathbb{G}_m) = \mathbb{Z}$, with the character $x \mapsto x^k$ denoted by $[k]$.

Let $G_0^{(1)}$ stand for the group SL_n/F and $G^{(1)} = \mathrm{Res}_{F/\mathbb{Q}}(G_0^{(1)})$. The superscript (1) will mean that we have intersected with SL_n; for example, $T^{(1)} = T \cap \mathrm{SL}_n$.

2.1.3 The symmetric space

For any topological group $\mathcal{G}$, let $\mathcal{G}^0$ stand for the connected component of the identity, and $\pi_0(\mathcal{G}) = \mathcal{G}/\mathcal{G}^0$ stand for the group of connected components. Note that $G(\mathbb{R}) = \prod_{v \in \mathsf{S}_\infty} \mathrm{GL}_n(\mathbb{R})$. Similarly, $Z(\mathbb{R}) \cong \prod_{v \in \mathsf{S}_\infty} \mathbb{R}^\times$, where each copy of $\mathbb{R}^\times$ consists of nonzero scalar matrices in the corresponding copy of $\mathrm{GL}_n(\mathbb{R})$. The subgroup $S(\mathbb{R})$ of $Z(\mathbb{R})$ consists of $\mathbb{R}^\times$ diagonally embedded in $\prod_{v \in \mathsf{S}_\infty} \mathbb{R}^\times$. The group $K_\infty^{(1)} := \prod_{v \in S_\infty} \mathrm{SO}(n)$ is a maximal compact subgroup of $G^{(1)}(\mathbb{R})$ and the corresponding Cartan involution Θ on $G^{(1)}(\mathbb{R})$ is given by $g \mapsto {}^t g^{-1}$ on each factor. Similarly, $C_\infty := \prod_{v \in S_\infty} \mathrm{O}(n)$ is a maximal compact subgroup of $G(\mathbb{R})$. Let

$$K_\infty := C_\infty S(\mathbb{R}) = C_\infty S(\mathbb{R})^0, \quad \text{hence} \quad K_\infty^0 = K_\infty^{(1)} \times S(\mathbb{R})^0.$$

The torus $T^{(1)} \times_{\mathbb{Q}} \mathbb{R}$ is the maximal split torus which is invariant under Θ. Let $T[2]$ denote the 2-torsion subgroup of $T(\mathbb{R})$, then $K_\infty = K_\infty^0 \cdot T[2]$. Inclusion induces a canonical identification $\pi_0(K_\infty) = \pi_0(G(\mathbb{R}))$ that is isomorphic to an r-fold product of $\mathbb{Z}/2\mathbb{Z}$. The (generalized) symmetric space is defined as $X := G(\mathbb{R})/K_\infty^0$. On this space we have an action of $T[2]$ that acts transitively on the set of connected components.

2.1.4 The adèlic locally symmetric space

Let $\mathbb{A}$ be the ring of adèles of $\mathbb{Q}$, which we decompose into its infinite and its finite part: $\mathbb{A} = \mathbb{R} \times \mathbb{A}_f$. The group of adèles is given by $G(\mathbb{A}) = G(\mathbb{R}) \times G(\mathbb{A}_f)$. Elements in the adèlic group are denoted by underlined letters $\underline{g}, \underline{h}$, etc., and the decomposition of an element $\underline{g}$ into its infinite and its finite part will be denoted $\underline{g} = g_\infty \times \underline{g}_f$. Let $K_f = \prod_p K_p \subset G(\mathbb{A}_f)$ be an open-compact subgroup. The *adèlic symmetric space* is

$$G(\mathbb{A})/K_\infty^0 K_f \;=\; X \times (G(\mathbb{A}_f)/K_f) \;=\; G(\mathbb{R})/K_\infty^0 \times (G(\mathbb{A}_f)/K_f).$$

It is a product of the symmetric space $X = G(\mathbb{R})/K_\infty^0$ and an infinite discrete set $G(\mathbb{A}_f)/K_f$. On this space $G(\mathbb{Q})$ acts properly discontinuously and we get a quotient

$$\begin{array}{c} G(\mathbb{R})/K_\infty^0 \times G(\mathbb{A}_f)/K_f \\ \Big\downarrow \pi \\ G(\mathbb{Q}) \backslash \left(G(\mathbb{R})/K_\infty^0 \times G(\mathbb{A}_f)/K_f\right). \end{array} \tag{2.1}$$

The target space, called the *adèlic locally symmetric space of* G *with level structure* K_f, is denoted

$$\mathcal{S}^G_{K_f} \;:=\; G(\mathbb{Q})\backslash G(\mathbb{A})/K^0_\infty K_f.$$

To get an idea of what this space looks like, consider the action of $G(\mathbb{Q})$ on the discrete space $G(\mathbb{A}_f)/K_f$. It follows from classical finiteness results that this quotient is finite; let $\{\underline{g}_f^{(i)}\}_{i=1}^{i=m}$ be a finite set of representatives of $G(\mathbb{Q})\backslash G(\mathbb{A}_f)/K_f$. The stabilizer of the coset $\underline{g}_f^{(i)} K_f$ in $G(\mathbb{Q})$ is equal to

$$\Gamma_i \;:=\; \Gamma^{\underline{g}_f^{(i)}} \;:=\; G(\mathbb{Q}) \,\cap\, \underline{g}_f^{(i)} K_f (\underline{g}_f^{(i)})^{-1},$$

which is an arithmetic subgroup of $G(\mathbb{Q})$ that acts properly discontinuously on X. We say K_f is *neat* if all the subgroups $\Gamma^{\underline{g}_f^{(i)}}$ are torsion free. For any choice of K_f we can pass to a subgroup K'_f of finite index in K_f that is neat; we may take K'_f to be normal in K_f. We have the following decomposition of the adelic locally symmetric space

$$\mathcal{S}^G_{K_f} \;\cong\; \coprod_{i=1}^{m} \Gamma_i\backslash G(\mathbb{R})/K^0_\infty.$$

Let $\underline{g} = g_\infty \times \underline{g}_f \in G(\mathbb{A})$; there is a unique index i such that $\underline{g}_f K_f = \gamma \underline{g}_f^{(i)} K_f$ for some $\gamma \in G(\mathbb{Q})$. We will leave it to the reader to check that the map that sends $G(\mathbb{Q})\,\underline{g}\,K^0_\infty K_f$ in $\mathcal{S}^G_{K_f}$ to $\Gamma_i \gamma^{-1} g_\infty K^0_\infty$ sets up the required homeomorphism.

2.2 HIGHEST WEIGHT MODULES $\mathcal{M}_\lambda$ AND THE SHEAVES $\widetilde{\mathcal{M}}_\lambda$

2.2.1 The character module of $T/\mathbb{Q}$

Let $E/\mathbb{Q}$ be a finite Galois extension and assume that $\mathrm{Hom}(F, E) \neq 0$. We denote this set of embeddings by $\{\tau : F \to E\}$, on which we have a transitive action of the Galois group $\mathrm{Gal}(E/\mathbb{Q})$. The base change $T \times_{\mathbb{Q}} E$ is a split torus; more precisely, dropping the subscript $\mathbb{Q}$ for simplicity, we have

$$T \times E \;=\; \prod_{\tau : F \to E} T_0 \times_{F,\tau} E \;=\; \prod_{\tau : F \to E} T_0.$$

Often, the field E will be taken to be large enough (so that, for example, some module can be split off over E) and we will analyze the behavior of objects as we change E to a field E' always subject to the condition of being Galois over $\mathbb{Q}$ and containing an isomorphic copy of F.

Consider the group of characters of the torus T over E

$$X^*(T \times E) \;=\; \mathrm{Hom}(T \times E, \mathbb{G}_m),$$

on which there is a natural action of $\mathrm{Gal}(E/\mathbb{Q})$. Since $T = R_{F/\mathbb{Q}}(T_0)$, we have

$$X^*(T \times E) = \bigoplus_{\tau : F \to E} X^*(T_0 \times_\tau E) = \bigoplus_{\tau : F \to E} X^*(T_0), \tag{2.2}$$

and any element $\lambda \in X^*(T \times E)$ is of the form $\lambda = (\lambda^\tau)_{\tau : F \to E}$, with $\lambda^\tau \in X^*(T_0 \times_\tau E)$ a character of the split torus $T_0 \times_\tau E$ (since, under our hypothesis on E, we have $F \otimes_{\mathbb{Q}} E = \prod_{\tau : F \to E} E$). Any $\eta \in \mathrm{Gal}(E/\mathbb{Q})$ acts on $\lambda \in X^*(T \times E)$ by permutations

$${}^\eta\lambda \;=\; (({}^\eta\lambda)^\tau)_{\tau : F \to E} \;=\; (\lambda^{\eta^{-1} \circ \tau})_{\tau : F \to E}. \tag{2.3}$$

It is easy to see that ${}^{\eta_1\eta_2}\lambda = {}^{\eta_1}({}^{\eta_2}\lambda)$ for all $\eta_1, \eta_2 \in \mathrm{Gal}(E/\mathbb{Q})$.

2.2.2 The rationality field of $\lambda \in X^*(T \times E)$

Define the rationality field $E(\lambda)$ of any $\lambda \in X^*(T \times E)$ as

$$E(\lambda) \;:=\; E^{\{\eta \in \mathrm{Gal}(E/\mathbb{Q}) \;:\; {}^\eta\lambda = \lambda\}};$$

i.e., it is the subfield of E fixed by the subgroup of $\mathrm{Gal}(E/\mathbb{Q})$ which fixes λ. We have the following proposition.

Proposition 2.1. *The field $E(\lambda)$ is the subfield of E generated by the values of λ on $T(\mathbb{Q})$.*

Proof. Let $t \in T(\mathbb{Q}) = T_0(F) \hookrightarrow T(E) = \prod_{\tau : F \to E} T_0(\tau(F))$. Suppose we write $t = \mathrm{diag}(t_1, \ldots, t_n) \in T_0(F)$; then $t \mapsto (\tau(t))_\tau \in T(E)$, where

$$\tau(t) = \mathrm{diag}(\tau(t_1), \ldots, \tau(t_n)) \in T_0(\tau(F)).$$

Given $\lambda \in X^*(T \times E)$ written as $\lambda = (\lambda^\tau)_\tau$, let $\eta \in \mathrm{Gal}(E/\mathbb{Q})$ be such that ${}^\eta\lambda = \lambda$. This means that $(\lambda^\tau)_\tau = (\lambda^{\eta^{-1} \circ \tau})_\tau$ or that $\lambda^\tau = \lambda^{\eta^{-1} \circ \tau}$ for all $\tau : F \to E$. For $t \in T(\mathbb{Q})$ as above we have

$$\begin{aligned} \eta(\lambda(t)) = \eta(\prod_\tau \lambda^\tau(\tau(t))) &= \prod_\tau \lambda^\tau(\eta(\tau(t))) \\ &= \prod_\mu \lambda^{\eta^{-1} \circ \mu}(\mu(t)) = \prod_\mu \lambda^\mu(\mu(t)) = \lambda(t). \end{aligned}$$

This gives

$$\mathbb{Q}(\lambda) := \mathbb{Q}(\{\lambda(t) \,:\, t \in T(\mathbb{Q})\}) \;\subset\; E^{\{\eta \in \mathrm{Gal}(E/\mathbb{Q}) \,:\, {}^{\eta}\lambda = \lambda\}},$$

or that $\{\eta \in \mathrm{Gal}(E/\mathbb{Q}) \,:\, {}^{\eta}\lambda = \lambda\} \subset \mathrm{Gal}(E/\mathbb{Q}(\lambda))$. To prove the reverse inclusion, let $\eta \in \mathrm{Gal}(E/\mathbb{Q}(\lambda))$. Then by definition $\eta(\lambda(t)) = \lambda(t)$ for all $t \in T(\mathbb{Q})$ or that the algebraic characters ${}^{\eta}\lambda$ and λ agree on $T(\mathbb{Q})$, which is Zariski dense in $T(F)$; hence ${}^{\eta}\lambda = \lambda$, whence $\{\eta \in \mathrm{Gal}(E/\mathbb{Q}) \,:\, {}^{\eta}\lambda = \lambda\} = \mathrm{Gal}(E/\mathbb{Q}(\lambda))$. □

The notation employed in the proof suggests that there is a canonical definition of the rationality field $\mathbb{Q}(\lambda)$ that is independent of the ambient splitting field E.

2.2.3 The Weyl group and the pairing on $X^*(T^{(1)})$

The normalizer $N(T)$ of the maximal torus is a subgroup of $G/\mathbb{Q}$ whose connected component of the identity is $T/\mathbb{Q}$. The quotient $N(T)/T = \mathcal{W}$ is a finite algebraic group scheme over $\mathbb{Q}$. For the base change $G \times E \supset N(T) \times E \supset T \times E$ we have $\mathcal{W}(E) = N(T)(E)/T(E)$ and $\mathcal{W}(E) =: W$ will be the absolute Weyl group, which is a product

$$W = \prod_{\tau : F \to E} W_0^{\tau}$$

with each W_0^{τ} isomorphic to W_0 the symmetric group in n letters, which is the Weyl group of G_0/F with respect to T_0/F. The Galois group $\mathrm{Gal}(E/\mathbb{Q})$ acts on W by permutations as in (2.3). There is a positive definite symmetric pairing

$$\langle\ ,\ \rangle : X^*(T^{(1)} \times E) \;\times\; X^*(T^{(1)} \times E) \;\longrightarrow\; \mathbb{R},$$

which is invariant under the action of W. The direct sum decomposition (2.2) for $T^{(1)}$ is orthogonal with respect to this pairing, and on each summand it is unique up to a scalar. Restriction of characters gives an inclusion

$$X^*(T \times E) \;\subset\; X^*(T^{(1)} \times E) \oplus X^*(Z \times E).$$

We extend the form $\langle\ ,\ \rangle$ trivially by zero on $X^*(Z \times E)$; and the Weyl group action is also trivial on this summand.

2.2.4 Standard basis and fundamental basis

For the moment we only consider T_0/F. The character module $X^*(T_0)$ is a free abelian group on $\{\mathbf{e}_1, \ldots, \mathbf{e}_n\}$, where, for any $t = \mathrm{diag}(t_1, \ldots, t_n) \in T_0(A)$, with A an F-algebra, we have $\mathbf{e}_i(t) = t_i \in A^{\times}$. For integers $b_1, \ldots, b_n$, the character $\lambda = (b_1, \ldots, b_n) := \sum b_i \mathbf{e}_i$ is given by $\lambda(t) = \prod_i t_i^{b_i}$. We will call

$$\{\mathbf{e}_1, \ldots, \mathbf{e}_n\}$$

the *standard basis* for $X^*(T_0)$. For example, the determinant character is given by

$$\boldsymbol{\delta}_n := \det = (1, \ldots, 1) = \sum_{i=1}^{n} \mathbf{e}_i.$$

The simple roots for SL_n or GL_n are given by

$$\{\boldsymbol{\alpha}_1, \ldots, \boldsymbol{\alpha}_{n-1}\}, \quad \boldsymbol{\alpha}_i := \mathbf{e}_i - \mathbf{e}_{i+1}.$$

The fundamental weights $\{\boldsymbol{\gamma}_1, \ldots, \boldsymbol{\gamma}_{n-1}\}$ in $X^*_{\mathbb{Q}}(T_0) := X^*(T_0) \otimes \mathbb{Q}$ are defined by

$$\frac{2\langle \boldsymbol{\gamma}_i, \boldsymbol{\alpha}_j \rangle}{\langle \boldsymbol{\alpha}_j, \boldsymbol{\alpha}_j \rangle} = \delta_{ij}, \quad \text{and} \quad \boldsymbol{\gamma}_i|_{Z_0} = 0.$$

(This makes sense only when $\boldsymbol{\gamma}_i$ is in $X^*_{\mathbb{Q}}(T_0)$.) In terms of the standard basis the fundamental weights are given by

$$\boldsymbol{\gamma}_i = (\mathbf{e}_1 + \cdots + \mathbf{e}_i) - \frac{i}{n}\boldsymbol{\delta}_n = \left(1 - \frac{i}{n}, \cdots, 1 - \frac{i}{n}, -\frac{i}{n}, \cdots, -\frac{i}{n}\right).$$

The basis for $X^*_{\mathbb{Q}}(T_0)$ given by

$$\{\boldsymbol{\gamma}_1, \ldots, \boldsymbol{\gamma}_i, \ldots, \boldsymbol{\gamma}_{n-1}, \boldsymbol{\delta}_n\}$$

will be called the *fundamental basis*. This basis respects the decomposition

$$T_0 = T_0^{(1)} \cdot Z_0, \quad \text{(up to isogeny)};$$

i.e., restriction of characters gives an inclusion $X^*(T_0) \subset X^*(T_0^{(1)}) \oplus X^*(Z_0)$, which, after tensoring by $\mathbb{Q}$, becomes an isomorphism:

$$X^*_{\mathbb{Q}}(T_0) \cong X^*_{\mathbb{Q}}({T_0}^{(1)}) \oplus X^*_{\mathbb{Q}}(Z_0).$$

The restriction to $T_0^{(1)}$ of $\{\boldsymbol{\gamma}_1, \ldots, \boldsymbol{\gamma}_{n-1}\}$ spans $X^*(T_0^{(1)})$, and the determinant character $\boldsymbol{\delta}_n$ spans $X^*_{\mathbb{Q}}(Z_0)$.

For example, half the sum of positive roots for G_0 or $G_0^{(1)}$ is in $X^*_{\mathbb{Q}}(T_0^{(1)})$, and is given by

$$\boldsymbol{\rho}_n := \sum_{i=1}^{n-1} \gamma_i = \left(\frac{n-1}{2}, \frac{n-3}{2}, \cdots, \frac{-(n-1)}{2}\right).$$

Given $\lambda \in X^*_{\mathbb{Q}}(T_0)$ we will write

$$\lambda = \sum_{i=1}^{n-1}(a_i - 1)\boldsymbol{\gamma}_i + d \cdot \boldsymbol{\delta}_n.$$

The $(a_i - 1)$ might seem strange here, but has the virtue that it will simplify some expressions later on. (See, for example, the discussion that follows on motivic weight.) The above expression for λ is the same as writing $\lambda + \boldsymbol{\rho}_n = \sum_{i=1}^{n-1} a_i \boldsymbol{\gamma}_i + d \cdot \boldsymbol{\delta}_n$. Let's write

$$\lambda = \lambda^{(1)} + \lambda_{\mathrm{ab}}, \quad \lambda^{(1)} := \sum_{i=1}^{n-1}(a_i - 1)\boldsymbol{\gamma}_i, \quad \lambda_{ab} := d \cdot \boldsymbol{\delta}_n, \tag{2.4}$$

and call $\lambda^{(1)}$ the semisimple part of λ, and λ_{ab} the abelian part of λ.

Let's describe the dictionary between the standard and the fundamental bases. Let $\lambda \in X^*_{\mathbb{Q}}(T_0)$ be written as $\lambda = \sum_{i=1}^{n-1}(a_i - 1)\boldsymbol{\gamma}_i + d \cdot \boldsymbol{\delta}_n$. Formally, as a *character* of T_0, it may be written as

$$\begin{aligned} t &= \mathrm{diag}(t_1, \ldots, t_n) \mapsto \lambda(t) \\ &= t_1^{a_1+a_2+\cdots+a_{n-1}-(n-1)} \cdot t_2^{a_2+\cdots+a_{n-1}-(n-2)} \cdots t_{n-1}^{a_{n-1}-1} \cdot (t_1 \cdots t_n)^{r_\lambda}, \end{aligned}$$

where

$$r_\lambda := (nd - \sum_{i=1}^{n-1} i(a_i - 1))/n.$$

It's clear then that in the standard basis, if $\lambda = (b_1, \ldots, b_n)$ then

$$\begin{aligned} b_1 &= a_1 + a_2 + \cdots + a_{n-1} - (n-1) + r_\lambda, \\ b_2 &= a_2 + \cdots + a_{n-1} - (n-2) + r_\lambda, \\ &\vdots \\ b_{n-1} &= a_{n-1} - 1 + r_\lambda, \\ b_n &= r_\lambda. \end{aligned} \tag{2.5}$$

Conversely,

$$\begin{aligned} a_i - 1 &= b_i - b_{i+1}, \quad \textit{for } 1 \le i \le n-1, \text{ and} \\ d &= (b_1 + \cdots + b_n)/n. \end{aligned} \tag{2.6}$$

All the above notations are adapted for characters of $T/\mathbb{Q} = R_{F/\mathbb{Q}}(T_0)$.

Hence, for $\lambda = (\lambda^\tau)_{\tau:F\to E}$ in $X^*_{\mathbb{Q}}(T \times E)$, we have:

$$\lambda^\tau = \sum_{i=1}^{n-1}(a_i^\tau - 1)\boldsymbol{\gamma}_i + d^\tau \cdot \boldsymbol{\delta}_n = (b_1^\tau, \ldots, b_n^\tau). \tag{2.7}$$

If we define $\rho_n = (\boldsymbol{\rho}_n)_{\tau:F\to E}$ then we can form $\lambda + \rho_n$ and get

$$(\lambda + \rho_n)^\tau = \sum_{i=1}^{n-1} a_i^\tau \boldsymbol{\gamma}_i + d^\tau \cdot \boldsymbol{\delta}_n. \tag{2.8}$$

We also put $\delta_n = (\boldsymbol{\delta}_n)_{\tau:F\to E}$; actually we are interested only in the case that $d^\tau = d$ for all τ, and in this case we write $\lambda = \lambda^{(1)} + d\delta_n$; see Lem. 2.3.

2.2.5 Integral weights

Let $\lambda = \sum_{i=1}^{n-1}(a_i - 1)\boldsymbol{\gamma}_i + d \cdot \boldsymbol{\delta}_n = (b_1, \ldots, b_n) \in X^*_{\mathbb{Q}}(T_0)$. We say that λ is an integral weight if and only if $\lambda \in X^*(T_0)$, which is the same as saying that $b_i \in \mathbb{Z}$ for all i. In terms of the fundamental basis: $\lambda \in X^*_{\mathbb{Q}}(T_0)$ is integral if and only if

$$\lambda \in X^*(T_0) \iff \begin{cases} a_i \in \mathbb{Z}, \quad 1 \le i \le n-1, \\ nd \in \mathbb{Z}, \\ nd \equiv \sum_{i=1}^{n-1} i(a_i - 1) \pmod{n}. \end{cases} \tag{2.9}$$

The last congruence condition is the same as saying that $r_\lambda \in \mathbb{Z}$.

A weight $\lambda = (\lambda^\tau)_{\tau:F\to E} \in X^*_{\mathbb{Q}}(T \times E)$ is integral if and only if each λ^τ is integral.

2.2.6 Dominant integral weights

Let $\lambda = (b_1, \ldots, b_n) = \sum_{i=1}^{n-1}(a_i - 1)\boldsymbol{\gamma}_i + d\cdot\boldsymbol{\delta} \in X^*(T_0)$ be an integral weight. For the choice of the Borel subgroup being the standard upper triangular subgroup B_0, λ is dominant if and only if

$$a_i \ge 1 \text{ for } 1 \le i \le n-1 \text{ (no condition on } d) \iff b_1 \ge b_2 \ge \cdots \ge b_n. \tag{2.10}$$

A weight $\lambda = (\lambda^\tau)_{\tau:F\to E} \in X^*_{\mathbb{Q}}(T \times E)$ is dominant-integral if and only if each λ^τ is dominant-integral. Denote the set of dominant-integral weights for T_0 (resp., T) by $X^+(T_0)$ (resp., $X^+(T \times E)$).

Remark 2.2. If λ is a dominant-integral weight then the *relevant* information is given by its semisimple part $\lambda^{(1)}$. The number d is uninteresting, the class of nd

mod $\mathbb{Z}$ is determined by $\lambda^{(1)}$, and if we modify $d \to d+m$ we have a canonical identification of the resulting cohomology groups; see Sect. 5.2.4.

2.2.7 The representation $(\rho_\lambda, \mathcal{M}_\lambda)$

Let $\lambda \in X^+(T \times E)$ be a dominant-integral weight and $(\rho_\lambda, \mathcal{M}_\lambda)$ be the absolutely irreducible finite-dimensional representation of $G \times E(\lambda)$ of highest weight λ. We can get hold of $\mathcal{M}_\lambda$ after going to a large enough Galois extension $E/\mathbb{Q}$ and descend to $E(\lambda)$. Put

$$\mathcal{M}_{\lambda,E} \;=\; \bigotimes_{\tau:F\to E} \mathcal{M}_{\lambda^\tau},$$

where $\mathcal{M}_{\lambda^\tau}/E$ is the absolutely irreducible finite-dimensional representation of $G_0 \times_\tau E = \mathrm{GL}_n/F \times_\tau E$ with highest weight λ^τ. If necessary, we will write $(\rho_{\lambda^\tau}, \mathcal{M}_{\lambda^\tau})$ for this representation. The group $G(\mathbb{Q}) = \mathrm{GL}_n(F)$ acts on $\mathcal{M}_{\lambda,E}$ diagonally; i.e., $\gamma \in G(\mathbb{Q})$ acts on a pure tensor $\otimes_\tau m_\tau$ via

$$\gamma \cdot (\otimes_\tau m_\tau) \;=\; \otimes_\tau(\tau(\gamma)\cdot m_\tau). \tag{2.11}$$

We can produce a descent datum and show that $\mathcal{M}_{\lambda,E}$ is defined over $E(\lambda)$, and an $E(\lambda)$-structure, which will be unique up to homotheties by $E^\times$, will be denoted $\mathcal{M}_\lambda$.

2.2.8 The sheaf $\widetilde{\mathcal{M}}_\lambda$ and algebraic dominant-integral weights

For the moment, let $\mathcal{M}$ be a finite-dimensional $\mathbb{Q}$-vector space and $\rho : G/\mathbb{Q} \to \mathrm{GL}(\mathcal{M})$ be a rational representation of the algebraic group $G/\mathbb{Q}$. This representation ρ provides a sheaf $\widetilde{\mathcal{M}}$ of $\mathbb{Q}$-vector spaces on $\mathcal{S}^G_{K_f}$: the sections over an open subset $V \subset \mathcal{S}^G_{K_f}$ are given by

$$\widetilde{\mathcal{M}}(V) = \{s : \pi^{-1}(V) \to \mathcal{M} \mid s \text{ locally constant and}$$
$$s(\gamma v) = \rho(\gamma)s(v),\ \gamma \in G(\mathbb{Q}),\ v \in \pi^{-1}(V)\}, \tag{2.12}$$

where π is as in (2.1). Take $\mathcal{M} = \mathcal{M}_\lambda$ as in §2.2.7; if the central character of ρ_λ is not the type of an algebraic Hecke character of F then the sheaf $\widetilde{\mathcal{M}}_\lambda$ is the zero sheaf. (See [24, 1.1.3].) We record this as the following lemma.

Lemma 2.3. *Let $\lambda \in X^+(T \times E)$; $\lambda = (\lambda^\tau)_{\tau:F\to E}$ with $\lambda^\tau = \sum_{i=1}^{n-1}(a_i^\tau - 1)\boldsymbol{\gamma}_i + d^\tau \cdot \boldsymbol{\delta}$. If there exist τ_1 and τ_2 in $\mathrm{Hom}(F,E)$ such that $d^{\tau_1} \neq d^{\tau_2}$ then $\widetilde{\mathcal{M}}_\lambda = 0$. Equivalently, if $\widetilde{\mathcal{M}}_\lambda \neq 0$ then $\lambda^{\tau_1}_{\mathrm{ab}} = \lambda^{\tau_2}_{\mathrm{ab}}$.*

Proof. If some $d^{\tau_1} \neq d^{\tau_2}$ then consider the central character of ρ_λ on a suitable congruence subgroup of the units in the diagonal center: $\mathcal{O}_F^\times \simeq S(\mathcal{O}_F) \subset \mathrm{GL}_n(F) = G(\mathbb{Q})$ to see that every stalk is zero, and hence the sheaf is the zero sheaf. □

Define $X^+_{\mathrm{alg}}(T \times E)$ to be the subset of those dominant-integral weights that satisfy the *algebraicity* condition that $d^{\tau_1} = d^{\tau_2}$ for all τ_1 and τ_2 in $\mathrm{Hom}(F, E)$; i.e.,

$$X^+_{\mathrm{alg}}(T \times E) := \{\lambda \in X^+(T \times E) \ : \ \lambda^{\tau_1}_{ab} = \lambda^{\tau_2}_{ab}, \ \forall \tau_1, \tau_2 \in \mathrm{Hom}(F, E)\}.$$

Henceforth, we will consider only such algebraic, dominant, and integral highest weights λ. Algebraicity means that the central character $\omega_\lambda \in X^*(T \times E)$ factors via the norm character $N_{F/\mathbb{Q}}$; i.e.,

$$\omega_\lambda(x) \ = \ N_{F/\mathbb{Q}}(x)^d, \tag{2.13}$$

and its restriction to $S = \mathbb{G}_m$ is given by

$$\omega_\lambda(y) \ = \ y^{rnd}. \tag{2.14}$$

If K_f is neat then every stalk of $\widetilde{\mathcal{M}}_\lambda$ is isomorphic to $\mathcal{M}_\lambda$ and the sheaf $\widetilde{\mathcal{M}}_\lambda$ is a local system. Note that $\widetilde{\mathcal{M}}_\lambda$ is a sheaf of $E(\lambda)$-vector spaces, and the base change $\mathcal{M}_{\lambda,E} = \mathcal{M}_\lambda \otimes_{E(\lambda)} E$ gives a sheaf $\widetilde{\mathcal{M}}_{\lambda,E}$ of E-vector spaces on $\mathcal{S}^G_{K_f}$.

2.3 COHOMOLOGY OF THE SHEAVES $\widetilde{\mathcal{M}}_\lambda$

A fundamental problem at the heart of this monograph is to understand the arithmetic information contained in the sheaf-theoretically defined cohomology groups $H^\bullet(\mathcal{S}^G_{K_f}, \widetilde{\mathcal{M}}_{\lambda,E})$.

2.3.1 Functorial properties upon changing the level structure

An inclusion $K_{1,f} \subset K_{2,f}$ of open-compact subgroups gives a canonical surjective map $\mathcal{S}^G_{K_{1,f}} \twoheadrightarrow \mathcal{S}^G_{K_{2,f}}$, which in turn induces a canonical map

$$H^\bullet(\mathcal{S}^G_{K_{2,f}}, \widetilde{\mathcal{M}}_{\lambda,E}) \to H^\bullet(\mathcal{S}^G_{K_{1,f}}, \widetilde{\mathcal{M}}_{\lambda,E}).$$

Pass to the limit over all open-compact subgroups K_f and define

$$H^\bullet(\mathcal{S}^G, \widetilde{\mathcal{M}}_{\lambda,E}) \ := \ \varinjlim_{K_f} H^\bullet(\mathcal{S}^G_{K_f}, \widetilde{\mathcal{M}}_{\lambda,E}).$$

On this limit, there is an action of $\pi_0(G(\mathbb{R})) \times G(\mathbb{A}_f)$, which we now describe. An element $x_\infty \in \pi_0(G(\mathbb{R})) = \pi_0(K_\infty)$ is represented by an element of $T[2]$ and so normalizes K_∞. Let $\underline{x}_f \in G(\mathbb{A}_f)$ and put $\underline{x} = x_\infty \times \underline{x}_f$. Right multiplication by $\underline{x}$, i.e., $g \mapsto g\underline{x}$, induces a map $m_{\underline{x}} : \mathcal{S}^G_{K_f} \longrightarrow \mathcal{S}^G_{\underline{x}_f^{-1} K_f \underline{x}_f}$. This gives a canonical map in cohomology:

$$m^\bullet_{\underline{x}} : H^\bullet(\mathcal{S}^G_{\underline{x}_f^{-1} K_f \underline{x}_f}, m_{\underline{x},*}\widetilde{\mathcal{M}}_{\lambda,E}) \longrightarrow H^\bullet(\mathcal{S}^G_{K_f}, \widetilde{\mathcal{M}}_{\lambda,E}).$$

(This is because, for any continuous map $f : X \to Y$ of topological spaces, and for any sheaf $\mathcal{F}$ of abelian groups (say) on X, we have a canonical map in sheaf cohomology $H^\bullet(Y, f_*\mathcal{F}) \to H^\bullet(X, \mathcal{F})$.) Next, we have $m_{\underline{x},*}\widetilde{\mathcal{M}}_{\lambda,E} = \widetilde{\mathcal{M}}_{\lambda,E}$ as sheaves on $\mathcal{S}^G_{\underline{x}_f^{-1} K_f \underline{x}_f}$; this is an easy verification using the definition of the sheaves. Hence, any $\underline{x}$ gives a canonical map

$$m^\bullet_{\underline{x}} : H^\bullet(\mathcal{S}^G_{\underline{x}_f^{-1} K_f \underline{x}_f}, \widetilde{\mathcal{M}}_{\lambda,E}) \longrightarrow H^\bullet(\mathcal{S}^G_{K_f}, \widetilde{\mathcal{M}}_{\lambda,E}).$$

Passing to the limit over all K_f gives the action of $\underline{x}$ on $H^\bullet(\mathcal{S}^G, \widetilde{\mathcal{M}}_{\lambda,E})$. The cohomology for the spaces with level structure can then be retrieved by taking invariants:

$$H^\bullet(\mathcal{S}^G_{K_f}, \widetilde{\mathcal{M}}_{\lambda,E}) \cong H^\bullet(\mathcal{S}^G, \widetilde{\mathcal{M}}_{\lambda,E})^{K_f}.$$

2.3.2 Hecke action

We let

$$\mathcal{H}^G_{K_f} = C_c^\infty(G(\mathbb{A}_f)/\!\!/K_f)$$

be the set of all locally constant and compactly supported bi-K_f-invariant $\mathbb{Q}$-valued functions on $G(\mathbb{A}_f)$. Take the Haar measure on $G(\mathbb{A}_f)$ to be the product of local Haar measures, and for every prime p, the local measure is normalized so that $\mathrm{vol}(G(\mathbb{Z}_p)) = 1$. Then $\mathcal{H}^G_{K_f}$ is a $\mathbb{Q}$-algebra under convolution of functions.

The $\pi_0(G(\mathbb{R}))\times G(\mathbb{A}_f)$ action on $H^\bullet(\mathcal{S}^G_{K_f}, \widetilde{\mathcal{M}}_{\lambda,E})$, after taking K_f-invariants, induces an action of $\pi_0(G(\mathbb{R})) \times \mathcal{H}^G_{K_f}$ on $H^\bullet(\mathcal{S}^G_{K_f}, \widetilde{\mathcal{M}}_{\lambda,E})$.

2.3.3 Functorial properties upon changing the field E

Consider another field E', also Galois over $\mathbb{Q}$ with an injection $\iota : E \to E'$. Then ι induces an isomorphism $X^*(T \times E) \longrightarrow X^*(T \times E')$ written as $\lambda \mapsto {}^\iota\lambda$. The map $\tau \mapsto \iota \circ \tau$ is a bijection from $\mathrm{Hom}(F, E)$ onto $\mathrm{Hom}(F, E')$. Hence any $\tau' \in \mathrm{Hom}(F, E')$ is of the form $\tau' = \iota \circ \tau$ for a unique $\tau \in \mathrm{Hom}(F, E)$ and we put $\tau = \iota^{-1} \circ \tau'$. If $\lambda = (\lambda^\tau)_{\tau : F \to E}$ then

$${}^\iota\lambda = (\lambda^\tau)_{\iota\circ\tau : F \to E'} = (\lambda^{\iota^{-1}\circ\tau'})_{\tau' : F \to E'}.$$

We get an identification $\iota^* : \mathcal{M}_{\lambda,E} \otimes_{E,\iota} E' \xrightarrow{\sim} \mathcal{M}_{{}^\iota\lambda,E'}$. This yields an identification $\widetilde{\mathcal{M}}_{\lambda,E} \otimes_{E,\iota} E' \xrightarrow{\sim} \widetilde{\mathcal{M}}_{\lambda,E'}$ of sheaves that induces an isomorphism

$$\iota^\bullet : H^\bullet(\mathcal{S}^G_{K_f}, \widetilde{\mathcal{M}}_{\lambda,E}) \otimes_{E,\iota} E' \xrightarrow{\sim} H^\bullet(\mathcal{S}^G_{K_f}, \widetilde{\mathcal{M}}_{{}^\iota\lambda,E'}).$$

The map $\iota^\bullet$ is $\pi_0(G(\mathbb{R})) \times \mathcal{H}^G_{K_f}$-equivariant, since it came from a morphism of sheaves whereas the Hecke action on these cohomology groups was intrinsic to the space $\mathcal{S}^G_{K_f}$.

We may assume $E = E'$; then ι is an element of the Galois group. If $\eta_1, \eta_2 \in \mathrm{Gal}(E/\mathbb{Q})$ then it is clear that $(\eta_1 \circ \eta_2)^* = \eta_1^* \circ \eta_2^*$, and hence $(\eta_1 \circ \eta_2)^\bullet = \eta_1^\bullet \circ \eta_2^\bullet$, i.e.,

$$\begin{array}{ccc}
H^\bullet(\mathcal{S}^G_{K_f}, \widetilde{\mathcal{M}}_{\lambda,E}) & \xrightarrow{\ \eta_2^\bullet\ } & H^\bullet(\mathcal{S}^G_{K_f}, \widetilde{\mathcal{M}}_{{}^{\eta_2}\lambda,E}) \\
& \searrow^{(\eta_1\circ\eta_2)^\bullet} & \downarrow{\scriptstyle \eta_1^\bullet} \\
& & H^\bullet(\mathcal{S}^G_{K_f}, \widetilde{\mathcal{M}}_{{}^{\eta_1\circ\eta_2}\lambda,E})
\end{array}$$

We say that the system of cohomology groups

$$\{H^\bullet(\mathcal{S}^G_{K_f}, \widetilde{\mathcal{M}}_{{}^\eta\lambda,E})\}_{\eta\in\mathrm{Gal}(E/\mathbb{Q})}$$

is defined over $\mathbb{Q}$.

2.3.4 The fundamental long exact sequence

Let $\bar{\mathcal{S}}^G_{K_f}$ be the Borel–Serre compactification of $\mathcal{S}^G_{K_f}$; i.e., $\bar{\mathcal{S}}^G_{K_f} = \mathcal{S}^G_{K_f} \cup \partial\mathcal{S}^G_{K_f}$, where the boundary is stratified as $\partial\mathcal{S}^G_{K_f} = \cup_P \partial_P \mathcal{S}^G_{K_f}$ with P running through the $G(\mathbb{Q})$-conjugacy classes of proper parabolic subgroups defined over $\mathbb{Q}$. (See Borel–Serre [10].) The sheaf $\widetilde{\mathcal{M}}_{\lambda,E}$ on $\mathcal{S}^G_{K_f}$ naturally extends, using the definition of the Borel–Serre compactification, to a sheaf on $\bar{\mathcal{S}}^G_{K_f}$ that we also denote by $\widetilde{\mathcal{M}}_{\lambda,E}$. The inclusion $\mathcal{S}^G_{K_f} \hookrightarrow \bar{\mathcal{S}}^G_{K_f}$ is a homotopy equivalence, and hence the restriction from $\bar{\mathcal{S}}^G_{K_f}$ to $\mathcal{S}^G_{K_f}$ induces an isomorphism in cohomology

$$H^\bullet(\bar{\mathcal{S}}^G_{K_f}, \widetilde{\mathcal{M}}_{\lambda,E}) \xrightarrow{\sim} H^\bullet(\mathcal{S}^G_{K_f}, \widetilde{\mathcal{M}}_{\lambda,E}).$$

The cohomology of the boundary $H^\bullet(\partial\mathcal{S}^G_{K_f}, \widetilde{\mathcal{M}}_{\lambda,E})$ and the cohomology with compact supports $H^\bullet_c(\mathcal{S}^G_{K_f}, \widetilde{\mathcal{M}}_{\lambda,E})$ are naturally modules for $\pi_0(G(\mathbb{R})) \times \mathcal{H}^G_{K_f}$; the Hecke action on these other cohomology groups are described exactly as in Sect. 2.3.2. Our basic object of interest is the following long exact sequence of

$\pi_0(G(\mathbb{R})) \times \mathcal{H}^G_{K_f}$-modules:

$$\cdots \longrightarrow H^i_c(\mathcal{S}^G_{K_f}, \widetilde{\mathcal{M}}_{\lambda,E}) \xrightarrow{\mathfrak{i}^\bullet} H^i(\bar{\mathcal{S}}^G_{K_f}, \widetilde{\mathcal{M}}_{\lambda,E}) \xrightarrow{\mathfrak{r}^\bullet} H^i(\partial\mathcal{S}^G_{K_f}, \widetilde{\mathcal{M}}_{\lambda,E}) \longrightarrow H^{i+1}_c(\mathcal{S}^G_{K_f}, \widetilde{\mathcal{M}}_{\lambda,E}) \longrightarrow \cdots .$$

The image of cohomology with compact supports inside the full cohomology is called *inner* or *interior* cohomology and is denoted

$$H^\bullet_!(\mathcal{S}^G_{K_f}, \widetilde{\mathcal{M}}_{\lambda,E}) := \mathrm{Image}(\mathfrak{i}^\bullet) = \mathrm{Image}\left(H^\bullet_c(\mathcal{S}^G_{K_f}, \widetilde{\mathcal{M}}_{\lambda,E}) \to H^\bullet(\mathcal{S}^G_{K_f}, \widetilde{\mathcal{M}}_{\lambda,E})\right).$$

Complementary to inner cohomology is the theory of Eisenstein cohomology, which is designed to describe the image of the restriction map $\mathfrak{r}^\bullet$. Let's note a slight abuse of terminology: *inner cohomology* is not an honest-to-goodness cohomology theory; for example, a short exact sequence of sheaves would not give a long exact sequence in inner cohomology. However, the terminology is very convenient and helps with our geometric intuition of the nature of the cohomology classes therein.

Our considerations in Sect. 2.3.3 about functorial properties on changing the base field E apply verbatim to the cohomology groups $H^\bullet_?(\mathcal{S}^G_{K_f}, \widetilde{\mathcal{M}}_{\lambda,E})$, where $? \in \{empty, c, !, \partial\}$; by $H^\bullet_\partial(\mathcal{S}^G_{K_f}, \widetilde{\mathcal{M}}_{\lambda,E})$ we mean $H^\bullet(\partial\mathcal{S}^G_{K_f}, \widetilde{\mathcal{M}}_{\lambda,E})$.

2.3.5 Inner cohomology and the inner spectrum $\mathrm{Coh}_!(G, \lambda)$

Inner cohomology is a semisimple module for the Hecke algebra (see Sect. 3.2.3). After taking $E/\mathbb{Q}$ to be a sufficiently large finite Galois extension there is an isotypical decomposition:

$$H^\bullet_!(\mathcal{S}^G_{K_f}, \mathcal{M}_{\lambda,E}) = \bigoplus_{\pi_f \in \mathrm{Coh}_!(G,K_f,\lambda)} H^\bullet_!(\mathcal{S}^G_{K_f}, \mathcal{M}_{\lambda,E})(\pi_f), \tag{2.15}$$

where π_f is an isomorphism type of an absolutely irreducible $\mathcal{H}^G_{K_f}$-module; i.e., there is an E-vector space V_{π_f} with an absolutely irreducible action π_f of $\mathcal{H}^G_{K_f}$. Let $\mathcal{H}^G_{K_p} = C_c^\infty(G(\mathbb{Q}_p)/\!\!/K_p)$ be the local Hecke algebra consisting of all locally constant and compactly supported bi-K_p-invariant $\mathbb{Q}$-valued functions on $G(\mathbb{Q}_p)$. The local factors $\mathcal{H}^G_{K_p}$ are commutative outside a finite set $\mathsf{S} = \mathsf{S}_{K_f}$ of primes and the factors for two different primes commute with each other. For $p \notin \mathsf{S}$ the commutative algebra $\mathcal{H}^G_{K_p}$ acts on V_{π_f} by a homomorphism $\pi_p : \mathcal{H}^G_{K_p} \to E$. Let V_{π_p} be the one-dimensional E-vector space E with the distinguished basis element $1 \in E$ and with the action π_p on it. Then

$$V_{\pi_f} = V_{\pi_f,\mathsf{S}} \otimes'_{p \notin \mathsf{S}} V_{\pi_p} = \otimes_{p \in \mathsf{S}} V_{\pi_p} \otimes E, \tag{2.16}$$

where the absolutely irreducible $\mathcal{H}^G_{K_f,\mathsf{S}}$- module $V_{\pi_f,\mathsf{S}}$ module is decomposed as a tensor product $V_{\pi_f,\mathsf{S}} = \otimes_{p\in\mathsf{S}} V_{\pi_p}$ of absolutely irreducible $\mathcal{H}^G_{K_p}$-modules. We decompose the Hecke algebra

$$\mathcal{H}^G_{K_f} \;=\; \mathcal{H}^G_{K_f,\mathsf{S}} \times \otimes_{p\notin\mathsf{S}} \mathcal{H}^G_{K_p} \;=\; \mathcal{H}^G_{K_f,\mathsf{S}} \times \mathcal{H}^{G,\mathsf{S}}_{K_f},$$

where the first factor acts on the first factor $V_{\pi_f,\mathsf{S}}$ and the second factor acts via the homomorphism $\pi_f^{\mathsf{S}} : \mathcal{H}^{G,\mathsf{S}}_{K_f} \to E$. The set $\mathrm{Coh}_!(G, K_f, \lambda)$ of isomorphism classes that occur with strictly positive multiplicity in (2.15) is called the inner spectrum of G with λ-coefficients and level structure K_f. Taking the union over all K_f, the inner spectrum of G with λ-coefficients is defined to be

$$\mathrm{Coh}_!(G,\lambda) \;=\; \bigcup_{K_f} \mathrm{Coh}_!(G, K_f, \lambda).$$

Going back to functorial properties on changing the field E via $\iota : E \to E'$, note that the map $\iota^\bullet$, since it came from a morphism of sheaves, preserves inner cohomology:

$$\iota^\bullet : H^\bullet_!(\mathcal{S}^G_{K_f}, \widetilde{\mathcal{M}}_{\lambda,E}) \longrightarrow H^\bullet_!(\mathcal{S}^G_{K_f}, \widetilde{\mathcal{M}}_{{}^\iota\lambda,E'}).$$

Given $\pi_f \in \mathrm{Coh}_!(G, K_f, \lambda)$, we deduce ${}^\iota\pi_f \in \mathrm{Coh}_!(G, K_f, {}^\iota\lambda)$ and that the π_f-isotypic component is mapped by $\iota^\bullet$ onto the ${}^\iota\pi_f$-isotypic component. Furthermore, given any character ε of $\pi_0(G(\mathbb{R}))$ we have

$$\iota^\bullet\left(H^\bullet_!(\mathcal{S}^G_{K_f}, \widetilde{\mathcal{M}}_{\lambda,E})(\pi_f \times \varepsilon)\right) \;=\; H^\bullet_!(\mathcal{S}^G_{K_f}, \widetilde{\mathcal{M}}_{{}^\iota\lambda,E'})({}^\iota\pi_f \times \varepsilon),$$

which, via an abuse of notation, we may write as $\iota^\bullet(\pi_f \times \varepsilon) = {}^\iota\pi_f \times \varepsilon$.

If E is large enough so that the π_f-isotypic component can be split off from inner cohomology, then we can define the rationality field of π_f as

$$E(\pi_f) \;:=\; E^{\{\eta\in\mathrm{Gal}(E/\mathbb{Q})\;:\;{}^\eta\pi_f = \pi_f\}}.$$

One may see, using strong multiplicity one for GL_n/F, that the field $E(\pi_f)$ is the subfield of E generated by values of π_p for $p \notin \mathsf{S}$.

2.4 INTEGRAL COHOMOLOGY

Let's briefly discuss how to refine many of the foregoing considerations to talk about integral sheaves and their cohomology and why this is interesting. This aspect is not so relevant for the results in this book, but it will become relevant when we apply certain refinements of the results to arithmetic questions. The reader should consult [27] and [30] for more details.

Let $\mathcal{O}_F$ be the ring of integers of our field $F/\mathbb{Q}$. Instead of starting from algebraic groups over F we start from the split reductive group scheme $\mathrm{GL}_n/\mathcal{O}_F$ and consider the affine group scheme $G/\mathbb{Z} = R_{\mathcal{O}_F/\mathbb{Z}}(\mathrm{GL}_n/\mathcal{O}_F)$. Accordingly we define the Borel subgroup $B/\mathbb{Z}$ and the "torus" $T/\mathbb{Z}$. We take the flag variety of all Borel subgroups $\mathcal{X}/\mathbb{Z} = B\backslash G$. (See, for example, Demazure and Grothendieck [17, XXII, 5.8].) Let w_0 be the element of the Weyl group of longest length that is represented by an element of $G(\mathbb{Z})$. Let $\lambda \in X^*(T \times E)$; the weight $w_0(\lambda)$ gives a line bundle $\mathcal{L}_{w_0(\lambda)}$ on $\mathcal{X} \times \mathcal{O}_E$ and this line bundle $\mathcal{L}_{w_0(\lambda)}$ has global sections if and only if λ is dominant. Let $A(G)$ be the algebra of regular functions on $G/\mathbb{Z}$. Then

$$\begin{aligned}\mathcal{M}_{\lambda,\mathcal{O}} &= H^0(B\backslash G, \mathcal{L}_{w_0(\lambda)})\\ &= \{f \in A(G) \otimes \mathcal{O}_E \ : \ f(bg) = w_0(\lambda)(b)f(g), \ \forall b \in B(\mathcal{O}_E), \forall g \in G(\mathcal{O}_E)\}.\end{aligned}$$

This is a module for $G/\mathbb{Z}$, where the ρ_λ-action of $g \in G(\mathcal{O}_E)$ on any $f \in \mathcal{M}_{\lambda,\mathcal{O}}$ is by right-shifts:

$$(\rho_\lambda(g)(f))(g') = f(g'g), \quad \forall g, g' \in G(\mathcal{O}_E).$$

This is a finitely generated locally free $\mathcal{O}_E$ module and $\mathcal{M}_{\lambda,\mathcal{O}} \otimes E$ is our old module $\mathcal{M}_\lambda$ (the Borel–Weil theorem). The highest/lowest weight vector can be explicitly written down. See [27], and see also Chap. 8.

Starting from $\mathcal{M}_{\lambda,\mathcal{O}}$ we can construct an integral sheaf $\widetilde{\mathcal{M}}_{\lambda,\mathcal{O}}$ of $\mathcal{O}$-modules on $\mathcal{S}^G_{K_f}$. (The reader is referred to [30, Chap. 6] for a detailed description of the construction of this sheaf.) We may then study the integral cohomology groups $H^\bullet(\mathcal{S}^G_{K_f}, \widetilde{\mathcal{M}}_{\lambda,\mathcal{O}})$ and variations such as inner and boundary cohomology. If necessary, to ignore torsion in integral cohomology, one may pass to the image of integral cohomology in rational cohomology; i.e., we may consider

$$H^\bullet_{\mathrm{int}}(\mathcal{S}^G_{K_f}, \widetilde{\mathcal{M}}_{\lambda,\mathcal{O}})/\mathrm{Torsion} \subset H^\bullet(\mathcal{S}^G_{K_f}, \widetilde{\mathcal{M}}_\lambda).$$

The Hecke operators described earlier, in general, do not stabilize integral cohomology, but need to be modified; indeed at every finite prime p, a p-optimal modification is possible. This is reviewed in Sect. 7.1.2 on cohomological L-functions; see also [30, Sect. 6.3].

We may attempt to refine the rationality results on the ratios of critical L-values that are proven in this monograph by asking for integrality results. The periods that come up at various places are always numbers that are defined only up to an element in $E^\times$. If we work with integral cohomology we are able to fix these periods up to an element in $\mathcal{O}_E^\times$. Then it becomes possible to formulate refined versions of the results in Kazhdan, Mazur, and Schmidt [44], Kasten and Schmidt [43], and we may speak about estimates for denominators. We will discuss this partially in Sect. 8.4.7. But integrality results are subtler than rationality results. We do not attempt it here, as it would take us way outside of the original scope of the monograph. The reader should bear in mind that in principle one can go through the entire development of our tools, replacing at every step rational cohomology $H^\bullet(\mathcal{S}^G_{K_f}, \widetilde{\mathcal{M}}_{\lambda,E})$ by integral cohomology $H^\bullet(\mathcal{S}^G_{K_f}, \widetilde{\mathcal{M}}_{\lambda,\mathcal{O}})$. Such an exercise, which offers many challenges along the way, has been carried out in some GL_2 situations in [30], and also some foundational considerations in greater generality are explained in [27]. Delicate problems involving the denominators of Eisenstein classes crop up as alluded to in the Introduction and that are discussed in [30]. It would be a very interesting and fruitful exercise to study integrality properties of special values of L-functions and their bearing on congruences for automorphic forms.

Chapter Three

Analytic Tools

We will now go to the transcendental level, i.e., take an embedding $\iota : E \to \mathbb{C}$, and extend the ground field to $\mathbb{C}$. For all of Chap. 3, we will work over $\mathbb{C}$ and therefore we suppress the subscript $\mathbb{C}$. Starting from Chap. 4 we will return to working at an arithmetic level, i.e., over a Galois extension $E/\mathbb{Q}$ that contains a copy of the totally real base field F, but for now, we work over $\mathbb{C}$.

3.1 CUSPIDAL PARAMETERS AND THE REPRESENTATION $\mathbb{D}_\lambda$ AT INFINITY

3.1.1 Lie algebras and relative Lie algebra cohomology

Let $\mathfrak{g}$ denote the Lie algebra of $G/\mathbb{Q}$. Similarly, let $\mathfrak{g}^{(1)}$, $\mathfrak{b}$, $\mathfrak{t}$, $\mathfrak{z}$, and $\mathfrak{s}$ be the Lie algebras of $G^{(1)}/\mathbb{Q}$, $B/\mathbb{Q}$, $T/\mathbb{Q}$, $Z/\mathbb{Q}$, and S, respectively. Let $\mathfrak{g}_\infty$ be the Lie algebra of $G(\mathbb{R})$, and $\mathfrak{k}_\infty$ that of K^0_∞.

We consider Harish-Chandra modules, also called $(\mathfrak{g}_\infty, K^0_\infty)$-modules, (π, V), where V is the space of smooth K^0_∞-finite vectors of a reasonable representation $(\pi, \mathcal{V})$ of $G(\mathbb{R})$. For example, if $(\pi, \mathcal{V})$ is essentially unitary (i.e., unitary up to a twist) and irreducible, then the space of K^0_∞-finite vectors is automatically smooth. Given a $(\mathfrak{g}_\infty, K^0_\infty)$-module (π, V) by $H^\bullet(\mathfrak{g}_\infty, K^0_\infty; V)$ we mean the cohomology of the complex $\mathrm{Hom}_{K^0_\infty}(\Lambda^\bullet(\mathfrak{g}_\infty/\mathfrak{k}_\infty), V)$. (See Borel and Wallach [9].) If $(\pi, \mathcal{V})$ is a (reasonable) representation of $G(\mathbb{R})$, then by $H^\bullet(\mathfrak{g}_\infty, K^0_\infty; \mathcal{V})$, we mean $H^\bullet(\mathfrak{g}_\infty, K^0_\infty; V)$ with V as earlier. It will be important later on to stay as far as possible at a *rational* level. Note: $\mathfrak{g}_\infty = \mathfrak{g} \otimes_\mathbb{Q} \mathbb{R}$. Similarly, let $\mathfrak{k}^{(1)}$ be the Lie algebra of the $\mathbb{Q}$-group $\prod_{\tau : F \to \mathbb{R}} \mathrm{SO}(n)$; then $\mathfrak{k}_\infty = (\mathfrak{s} \oplus \mathfrak{k}^{(1)}) \otimes_\mathbb{Q} \mathbb{R}$. Given a $(\mathfrak{g}_\infty, K^0_\infty)$-module (π, V), using connectedness of K^0_∞, we have $\mathrm{Hom}_{K^0_\infty}(\Lambda^\bullet(\mathfrak{g}_\infty/\mathfrak{k}_\infty), V) = \mathrm{Hom}_{K^0_\infty}(\Lambda^\bullet(\mathfrak{g}/\mathfrak{k}), V) = \mathrm{Hom}_\mathfrak{k}(\Lambda^\bullet(\mathfrak{g}/\mathfrak{k}), V)$. We will consider modules (π, V) with a $\mathbb{Q}$-structure, i.e., $V = V_0 \otimes_\mathbb{Q} \mathbb{C}$, and then

$$H^\bullet(\mathfrak{g}_\infty, K^0_\infty; V) = H^\bullet(\mathfrak{g}, K^0_\infty; V) = H^\bullet(\mathfrak{g}, \mathfrak{k}; V_0) \otimes \mathbb{C}. \tag{3.1}$$

See Chap. 8.

3.1.2 Pure dominant-integral weights

Not all algebraic dominant-integral weights can support inner cohomology. We have the following purity condition.

Lemma 3.1 (Purity Lemma). *Let $\lambda \in X^+_{\mathrm{alg}}(T \times \mathbb{C})$ be an algebraic dominant-integral weight; and say $\lambda = (\lambda^\nu)_{\nu:F\to\mathbb{C}}$, with $\lambda^\nu = \sum_{i=1}^{n-1}(a_i^\nu - 1)\boldsymbol{\gamma}_i + d\cdot\boldsymbol{\delta}$. Suppose there is an irreducible essentially unitary representation $\mathcal{V}$ of $G(\mathbb{R})$ such that $H^\bullet(\mathfrak{g}, K^0_\infty; \mathcal{V}\otimes\mathcal{M}_\lambda) \neq 0$. Then λ is essentially self-dual; i.e., for each ν, we have $a_i^\nu = a_{n-i}^\nu$.*

This is a well-known consequence of Wigner's Lemma. For a proof, the interested reader can see the discussion in Wallach [71, Sect. 9.4.1–9.4.6]. An algebraic dominant-integral essentially self-dual weight λ will be called a *pure* weight. Let's summarize the restrictions on all the ingredients going into λ if it is a pure weight.

Lemma 3.2. *Let $\lambda \in X^*_{\mathrm{alg}}(T\times E)$ be an algebraic weight. Suppose we write*

$$\lambda = (\lambda^\tau)_{\tau:F\to E}, \quad \lambda^\tau \;=\; \sum_{i=1}^{n-1}(a_i^\tau - 1)\boldsymbol{\gamma}_i + d\cdot\boldsymbol{\delta} \;=\; (b_1^\tau,\ldots,b_n^\tau).$$

If λ is dominant, integral, and pure, then:

1. *Integrality:*
   - $a_i^\tau \in \mathbb{Z}$ *for all* τ *and all* i;
   - $nd \in \mathbb{Z}$; *and*
   - $r_{\lambda^\tau} \in \mathbb{Z}$, *i.e.,* $nd \equiv \sum i(a_i^\tau - 1) \pmod{n}$;
   - $b_i^\tau \in \mathbb{Z}$ *for all* τ *and all* i.
2. *Dominance:*
   - $a_i^\tau \geq 1$ *for all* τ *and all* i;
   - $b_1^\tau \geq b_2^\tau \geq \cdots \geq b_n^\tau$.
3. *Purity:*
   - $a_i^\tau = a_{n-i}^\tau$ *and* $2d \in \mathbb{Z}$. *In particular, if* n *is odd, then* $d \in \mathbb{Z}$;
   - *there exists* $w \in \mathbb{Z}$ *such that* $b_i^\tau + b_{n-i+1}^\tau = w$ *for all* τ *and all* i. *Indeed,* $w = 2d$.

Proof. For integrality and dominance see Sect. 2.2.5 and Sect. 2.2.6, respectively. Under purity, we need to show that $2d \in \mathbb{Z}$. Recall: by integrality $r_{\lambda^\tau} \in \mathbb{Z}$ and, by definition, $nd = \sum_{i=1}^{n-1} i(a_i^\tau - 1) + nr_{\lambda^\tau}$. The purity condition implies

$$\begin{aligned} 2nd = nd + nd &= \sum_{i=1}^{n-1} i(a_i^\tau - 1) + \sum_{i=1}^{n-1}(n-i)(a_i^\tau - 1) + 2nr_{\lambda^\tau} \\ &= n\sum_{i=1}^{n-1}(a_i^\tau - 1) + 2nr_{\lambda^\tau}. \end{aligned}$$

Hence we have

$$2d = \sum_{i=1}^{n-1}(a_i^\tau - 1) + 2r_{\lambda^\tau}. \tag{3.2}$$

This implies that $2d \in \mathbb{Z}$. Furthermore, if n is odd then $d \in \mathbb{Z}$, since we already had $nd \in \mathbb{Z}$ by integrality. In terms of the standard basis, the condition $a_i^\tau = a_{n-i}^\tau$ translates to $b_i^\tau - b_{i+1}^\tau + 1 = b_{n-i}^\tau - b_{n-i+1}^\tau + 1$, which is the same as $b_i^\tau + b_{n-i+1}^\tau = b_{i+1}^\tau + b_{n-i}^\tau$. In other words, $b_i^\tau + b_{n-i+1}^\tau$ is independent of i; say, $b_i^\tau + b_{n-i+1}^\tau = w^\tau$. Then

$$2nd = nd + nd = \sum_{i=1}^{n} b_i^\tau + b_{n-i+1}^\tau = nw^\tau,$$

or that $w^\tau = 2d$ is independent of τ; hence $b_i^\tau + b_{n-i+1}^\tau = 2d$ for all i and τ. □

Denote by $X_0^*(T \times E)$ the set of pure weights. For $\lambda \in X_0^*(T \times E)$, with the notations as earlier, we shall call the integer $2d$ the *purity weight* of λ; furthermore, if we write $\lambda = \lambda^{(1)} + d\boldsymbol{\delta}$, then note that the dual weight of λ is given by $\lambda^\vee = \lambda^{(1)} - d\boldsymbol{\delta}$; i.e., $\lambda^\vee = \lambda - 2d\boldsymbol{\delta}$, which implies $\mathcal{M}_\lambda^\vee = \mathcal{M}_\lambda \otimes (\det)^{-2d}$.

3.1.3 Motivic weight

Let $\lambda \in X_0^*(T \times E)$ be as earlier. The *motivic weight* of λ is defined to be the integer

$$\mathbf{w}_\lambda := \max\{\mathbf{w}_{\lambda^\tau} \mid \tau : F \to E\}, \quad \text{where} \quad \mathbf{w}_{\lambda^\tau} = \sum_{i=1}^{n-1} a_i^\tau.$$

From (3.2) we get the parity conditions

$$\mathbf{w}_{\lambda^\tau} \equiv 2d + n - 1 \pmod 2, \ \forall \tau; \quad \text{hence} \quad \mathbf{w}_\lambda \equiv 2d + n - 1 \pmod 2. \tag{3.3}$$

In particular, if n is odd then $\mathbf{w}_\lambda \equiv 0 \pmod 2$. These parity conditions play an important role in the numerology concerning cuspidal parameters, Hodge pairs, critical points, etc.

3.1.4 Cuspidal parameters and the representation at infinity

Let $\lambda \in X_0^*(T \times \mathbb{C})$ with $\lambda = (\lambda^\nu)_{\nu:F\to\mathbb{C}}$. The infinitesimal character of an irreducible admissible $(\mathfrak{g}, K_\infty^0)$-module π_∞ such that $H^\bullet(\mathfrak{g}, K_\infty^0; \pi_\infty \otimes \mathcal{M}_\lambda) \neq 0$, by Wigner's Lemma, is uniquely determined by λ. Hence, up to isomorphism, there are only finitely many such π_∞ (this is a consequence of the Langlands classification; the reader is referred to Wallach [71, Thm. 5.5.6]). We denote this finite set by $\mathrm{Coh}_\infty(G, \lambda)$. Among this finite set of possible $(\mathfrak{g}, K_\infty^0)$-modules, exactly one, up to twisting by sign characters (see later), is generic, i.e., admits a Whittaker model. Only this representation can appear as the representation at infinity of a global cohomological cuspidal representation.

Let $\lambda^\nu = \sum_{i=1}^{n-1}(a_i^\nu - 1)\boldsymbol{\gamma}_i + d\cdot\boldsymbol{\delta}_n$. The *cuspidal parameter of* λ is defined as $\ell = (\ell^\nu)_{\nu:F\to\mathbb{C}}$, where $\ell^\nu = (\ell_1^\nu, \ldots, \ell_n^\nu)$ and

$$
\begin{aligned}
\ell_1^\nu &:= a_1^\nu + a_2^\nu + a_3^\nu + \cdots + a_{n-1}^\nu && = a_1^\nu + a_2^\nu + a_3^\nu + \cdots + a_{n-1}^\nu = \mathbf{w}_{\lambda^\nu} \\
\ell_2^\nu &:= -a_1^\nu + a_2^\nu + a_3^\nu + \cdots + a_{n-1}^\nu && = a_2^\nu + a_3^\nu + \cdots + a_{n-2}^\nu \\
\ell_3^\nu &:= -a_1^\nu - a_2^\nu + a_3^\nu + \cdots + a_{n-1}^\nu && = a_3^\nu + \cdots + a_{n-3}^\nu \\
& && \vdots \\
\ell_n^\nu &:= -a_1^\nu - a_2^\nu - \cdots - a_{n-1}^\nu = -\mathbf{w}_{\lambda^\nu}
\end{aligned}
\tag{3.4}
$$

The integers ℓ_j^ν, are also called as the cuspidal parameters of λ. Observe that

- ℓ depends only on the semisimple part $\lambda^{(1)}$ and not on the abelian part λ_{ab} of λ; i.e., ℓ depends only on the a_i^ν's, and not on the purity weight $2d$.
- $\ell_1^\nu > \ell_2^\nu > \cdots > \ell_{[n/2]}^\nu > 0$ and $\ell_i^\nu = -\ell_{n-i+1}^\nu$.
- $\ell^\nu = (\mathbf{w}_{\lambda^\nu}, \mathbf{w}_{\lambda^\nu} - 2a_1^\nu, \mathbf{w}_{\lambda^\nu} - 2a_1^\nu - 2a_2^\nu, \ldots, -\mathbf{w}_{\lambda^\nu})$. It readily follows that
$$\ell_i^\nu \equiv \mathbf{w}_\lambda \equiv 2d + n - 1 \pmod 2; \tag{3.5}$$
i.e., every cuspidal parameter has the same parity as the motivic weight. Furthermore, if n is odd, then all the cuspidal parameters, the motivic weight and the purity weight are even; however, if n is even, then all the cuspidal parameters and the motivic weight have the same parity which is opposite to the parity of the purity weight.

For any integer $\ell \geq 1$, let D_ℓ be the discrete series representation of $\mathrm{GL}_2(\mathbb{R})$ of lowest nonnegative K-type corresponding to $\ell + 1$ and with central character $\mathrm{sgn}^{\ell+1}$. The Langlands parameter of D_ℓ is $\mathrm{Ind}_{\mathbb{C}^\times}^{W_\mathbb{R}}(\chi_\ell)$, where $W_\mathbb{R}$ is the Weil group of $\mathbb{R}$, and χ_ℓ is the character of $\mathbb{C}^\times$ sending z to $(z/\bar{z})^{\ell/2}$, or $\chi_\ell(re^{it}) = e^{i\ell t}$. The infinitesimal character of D_ℓ is $(\ell/2, -\ell/2)$. For example, a holomorphic cuspidal modular form of weight k generates D_{k-1} as the representation at infinity. (For more details, the reader is referred to [57, Sects. 3.1.2–3.1.5].)

Let R_n stand for the standard parabolic subgroup of GL_n of type $(2, 2, \ldots, 2)$ if n is even, and of type $(2, 2, \ldots, 2, 1)$ if n is odd. For $\lambda = (\lambda^\nu)_{\nu:F\to\mathbb{C}} \in X_0^*(T \times \mathbb{C})$, define $\mathbb{D}_{\lambda^\nu}$ as

$$\begin{cases} \mathrm{Ind}_{R_n(\mathbb{R})}^{\mathrm{GL}_n(\mathbb{R})}\left((D_{\ell_1^\nu} \otimes |\cdot|_{\mathbb{R}}^{-d}) \otimes \cdots \otimes (D_{\ell_{n/2}^\nu} \otimes |\cdot|_{\mathbb{R}}^{-d})\right), & \text{if } n \text{ is even;} \\ \mathrm{Ind}_{R_n(\mathbb{R})}^{\mathrm{GL}_n(\mathbb{R})}\left((D_{\ell_1^\nu} \otimes |\cdot|_{\mathbb{R}}^{-d}) \otimes \cdots \otimes (D_{\ell_{(n-1)/2}^\nu} \otimes |\cdot|_{\mathbb{R}}^{-d}) \otimes |\cdot|_{\mathbb{R}}^{-d}\right), & \text{if } n \text{ is odd,} \end{cases}$$

where $\mathrm{Ind}_{R_n(\mathbb{R})}^{\mathrm{GL}_n(\mathbb{R})}$ denotes normalized parabolic induction. Identify the set S_∞ of infinite places with the set Σ_F; say, $v \in \mathsf{S}_\infty$ corresponds to $\nu_v \in \Sigma_F$. Define

$$\mathbb{D}_\lambda = \bigotimes_{v \in \mathsf{S}_\infty} \mathbb{D}_{\lambda^{\nu_v}}. \tag{3.6}$$

We will now discuss twisting by sign characters. Identify the sign-character $\mathrm{sgn} : \mathbb{R}^\times \to \{\pm 1\}$ with -1, and the trivial character of $\mathbb{R}^\times$ with $+1$. Let $\varepsilon = (\varepsilon_v)_{v\in\mathsf{S}_\infty}$ be an r-tuple of signs; i.e., $\varepsilon_v \in \{\pm 1\}$. Then ε gives a character of order 2 on $\prod_{v\in\mathsf{S}_\infty} \mathbb{R}^\times$, and via the determinant map, ε gives of a character of $G(\mathbb{R})/G(\mathbb{R})^0 = \pi_0(G(\mathbb{R}))$.

1. If n is even then $\mathbb{D}_\lambda \otimes \varepsilon \cong \mathbb{D}_\lambda$ for every ε. However, if
2. n is odd, then for every nontrivial ε we have $\mathbb{D}_\lambda \otimes \varepsilon \not\cong \mathbb{D}_\lambda$. This may be seen by computing central characters. The central character $\omega_{\mathbb{D}_{\lambda^{\nu_v}} \otimes \varepsilon_v}$ of the v-th component of $\mathbb{D}_\lambda \otimes \varepsilon$ is

$$\omega_{\mathbb{D}_{\lambda^{\nu_v}} \otimes \varepsilon_v} = \mathrm{sgn}^{\frac{(n-1)}{2}} \cdot |\ |^{-nd} \cdot \varepsilon_v.$$

(We have used the fact that ℓ_j^ν is even when n is odd.)

We collect some basic properties of $\mathbb{D}_\lambda$ in the following proposition.

Proposition 3.3. *Let $\lambda \in X_0^*(T \times \mathbb{C})$ and $\mathbb{D}_\lambda$ be as in (3.6). Let $\varepsilon = (\varepsilon_v)_{v\in\mathsf{S}_\infty}$ be any r-tuple of signs. Then*

1. *$\mathbb{D}_\lambda \otimes \varepsilon$ is an irreducible essentially tempered representation.*
2. *$\mathbb{D}_\lambda \otimes \varepsilon$ admits a Whittaker model.*
3. *$H^\bullet(\mathfrak{g}, K_\infty^0; (\mathbb{D}_\lambda \otimes \varepsilon) \otimes \mathcal{M}_\lambda) \neq 0$.*
4. *Let π be an irreducible essentially tempered representation of $G(\mathbb{R})$. Suppose $H^\bullet(\mathfrak{g}, K_\infty^0; \pi \otimes \mathcal{M}_\lambda) \neq 0$, then $\pi = \mathbb{D}_\lambda \otimes \varepsilon$ for some ε.*

Proof. These are well-known results for $\mathrm{GL}_n(\mathbb{R})$; we refer the reader to the very useful survey articles of Knapp [47] and Mœglin [51] and to the references therein. (Irreduciblity follows from Speh's results. A representation irreducibly induced from essentially discrete series representation is essentially tempered. Admitting a Whittaker model is a hereditary property, and an essentially discrete series representation of $\mathrm{GL}_2(\mathbb{R})$ has a Whittaker model. Nonvanishing of cohomology follows from Delorme's Lemma; see later.) □

3.1.5 The cohomology degrees

The relative Lie algebra cohomology group in Prop. 3.3 can be computed using Delorme's Lemma; see, for example, Borel and Wallach [9, Thm. III.3.3]. We summarize what we need about these cohomology groups in the following proposition.

Proposition 3.4. *Let $\lambda \in X_0^*(T \times \mathbb{C})$, $\mathbb{D}_\lambda$ and ε be as earlier. Then*

$$H^\bullet(\mathfrak{g}, K_\infty^0; (\mathbb{D}_\lambda \otimes \varepsilon) \otimes \mathcal{M}_\lambda) \neq 0 \iff b_n^F \leq \bullet \leq \tilde{t}_n^F,$$

where the bottom degree b_n^F and the top degree $\tilde{t}_n^F$ are defined as:

$$b_n^{\mathbb{Q}} := \lfloor n^2/4 \rfloor, \qquad b_n^F := \mathsf{r}_F\, b_n^{\mathbb{Q}},$$

$$t_n^{\mathbb{Q}} := b_n^{\mathbb{Q}} + \lceil n/2 \rceil - 1, \quad t_n^F := \mathsf{r}_F\, t_n^{\mathbb{Q}}, \quad \tilde{t}_n^F := t_n^F + \mathsf{r}_F - 1,$$

where, recall that $\mathsf{r}_F = [F : \mathbb{Q}]$, and for any real number x, $\lfloor x \rfloor$ denotes the greatest integer less than or equal to x, and $\lceil x \rceil = -\lfloor -x \rfloor$ is the least integer greater than or equal to x.

Furthermore, the cohomology group $H^q(\mathfrak{g}, K_\infty^0; (\mathbb{D}_\lambda \otimes \varepsilon) \otimes \mathcal{M}_\lambda)$, as a module over $K_\infty/K_\infty^0 = \pi_0(K_\infty) = \pi_0(G_n(\mathbb{R}))$, in the extreme degrees $q = b_n^F$ or $q = \tilde{t}_n^F$ is given by

1. *If n is even, then $\mathbb{D}_\lambda \otimes \varepsilon = \mathbb{D}_\lambda$ for all ε, and*

$$H^q(\mathfrak{g}, K_\infty^0; \mathbb{D}_\lambda \otimes \mathcal{M}_\lambda) = \bigoplus_{\varepsilon \in \widehat{\pi_0(G_n(\mathbb{R}))}} \varepsilon.$$

2. *If n is odd then*

$$H^q(\mathfrak{g}, K_\infty^0; (\mathbb{D}_\lambda \otimes \varepsilon) \otimes \mathcal{M}_\lambda) = (-1)^{\mathsf{r}_F(d+(n-1)/2)}\varepsilon.$$

Proof. We just make a few comments, as all of this is well known. First of all, observe that $\mathfrak{g}/\mathfrak{k} = \mathfrak{g}^{(1)}/\mathfrak{k}^{(1)} \oplus \mathfrak{z}/\mathfrak{s}$, and hence $\Lambda^\bullet(\mathfrak{g}/\mathfrak{k}) = \Lambda^\bullet(\mathfrak{g}^{(1)}/\mathfrak{k}^{(1)}) \otimes \Lambda^\bullet(\mathfrak{z}/\mathfrak{s})$. This implies:

$$\begin{aligned}&\mathrm{Hom}_{K_\infty^0}(\Lambda^\bullet(\mathfrak{g}/\mathfrak{k}), (\mathbb{D}_\lambda \otimes \varepsilon) \otimes \mathcal{M}_\lambda)\\ &\quad = \mathrm{Hom}_{K_\infty^{(1)}}(\Lambda^\bullet(\mathfrak{g}^{(1)}/\mathfrak{k}^{(1)}), (\mathbb{D}_\lambda \otimes \varepsilon) \otimes \mathcal{M}_\lambda) \otimes \mathrm{Hom}(\Lambda^\bullet(\mathfrak{z}/\mathfrak{s}), \mathbb{C}),\end{aligned}$$

since the Lie algebra $\mathfrak{z}$ acts trivially on $(\mathbb{D}_\lambda \otimes \varepsilon) \otimes \mathcal{M}_\lambda$. The cohomology group $H^\bullet(\mathfrak{g}^{(1)}, K_\infty^{(1)}; (\mathbb{D}_\lambda \otimes \varepsilon) \otimes \mathcal{M}_\lambda)$ is calculated in Clozel [14, Lem. 3.14]; in particular, one sees that it is nonvanishing if and only if $b_n^F \leq \bullet \leq t_n^F$. For $(\mathfrak{g}, K_\infty^0)$-cohomology we need to go up to $\tilde{t}_n^F = t_n^F + \mathsf{r}_F - 1$ since the dimension of $\mathfrak{z}/\mathfrak{s}$ is $\mathsf{r}_F - 1$. If n is odd then $\mathrm{O}(n)/\mathrm{SO}(n)$ is represented by $\{\pm 1_n\}$, where 1_n is the identity $n \times n$-matrix. So, in case (2), it simply boils down to computing the central character of $(\mathbb{D}_\lambda \otimes \varepsilon) \otimes \mathcal{M}_\lambda$. □

3.2 SQUARE-INTEGRABLE COHOMOLOGY

3.2.1 de Rham complex

For a level structure $K_f \subset G(\mathbb{A}_f)$, in what follows, we consider various spaces of functions on $G(\mathbb{Q})\backslash G(\mathbb{A})/K_f$ that are naturally $G(\mathbb{R})$-modules. Assume for the moment that K_f is a neat subgroup. Consider a pure weight $\lambda \in X_0^*(T \times \mathbb{C})$. The sheaf $\widetilde{\mathcal{M}}_\lambda$ on $\mathcal{S}^G_{K_f}$, as constructed in Sect. 2.2.8, is a local system of $\mathbb{C}$-vector spaces. The cohomology $H^\bullet(\mathcal{S}^G_{K_f}, \widetilde{\mathcal{M}}_\lambda)$ is the cohomology of the de Rham complex $\Omega^\bullet(\mathcal{S}^G_{K_f}, \widetilde{\mathcal{M}}_\lambda)$. We have the isomorphism between the de Rham complex and the relative Lie algebra complex

$$\Omega^\bullet(\mathcal{S}^G_{K_f}, \widetilde{\mathcal{M}}_\lambda) \;=\; \mathrm{Hom}_{K^0_\infty}(\Lambda^\bullet(\mathfrak{g}/\mathfrak{k}), \mathcal{C}^\infty(G(\mathbb{Q})\backslash G(\mathbb{A})/K_f, \omega_\lambda^{-1}|_{S(\mathbb{R})^0}) \otimes \mathcal{M}_\lambda),$$

where $\mathcal{C}^\infty(G(\mathbb{Q})\backslash G(\mathbb{A})/K_f, \omega_\lambda^{-1}|_{S(\mathbb{R})^0})$ is the set all functions $\phi : G(\mathbb{A}) \to \mathbb{C}$ such that ϕ is smooth (which means smooth in the usual sense at the infinite places and locally constant at the finite places) and

$$\phi(\gamma\, \underline{g}\, \underline{k}_f\, a_\infty) = \omega_\lambda^{-1}(a_\infty)\phi(\underline{g}), \quad \forall \underline{g} \in G(\mathbb{A}),\ \gamma \in G(\mathbb{Q}),\ \underline{k}_f \in K_f,\ a_\infty \in S(\mathbb{R})^0.$$

We will abbreviate $\omega_\lambda^{-1}|_{S(\mathbb{R})^0} = \omega_\infty^{-1}$. Differentiating the action of $G(\mathbb{R})$ provides an action of the Lie algebra $\mathfrak{g}$ and hence an action of the universal enveloping algebra $\mathfrak{U}(\mathfrak{g})$ that will be denoted by $f \mapsto Uf$ for $U \in \mathfrak{U}(\mathfrak{g})$.

3.2.2 Definition of square-integrable cohomology

Inside $\mathcal{C}^\infty(G(\mathbb{Q})\backslash G(\mathbb{A})/K_f, \omega_\infty^{-1})$ lies the subspace $\mathcal{C}_2^\infty(G(\mathbb{Q})\backslash G(\mathbb{A})/K_f, \omega_\infty^{-1})$ of smooth functions ϕ satisfying the square-integrability condition:

$$\int_{S(\mathbb{R})^0 G(\mathbb{Q})\backslash G(\mathbb{A})} |(U\phi)(\underline{g})|^2\, |\det(g)|_F^{2d}\, dg < \infty,$$

for all $U \in \mathfrak{U}(\mathfrak{g})$. The integrand is trivial on $S(\mathbb{R})^0$ by (2.14). Now we define $H^\bullet_{(2)}(\mathcal{S}^G_{K_f}, \widetilde{\mathcal{M}}_\lambda)$ as the submodule of $H^\bullet(\mathcal{S}^G_{K_f}, \widetilde{\mathcal{M}}_\lambda)$ consisting of those cohomology classes that can be represented by square-integrable forms, i.e., by closed forms in

$$\mathrm{Hom}_{K^0_\infty}(\Lambda^\bullet(\mathfrak{g}/\mathfrak{k}), \mathcal{C}^\infty_2(G(\mathbb{Q})\backslash G(\mathbb{A})/K_f, \omega_\infty^{-1}) \otimes \mathcal{M}_\lambda).$$

3.2.3 A result of Borel and Garland

We know from the fundamental work of Langlands [50] that we have a decomposition into essentially unitary $G(\mathbb{R})$-modules:

$$\begin{aligned} &L^2(G(\mathbb{Q})\backslash G(\mathbb{A})/K_f, \omega_\infty^{-1}) \\ &\qquad = L^2_{\mathrm{disc}}(G(\mathbb{Q})\backslash G(\mathbb{A})/K_f, \omega_\infty^{-1}) \oplus L^2_{\mathrm{cont}}(G(\mathbb{Q})\backslash G(\mathbb{A})/K_f, \omega_\infty^{-1}), \end{aligned}$$

where the first summand in the right-hand side is the closure of the direct sum of irreducible subspaces. Hence, any square-integrable form

$$\omega \in \mathrm{Hom}_{K^0_\infty}(\Lambda^\bullet(\mathfrak{g}/\mathfrak{k}), \mathcal{C}^\infty_2(G(\mathbb{Q})\backslash G(\mathbb{A})/K_f, \omega_\infty^{-1}) \otimes \mathcal{M}_\lambda)$$

may be written as $\omega = \omega_{\mathrm{disc}} \oplus \omega_{\mathrm{cont}}$. The two projections given by the above decomposition commute with the action of $G(\mathbb{A})$; hence they commute with the action of $\mathfrak{U}(\mathfrak{g})$. This implies that both ω_{disc} and ω_{cont} are smooth; i.e.,

$$\omega_{\mathrm{disc}}, \omega_{\mathrm{cont}} \in \mathrm{Hom}_{K_\infty}(\Lambda^\bullet(\mathfrak{g}/\mathfrak{k}), \mathcal{C}^\infty_2(G(\mathbb{Q})\backslash G(\mathbb{A})/K_f, \omega_\infty^{-1}) \otimes \mathcal{M}_\lambda).$$

If ω is closed then $d\omega_{\mathrm{disc}} = d\omega_{\mathrm{cont}} = 0$.

Proposition 3.5. *The cohomology class* $[\omega_{\mathrm{cont}}] = 0$.

Proof. Let q be the degree of ω_{cont}. Since the $\mathcal{C}^\infty$ functions are dense in L^2, Lemma 5.5 in [7] implies that we can find a

$$\psi \in \mathrm{Hom}_{K_\infty}(\Lambda^{q-1}(\mathfrak{g}/\mathfrak{k}), \mathcal{C}^\infty_2(G(\mathbb{Q})\backslash G(\mathbb{A})/K_f, \omega_\infty^{-1}) \otimes \mathcal{M}_\lambda)$$

such that the L^2-norm of $\omega_{\mathrm{cont}} - d\psi$ becomes arbitrarily small (see also [30]). This fact can also be obtained by more general methods using the theory of distributions ([18, Thm. 24, 26]) or Hilbert space techniques ([1, Thm. C. 31]). We invoke Poincaré duality ([26, Thm. 4.8.9]; the finiteness assumptions therein are easily verified): The cohomology class $[\omega_{\mathrm{cont}}] \in H^p(\mathcal{S}^G_{K_f}, \widetilde{\mathcal{M}}_\lambda)$ is zero if and only if the value of the cup product pairing with any class $[\omega_2] \in H^{d-p}_c(\mathcal{S}^G_{K_f}, \widetilde{\mathcal{M}}^\vee_\lambda)$ is zero. But the absolute value $|[\omega_{\mathrm{cont}}] \cup [\omega_2]|$ of the cup product can be given by a scalar product integral. Therefore it can be estimated by the product of the two L^2-norms

$$|[\omega_{\text{cont}}] \cup [\omega_2]| = |[\omega_{\text{cont}} - d\psi] \cup [\omega_2]| \leq ||\omega - d\psi||_2 \, ||\omega_2||_2$$

(Cauchy–Schwarz inequality) and hence must be zero. □

We define $\mathrm{Coh}_\infty^{(2)}(G,\lambda)$ to be the finite set of isomorphism classes of essentially unitary $G(\mathbb{R})$-modules having nontrivial cohomology with $\mathcal{M}_\lambda$-coefficients. For each $\pi_\infty \in \mathrm{Coh}_\infty^{(2)}(G,\lambda)$ choose a representative $\mathcal{V}_{\pi_\infty}$, and let V_{π_∞} be the resulting Harish-Chandra module of K_∞-finite vectors. Then, we put

$$\begin{aligned} W_{\pi_\infty}^{(2)} &= \mathrm{Hom}_{G(\mathbb{R})}\left(\mathcal{V}_{\pi_\infty}, L^2(G(\mathbb{Q})\backslash G(\mathbb{A})/K_f, \omega_\infty^{-1})\right) \\ &= \mathrm{Hom}_{G(\mathbb{R})}\left(\mathcal{V}_{\pi_\infty}, L^2_{\text{disc}}(G(\mathbb{Q})\backslash G(\mathbb{A})/K_f, \omega_\infty^{-1})\right). \end{aligned}$$

It is clear that any $\Phi \in W_{\pi_\infty}^{(2)}$ sends V_{π_∞} to $\mathcal{C}_2^\infty(G(\mathbb{Q})\backslash G(\mathbb{A})/K_f, \omega_\infty^{-1})$. It follows from Borel and Garland [8] that the induced map

$$\bigoplus_{\pi_\infty \in \mathrm{Coh}_\infty^{(2)}(G,\lambda)} W_{\pi_\infty}^{(2)} \otimes H^\bullet(\mathfrak{g}, K_\infty^0; V_{\pi_\infty} \otimes \mathcal{M}_\lambda) \longrightarrow H^\bullet_{(2)}(\mathcal{S}_{K_f}^G, \widetilde{\mathcal{M}}_\lambda) \tag{3.7}$$

is surjective. In [8] it is proved that only finitely many summands in the above decomposition contribute to cohomology. We can also take the action of the Hecke algebra into account; its action via convolution on the discrete spectrum $L^2_{\text{disc}}(G(\mathbb{Q})\backslash G(\mathbb{A})/K_f, \omega_\infty^{-1})$ is self-adjoint. For each isomorphism class $\pi_\infty \in \mathrm{Coh}_\infty^{(2)}(G,\lambda)$, take $\mathcal{V}_{\pi_\infty}$ as earlier, and similarly, for each isomorphism class π_f of absolutely irreducible $\mathcal{H}_{K_f}^G$-module choose a representative V_{π_f} for π_f, and define

$$W_{\pi_\infty \otimes \pi_f}^{(2)} = \mathrm{Hom}_{G(\mathbb{R}) \times \mathcal{H}_{K_f}^G}\left(\mathcal{V}_{\pi_\infty} \otimes V_{\pi_f}, L^2_{\text{disc}}(G(\mathbb{Q})\backslash G(\mathbb{A})/K_f, \omega_\infty^{-1})\right).$$

Define $\mathrm{Coh}_{(2)}(G,\lambda,K_f)$ to be the set of those π_f for which there exists a π_∞ such that $W_{\pi_\infty \otimes \pi_f}^{(2)} \neq 0$. Then we get the refined decomposition

$$\begin{aligned} &\bigoplus_{\pi_\infty \times \pi_f} W_{\pi_\infty \otimes \pi_f}^{(2)} \otimes H^\bullet(\mathfrak{g}, K_\infty^0; V_{\pi_\infty} \otimes \mathcal{M}_\lambda) \otimes V_{\pi_f} \\ &\qquad \longrightarrow \bigoplus_{\pi_f \in \mathrm{Coh}_{(2)}(G,\lambda,K_f)} H^\bullet_{(2)}(\mathcal{S}_{K_f}^G, \widetilde{\mathcal{M}}_\lambda)(\pi_f), \end{aligned} \tag{3.8}$$

that is a surjective map onto $H^\bullet_{(2)}(\mathcal{S}_{K_f}^G, \widetilde{\mathcal{M}}_\lambda)$, and the summand on the right-hand side indexed by π_f is in fact the π_f-isotypic component. This implies that square-integrable cohomology is a semisimple module under the action of the Hecke algebra, hence so is inner cohomology. It is a delicate question to compute the kernel of the above map. Note that, by the multiplicity one theorem for the discrete spectrum for GL_n (which was proved in a special case by Jacquet [37], and in general by Mœglin and Waldspurger [52]), we have $\dim(W^{(2)}_{\pi_\infty\otimes\pi_f}) = 1$.

3.3 CUSPIDAL COHOMOLOGY

3.3.1 The cohomological cuspidal spectrum

The space of square-integrable functions contains the space of smooth cusp forms

$$\mathcal{C}^\infty_{\mathrm{cusp}}(G(\mathbb{Q})\backslash G(\mathbb{A})/K_f, \omega_\infty^{-1}) \;\subset\; \mathcal{C}^\infty_{(2)}(G(\mathbb{Q})\backslash G(\mathbb{A})/K_f, \omega_\infty^{-1}).$$

This inclusion induces an inclusion in cohomology (see Borel [6, Cor. 5.5]) and we define cuspidal cohomology by

$$H^\bullet_{\mathrm{cusp}}(\mathcal{S}^G_{K_f}, \widetilde{\mathcal{M}}_\lambda) \;:=\; H^\bullet\left(\mathfrak{g}, K^0_\infty; \mathcal{C}^\infty_{\mathrm{cusp}}(G(\mathbb{Q})\backslash G(\mathbb{A})/K_f, \omega_\infty^{-1}) \otimes \mathcal{M}_\lambda\right).$$

For $\pi_\infty \in \mathrm{Coh}_\infty(G,\lambda)$ and $\pi_f \in \mathrm{Coh}(G,\lambda,K_f)$, define

$$W^{\mathrm{cusp}}_{\pi_\infty\otimes\pi_f} \;:=\; \mathrm{Hom}_{(\mathfrak{g},K_\infty)\otimes\mathcal{H}^G_{K_f}}\left(V_{\pi_\infty}\otimes V_{\pi_f}, \mathcal{C}^\infty_{\mathrm{cusp}}(G(\mathbb{Q})\backslash G(\mathbb{A})/K_f, \omega_\infty^{-1})\right).$$

Define $\mathrm{Coh}_{\mathrm{cusp}}(G,\lambda,K_f)$ as the set of those $\pi_f \in \mathrm{Coh}(G,\lambda,K_f)$ for which we can find a $\pi_\infty \in \mathrm{Coh}_\infty(G,\lambda)$ such that $W^{\mathrm{cusp}}_{\pi_\infty\otimes\pi_f} \neq 0$. We have a surjective map

$$\bigoplus_{\pi_f\in\mathrm{Coh}_{\mathrm{cusp}}(G,\lambda,K_f)} \;\bigoplus_{\pi_\infty\in\mathrm{Coh}_\infty(G,\lambda)} W^{\mathrm{cusp}}_{\pi_\infty\otimes\pi_f} \otimes H^\bullet(\mathfrak{g}, K^0_\infty; V_{\pi_\infty}\otimes\mathcal{M}_\lambda)\otimes V_{\pi_f}$$
$$\longrightarrow H^\bullet_{\mathrm{cusp}}(\mathcal{S}^G_{K_f}, \mathcal{M}_\lambda),$$

which is in fact an isomorphism by a theorem of Borel, loc. cit.; this also implies that $H^\bullet_{\mathrm{cusp}}(\mathcal{S}^G_{K_f}, \mathcal{M}_\lambda) \subset H^\bullet_!(\mathcal{S}^G_{K_f}, \mathcal{M}_\lambda)$.

3.3.2 Consequence of multiplicity one and strong multiplicity one

By multiplicity one for the cuspidal spectrum for $G = R_{F/\mathbb{Q}}(\mathrm{GL}_n/F)$ (Jacquet [37] and Mœglin and Waldspurger [52]), if $W^{(\mathrm{cusp})}_{\pi_\infty\otimes\pi_f}$ is nonzero then it is of dimension 1. Furthermore, from strong multiplicity one for cuspidal representations due to Jacquet and Shalika [39], it follows that π_f is determined by its restriction π_f^{S} to the central subalgebra $\mathcal{H}^{G,\mathsf{S}} = \otimes_{v\notin\mathsf{S}}\mathcal{H}^G_{K_p}$ of $\mathcal{H}^G_{K_f}$.

3.3.3 The character of the component group I

Since a cuspidal automorphic representation has a global Whittaker model, it follows that the representation at infinity π_∞ is isomorphic to $\mathbb{D}_\lambda \otimes \varepsilon$ for some sign character ε, because these are the only $\pi_\infty \in \mathrm{Coh}_\infty(G,\lambda)$ that have a local Whittaker model. Applying the isomorphism mentioned in Sect. 3.3.1, and the cohomology of $\mathbb{D}_\lambda \otimes \varepsilon$ as described in Prop. 3.4 for extreme degrees $q \in \{b_n^F, \tilde{t}_n^F\}$, we conclude the following decomposition of cuspidal cohomology into nonzero absolutely irreducible $\pi_0(G(\mathbb{R})) \times \mathcal{H}^G_{K_f}$-modules:

$$H^q_{\mathrm{cusp}}(\mathcal{S}^G_{K_f}, \widetilde{\mathcal{M}}_\lambda)$$
$$= \begin{cases} \displaystyle\bigoplus_{\pi_f \in \mathrm{Coh}_{\mathrm{cusp}}(G,\lambda,K_f)} \bigoplus_{\varepsilon \in \widehat{\pi_0(G_n(\mathbb{R}))}} H^\bullet_{\mathrm{cusp}}(\mathcal{S}^G_{K_f}, \mathcal{M}_\lambda)(\pi_f \times \varepsilon), & \text{if } n \text{ is even,} \\ \displaystyle\bigoplus_{\pi_f \in \mathrm{Coh}_{\mathrm{cusp}}(G,\lambda,K_f)} H^\bullet_{\mathrm{cusp}}(\mathcal{S}^G_{K_f}, \mathcal{M}_\lambda)(\pi_f \times \varepsilon(\pi_f)), & \text{if } n \text{ is odd,} \end{cases} \tag{3.9}$$

where, when n is odd the canonical character $\varepsilon(\pi_f)$ that π_f can pair with is given by

$$\varepsilon(\pi_f) = (\varepsilon(\pi_f)_v)_{v \in \mathsf{S}_\infty}, \quad \varepsilon(\pi_f)_v(-1) = (-1)^d \omega_{\pi_v}(-1), \tag{3.10}$$

where ω_{π_v} is the central character of π_v.

This last assertion when n is odd may be seen as follows: Suppose we are given $\pi_f \in \mathrm{Coh}_{\mathrm{cusp}}(G,\lambda,K_f)$; then by strong multiplicity one, there is a unique π_∞ that it can pair with to give a cuspidal automorphic representation $\pi = \pi_\infty \otimes \pi_f$. Suppose, $\pi_\infty = \otimes_{v \in \mathsf{S}_\infty} \pi_v$. By Prop. 3.4, (2), we also see that $H^q(\mathfrak{g}, K_v^0; \pi_v \otimes \mathcal{M}_{\lambda_v})$, as a $K_v/K_v^0 = \mathrm{O}(n)/\mathrm{SO}(n)$-module, is the sign character whose value at -1 is $\omega_{\pi_v}(-1)(-1)^d$. To justify the notation $\varepsilon(\pi_f)$, we note that ω_{π_v} is completely determined by ω_{π_f} by automorphy of the central character $\omega_\pi = \omega_{\pi_\infty} \otimes \omega_{\pi_f}$. To parse it further, observe that $\omega_\pi = |\ |^{-nd} \otimes \omega_\pi^0$, where ω_π^0 is a character of finite order. Fix $v \in \mathsf{S}_\infty$, and apply weak approximation to choose an $a \in F^\times$ (which will depend on v) such that $a < 0$ as an element of F_v and $a > 0$ as an element of F_w for all $w \in \mathsf{S}_\infty - \{v\}$; then $\omega_{\pi_v}(-1) = \omega^0_{\pi_f}(a)$. Similarly, we also have (when n is odd):

$$\begin{aligned} \pi_\infty &= \mathbb{D}_\lambda \otimes \varepsilon_\infty(\pi_f); \\ \varepsilon_\infty(\pi_f) &= (\varepsilon_\infty(\pi_f)_v)_{v \in \mathsf{S}_\infty}; \\ \varepsilon_\infty(\pi_f)_v(-1) &= (-1)^{(n-1)/2} \omega_{\pi_v}(-1). \end{aligned} \tag{3.11}$$

The reader is also referred to the discussion in Gan and Raghuram [20, Sect. 3].

3.3.4 A filtration

We now drop the assumption that we are working over $\mathbb{C}$ and go back to our coefficient system $\mathcal{M}_{\lambda,E}$ defined over E. An embedding $\iota : E \to \mathbb{C}$ gives a chain of subspaces

$$\begin{aligned} H^\bullet_{\text{cusp}}(\mathcal{S}^G_{K_f}, \widetilde{\mathcal{M}}_{{}^\iota\lambda,\mathbb{C}}) &\subset H^\bullet_!(\mathcal{S}^G_{K_f}, \widetilde{\mathcal{M}}_{{}^\iota\lambda,\mathbb{C}}) \\ &\subset H^\bullet_{(2)}(\mathcal{S}^G_{K_f}, \widetilde{\mathcal{M}}_{{}^\iota\lambda,\mathbb{C}}) \subset H^\bullet(\mathcal{S}^G_{K_f}, \widetilde{\mathcal{M}}_{\lambda,E}) \otimes_{E,\iota} \mathbb{C}. \end{aligned}$$

Since the Hecke module $H^\bullet_!(\mathcal{S}^G_{K_f}, \widetilde{\mathcal{M}}_{{}^\iota\lambda,\mathbb{C}}) = H^\bullet_!(\mathcal{S}^G_{K_f}, \widetilde{\mathcal{M}}_{\lambda,E}) \otimes_{E,\iota} \mathbb{C}$ is a submodule of a semisimple module it is also semisimple. But then already the E-module $H^\bullet_!(\mathcal{S}^G_{K_f}, \widetilde{\mathcal{M}}_{\lambda,E})$ is semisimple and if E is large enough we get an isotypical decomposition

$$H^\bullet_!(\mathcal{S}^G_{K_f}, \widetilde{\mathcal{M}}_{\lambda,E}) \otimes_{E,\iota} \mathbb{C} = \bigoplus_{\pi_f \in \mathrm{Coh}_!(G,\lambda,K_f)} H^\bullet_!(\mathcal{S}^G_{K_f}, \widetilde{\mathcal{M}}_{\lambda,E})(\pi_f) \otimes_{E,\iota} \mathbb{C},$$

where the π_f are absolutely irreducible.

We have to understand the discrepancies between these spaces, especially the difference between the cuspidal and the square-integrable cohomology. This issue may become very delicate for a general reductive group, but for GL_n the situation is relatively simple: Take π_f in $\mathrm{Coh}_!(G, \lambda, K_f)$; then for any $\iota : E \to \mathbb{C}$, the module ${}^\iota\pi_f = \pi_f \otimes_{E,\iota} \mathbb{C} \in \mathrm{Coh}_{(2)}(G, \lambda, K_f)$. By definition, there is a ${}^\iota\pi_\infty$ such that $W^{(2)}_{{}^\iota\pi_\infty \otimes {}^\iota\pi_f}$ is one-dimensional. We have to find criteria to decide whether

$$W^{\text{cusp}}_{{}^\iota\pi_\infty \otimes {}^\iota\pi_f} = 0 \quad \text{or} \quad W^{\text{cusp}}_{{}^\iota\pi_\infty \otimes {}^\iota\pi_f} = W^{(2)}_{{}^\iota\pi_\infty \otimes {}^\iota\pi_f}.$$

We will show later (Thm. 4.7) that

$$\begin{aligned} &\text{the isomorphism type } {}^\iota\pi_f \in \mathrm{Coh}_{(2)}(G, \lambda, K_f) \text{ is cuspidal} \iff \\ &\qquad H^\bullet_!(\mathcal{S}^G_{K_f}, \widetilde{\mathcal{M}}_{\lambda,E})(\pi_f) \otimes_{E,\iota} \mathbb{C} = H^\bullet_{(2)}(\mathcal{S}^G_{K_f}, \widetilde{\mathcal{M}}_{\lambda,E} \otimes_{E,\iota} \mathbb{C})({}^\iota\pi_f). \end{aligned} \tag{3.12}$$

This implies that if ${}^\iota\pi_f$ is cuspidal for one embedding ι_0, then it is cuspidal for every embedding ι, giving another proof of a result due to Clozel [14].

Chapter Four

Boundary Cohomology

We will discuss some relevant details of the cohomology of the boundary of the Borel–Serre compactification of the locally symmetric space $\mathcal{S}^G_{K_f}$. (See [25, Sect. 1.1] and [30] for more details and proofs.)

4.1 A SPECTRAL SEQUENCE CONVERGING TO BOUNDARY COHOMOLOGY

Recall from Sect. 2.3.4 the Borel–Serre compactification of $\mathcal{S}^G_{K_f}$ and the associated long exact sequence. There is a spectral sequence built out of the cohomology of the boundary strata $\partial_P \mathcal{S}^G_{K_f}$ that converges to the cohomology of the boundary.

4.1.1 The spectral sequence at an arithmetic level

For $1 \leq i \leq n-1$, let $P_{0,i}$ be the standard maximal parabolic subgroup of $G_0 = \mathrm{GL}_n/F$ obtained by deleting the simple root $\boldsymbol{\alpha}_i$. (See Sect. 2.2.4.) Let $P_i = R_{F/\mathbb{Q}}(P_{0,i})$. Then P_i is a standard maximal parabolic subgroup of G; note that P_i is not absolutely maximal unless $F = \mathbb{Q}$. The standard parabolic subgroups P correspond to nonempty subsets $I = \Delta_P \subset \{1, 2, \dots, n-1\}$ where $P_i \supset P \iff i \in I$. Define $d(P) := \#I$. If $P \supset Q$ then $\Delta_P \subset \Delta_Q$ and by the Borel–Serre construction we have an embedding $\partial_Q \mathcal{S}^G_{K_f} \hookrightarrow \overline{\partial_P \mathcal{S}^G_{K_f}}$ that gives a restriction map in cohomology

$$\mathfrak{r}_{P,Q} : H^\bullet(\partial_P \mathcal{S}^G_{K_f}, \widetilde{\mathcal{M}}_{\lambda,E}) \simeq H^\bullet(\overline{\partial_P \mathcal{S}^G_{K_f}}, \widetilde{\mathcal{M}}_{\lambda,E}) \longrightarrow H^\bullet(\partial_Q \mathcal{S}^G_{K_f}, \widetilde{\mathcal{M}}_{\lambda,E}).$$

From this we get a spectral sequence converging to $H^\bullet(\partial \mathcal{S}^G_{K_f}, \widetilde{\mathcal{M}}_{\lambda,E})$ whose $E_1^{p,q}$ term is

$$E_1^{p,q} := \bigoplus_{d(P)=p+1} H^q(\partial_P \mathcal{S}^G_{K_f}, \widetilde{\mathcal{M}}_{\lambda,E}).$$

The boundary map $d : E_1^{p,q} \to E_1^{p+1,q}$ is given by

$$d(\xi_P) \;=\; \sum_{\substack{Q\subset P \\ d(Q)=p+2}} (-1)^{\sigma(P,Q)}\mathfrak{r}_{P,Q}(\xi_P) \tag{4.1}$$

for any $\xi_P \in H^q(\partial_P \mathcal{S}^G_{K_f}, \widetilde{\mathcal{M}}_{\lambda,E})$ with $d(P) = p+1$, and $\sigma(P,Q)$ is the place of the vertex defining P in the ordered set of vertices defining Q.

4.1.2 The spectral sequence at a transcendental level

Take an $\iota : E \to \mathbb{C}$ and define

$$\Omega^q_P(\mathcal{M}_{\iota\lambda}) \;:=\; \mathrm{Hom}_{K^0_\infty}\left(\Lambda^q(\mathfrak{g}/\mathfrak{k}), \mathcal{C}^\infty(P(\mathbb{Q})\backslash G(\mathbb{A})/K_f, \omega_\infty^{-1}) \otimes \mathcal{M}_{\iota\lambda}\right).$$

Also, define

$$\Omega^q_P(\mathcal{M}_{\iota\lambda})^{(0)} \;:=\; \mathrm{Hom}_{K^0_\infty}\left(\Lambda^q(\mathfrak{g}/\mathfrak{k}), \mathcal{C}^\infty(P(\mathbb{Q})U_P(\mathbb{A})\backslash G(\mathbb{A})/K_f, \omega_\infty^{-1}) \otimes \mathcal{M}_{\iota\lambda}\right).$$

We have the inclusion $\Omega^q_P(\mathcal{M}_{\iota\lambda})^{(0)} \subset \Omega^q_P(\mathcal{M}_{\iota\lambda})$, and taking the constant term $\mathcal{F}^P$ along P gives a projection on to that subspace; recall that the constant term map $\mathcal{F}^P : \mathcal{C}^\infty(P(\mathbb{Q})\backslash G(\mathbb{A})/K_f, \omega_\infty^{-1}) \to \mathcal{C}^\infty(P(\mathbb{Q})U_P(\mathbb{A})\backslash G(\mathbb{A})/K_f, \omega_\infty^{-1})$ is defined as

$$\mathcal{F}^P(\phi)(\underline{g}) \;=\; \int_{U_P(\mathbb{Q})\backslash U_P(\mathbb{A})} \phi(\underline{u}\underline{g})\, d\underline{u}.$$

Hence $\Omega^q_P(\mathcal{M}_{\iota\lambda})^{(0)}$ is a direct summand of $\Omega^q_P(\mathcal{M}_{\iota\lambda})$. Consider the double-complex $\Omega^{\bullet\bullet} = \Omega^{\bullet\bullet}(\mathcal{M}_{\iota\lambda})$:

$$\begin{array}{ccccccccc}
 & & 0 & & 0 & & & & 0 & & \\
 & & \uparrow & & \uparrow & & & & \uparrow & & \\
0 & \longrightarrow & \Omega^{n-2,0} & \longrightarrow & \Omega^{n-2,1} & \longrightarrow & \cdots & \longrightarrow & \Omega^{n-2,\mathsf{d}} & \longrightarrow & 0 \\
 & & \uparrow & & \uparrow & & & & \uparrow & & \\
 & & \vdots & & \vdots & & & & \vdots & & \\
 & & \uparrow & & \uparrow & & & & \uparrow & & \\
0 & \longrightarrow & \Omega^{1,0} & \longrightarrow & \Omega^{1,1} & \longrightarrow & \cdots & \longrightarrow & \Omega^{1,\mathsf{d}} & \longrightarrow & 0 \\
 & & \uparrow & & \uparrow & & & & \uparrow & & \\
0 & \longrightarrow & \Omega^{0,0} & \longrightarrow & \Omega^{0,1} & \longrightarrow & \cdots & \longrightarrow & \Omega^{0,\mathsf{d}} & \longrightarrow & 0 \\
 & & \uparrow & & \uparrow & & & & \uparrow & & \\
 & & 0 & & 0 & & & & 0 & &
\end{array}$$

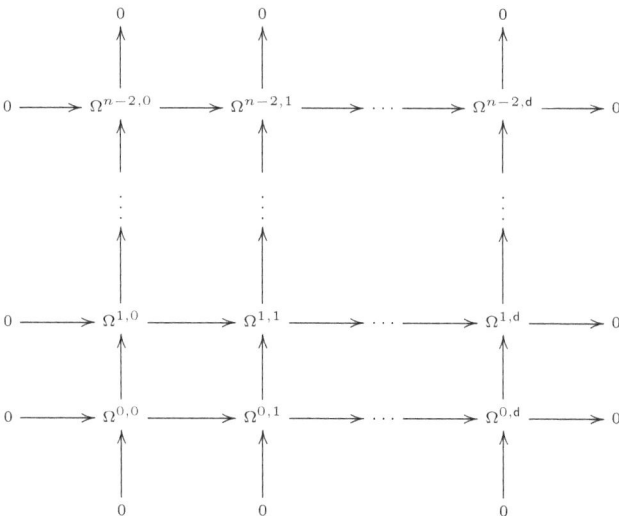

where

$$\Omega^{p,q} = \bigoplus_{d(P)=p} \Omega^q_P(\mathcal{M}_{\iota\lambda});$$

the horizontal arrow $\Omega^{p,q} \to \Omega^{p,q+1}$ is exterior differentiation; the number of columns d is the dimension of $\mathfrak{g}/\mathfrak{k}$; the vertical arrow $\Omega^{p,q} \to \Omega^{p+1,q}$ is defined exactly as in (4.1), i.e., as an alternating sum of maps $\Omega^q_P \to \Omega^q_Q$ for any $Q \subset P$ with $d(Q) = d(P) + 1$, while using the canonical map $\mathcal{C}^\infty(P(\mathbb{Q})\backslash G(\mathbb{A}), \omega_\infty^{-1}) \to \mathcal{C}^\infty(Q(\mathbb{Q})\backslash G(\mathbb{A}), \omega_\infty^{-1})$. The associated simple complex made from this double complex computes the cohomology of the boundary: $H^\bullet(\partial\mathcal{S}^G_{K_f}, \widetilde{\mathcal{M}}_{\iota\lambda})$. All of this remains true even if we work with the double subcomplex $\Omega^{\bullet\bullet}(\mathcal{M}_{\iota\lambda})^{(0)}$. For the vertical arrows, one uses a partial constant term map:

$$\mathcal{F}^Q_P : \mathcal{C}^\infty(P(\mathbb{Q})U_P(\mathbb{A})\backslash G(\mathbb{A}), \omega_\infty^{-1}) \to \mathcal{C}^\infty(Q(\mathbb{Q})U_Q(\mathbb{A})\backslash G(\mathbb{A}), \omega_\infty^{-1})$$

given by

$$\mathcal{F}^Q_P(\phi)(\underline{g}) = \int\limits_{U_Q(\mathbb{Q})U_P(\mathbb{A})\backslash U_Q(\mathbb{A})} \phi(\underline{u}\,\underline{g})\, d\underline{u}.$$

4.2 COHOMOLOGY OF $\partial_P\mathcal{S}^G$ AS AN INDUCED REPRESENTATION

It is clear from the $E_1^{p,q}$ term of the spectral sequence in Sect. 4.1.1 that to understand the cohomology of the boundary, we need to understand the cohomology of a single stratum $\partial_P\mathcal{S}^G_{K_f}$. In the following we do not need that P is maximal, i.e., that $\partial_P\mathcal{S}^G_{K_f}$ is open. It is well known that

$$H^\bullet(\partial_P\mathcal{S}^G_{K_f}, \widetilde{\mathcal{M}}_{\lambda,E}) = H^\bullet(P(\mathbb{Q})\backslash G(\mathbb{A})/K^0_\infty K_f, \widetilde{\mathcal{M}}_{\lambda,E}).$$

The space $P(\mathbb{Q})\backslash G(\mathbb{A})/K^0_\infty K_f$ fibers over locally symmetric spaces of M_P, as we now explain. Let Ξ_{K_f} be a complete set of representatives for $P(\mathbb{A}_f)\backslash G(\mathbb{A}_f)/K_f$. Let $K^P_\infty = K^0_\infty \cap P(\mathbb{R})$, and for $\xi_f \in \Xi_{K_f}$, let $K^P_f(\xi_f) = P(\mathbb{A}_f) \cap \xi_f K_f \xi_f^{-1}$. Then

$$P(\mathbb{Q})\backslash G(\mathbb{A})/K^0_\infty K_f = \coprod_{\xi_f \in \Xi_{K_f}} P(\mathbb{Q})\backslash P(\mathbb{A})/K^P_\infty K^P_f(\xi_f).$$

Let $\kappa_P : P \to P/U_P = M_P$ be the canonical map, and define $K^{M_P}_\infty = \kappa_P(K^P_\infty)$, and for $\xi_f \in \Xi_{K_f}$, let $K^{M_P}_f(\xi_f) = \kappa_P(K^P_f(\xi_f))$. Define

$$\underline{\mathcal{S}}^{M_P}_{K^{M_P}_f(\xi_f)} := M_P(\mathbb{Q})\backslash M_P(\mathbb{A})/K^{M_P}_\infty K^{M_P}_f(\xi_f).$$

The underline is to emphasize that we have divided out by $K_\infty^{M_P}$ which is not connected; see Sect. 4.2.1. Let $K_f^{U_P}(\xi_f) = U_P(\mathbb{A}_f) \cap \xi_f K_f \xi_f^{-1}$. We have the fibration:

$$U_P(\mathbb{Q})\backslash U_P(\mathbb{A})/K_f^{U_P}(\xi_f) \hookrightarrow P(\mathbb{Q})\backslash P(\mathbb{A})/K_\infty^P K_f^P(\xi_f) \twoheadrightarrow \underline{\mathcal{S}}^{M_P}_{K_f^{M_P}(\xi_f)}.$$

The corresponding Leray–Serre spectral sequence degenerates at E_2-level. See, for example, Schwermer's articles [60, Sect. 7] and [59, Thm. 2.7]. The cohomology of the total space is given in terms of the cohomology of the base with coefficients in the cohomology of the fiber. For the cohomology of the fiber, one uses a classical theorem due to van Est: if $\mathfrak{u}_P$ is the Lie algebra of U_P then the cohomology of the fiber is the same as the unipotent Lie algebra cohomology group $H^\bullet(\mathfrak{u}_P, \mathcal{M}_{\lambda,E})$, which is naturally an algebraic representation of M_P; the associated sheaf on $\underline{\mathcal{S}}^{M_P}_{K_f^{M_P}(\xi_f)}$ is denoted by putting a tilde on top. Putting all this together, we get

$$H^\bullet(\partial_P \mathcal{S}^G_{K_f}, \widetilde{\mathcal{M}}_{\lambda,E}) = \bigoplus_{\xi_f \in \Xi_{K_f}} H^\bullet\left(\underline{\mathcal{S}}^{M_P}_{K_f^{M_P}(\xi_f)}, \widetilde{H^\bullet(\mathfrak{u}_P, \mathcal{M}_{\lambda,E})}\right). \tag{4.2}$$

It is convenient to pass to the limit over all open-compact subgroups K_f and define

$$H^\bullet(\partial_P \mathcal{S}^G, \widetilde{\mathcal{M}}_{\lambda,E}) := \varinjlim_{K_f} H^\bullet(\partial_P \mathcal{S}^G_{K_f}, \widetilde{\mathcal{M}}_{\lambda,E}).$$

Let $\underline{\mathcal{S}}^{M_P} := M_P(\mathbb{Q})\backslash M_P(\mathbb{A})/K_\infty^{M_P}$. Now we can rewrite (4.2) as

$$H^\bullet(\partial_P \mathcal{S}^G, \widetilde{\mathcal{M}}_{\lambda,E})^{K_f} = \bigoplus_{\xi_f \in \Xi_{K_f}} H^\bullet\left(\underline{\mathcal{S}}^{M_P}, \widetilde{H^\bullet(\mathfrak{u}_P, \mathcal{M}_{\lambda,E})}\right)^{K_f^{M_P}(\xi_f)}. \tag{4.3}$$

Let's recall an exercise in basic Mackey theory for taking invariants in an induced representation and stated in a context immediately relevant to us.

Proposition 4.1. *Let V be a module for $M_P(\mathbb{A}_f)$ that is inflated up to a module for $P(\mathbb{A}_f)$ via the canonical projection $P(\mathbb{A}_f) \to M_P(\mathbb{A}_f)$. Let ${}^{\mathrm{a}}\mathrm{Ind}_{P(\mathbb{A}_f)}^{G(\mathbb{A}_f)}(V)$ stand for the algebraic (that is un-normalized) induction from $P(\mathbb{A}_f)$ to $G(\mathbb{A}_f)$ of V. Let $K_f \subset G(\mathbb{A}_f)$ be an open-compact subgroup. Then*

$$\left({}^{\mathrm{a}}\mathrm{Ind}_{P(\mathbb{A}_f)}^{G(\mathbb{A}_f)}(V)\right)^{K_f} = \bigoplus_{\xi_f \in \Xi_{K_f}} V^{K_f^{M_P}(\xi_f)}.$$

It is clear from Mackey theory as earlier that the right-hand side of (4.3) is the K_f-invariants of an algebraically induced representation. Passing to the limit over all K_f, we get the following important result.

Proposition 4.2. *The cohomology of $\partial_P \mathcal{S}^G$ is given by*

$$H^\bullet(\partial_P \mathcal{S}^G, \widetilde{\mathcal{M}}_{\lambda,E}) = {}^{\mathrm{a}}\mathrm{Ind}_{\pi_0(P(\mathbb{R}))\times P(\mathbb{A}_f)}^{\pi_0(G(\mathbb{R}))\times G(\mathbb{A}_f)} \left(H^\bullet(\underline{\mathcal{S}}^{M_P}, \widetilde{H^\bullet(\mathfrak{u}_P, \mathcal{M}_{\lambda,E})})\right).$$

The notation ${}^{\mathrm{a}}\mathrm{Ind}$ *stands for algebraic, or un-normalized, induction.*

The process of induction from $\pi_0(P(\mathbb{R}))$ to $\pi_0(G(\mathbb{R}))$ is important and needs some explanation.

4.2.1 Induction from $\pi_0(P(\mathbb{R}))$ to $\pi_0(G(\mathbb{R}))$

The parabolic subgroup P is of the form $P = R_{F/\mathbb{Q}}(P_0)$ for a parabolic subgroup P_0/F of $G_0 = \mathrm{GL}_n/F$. Since F is totally real, we identify the set of infinite places S_∞ with $\Sigma_F = \mathrm{Hom}(F, \mathbb{R})$, and $G(\mathbb{R}) = \prod_{\tau\in\Sigma_F} \mathrm{GL}_n(\mathbb{R})$ and $P(\mathbb{R}) = \prod_{\tau\in\Sigma_F} P_0^\tau(\mathbb{R})$, where $P_0^\tau = P_0 \times_{F,\tau} \mathbb{R}$. Furthermore, suppose, P_0 corresponds to the partition $n = n_1 + \cdots + n_k$ with $k \geq 2$ and $n_j \geq 1$, then $M_{P_0} = \mathrm{GL}_{n_1} \times \cdots \times \mathrm{GL}_{n_k}/F$. We have

$$\begin{aligned}
K_\infty^{M_P} &= \kappa_P(P(\mathbb{R}) \cap K_\infty^0) \\
&= \kappa_P\left(P(\mathbb{R}) \cap (S(\mathbb{R})^0 \cdot \prod_\tau \mathrm{SO}(n))\right) \\
&= \kappa_P\left(S(\mathbb{R})^0 \cdot \prod_\tau (P_0^\tau(\mathbb{R}) \cap \mathrm{SO}(n))\right) \\
&= \kappa_P\left(S(\mathbb{R})^0 \cdot \prod_\tau \mathrm{S}(\mathrm{O}(n_1) \times \cdots \times \mathrm{O}(n_k))\right) \\
&\simeq S(\mathbb{R})^0 \cdot \prod_\tau \mathrm{S}(\mathrm{O}(n_1) \times \cdots \times \mathrm{O}(n_k)).
\end{aligned}$$

Note that $\pi_0(P(\mathbb{R}))$ has order $2^{\mathsf{r}k}$ and $\pi_0(K_\infty^{M_P})$ has order $2^{\mathsf{r}(k-1)}$. (Recall, $\mathsf{r} = [F : \mathbb{Q}]$.) Inclusion of components induces a canonical surjective map $\pi_0(P(\mathbb{R})) \to \pi_0(G(\mathbb{R}))$ giving a short exact sequence:

$$1 \longrightarrow \pi_0(K_\infty^{M_P}) \longrightarrow \pi_0(P(\mathbb{R})) \longrightarrow \pi_0(G(\mathbb{R})) \longrightarrow 1. \tag{4.4}$$

In the definition of $\underline{\mathcal{S}}^{M_P}$ we have divided out by $K_\infty^{M_P}$. For $\mathcal{S}^{M_P}$ we divide only by $K_\infty^{M_P,0}$, the connected component of the identity in $K_\infty^{M_P}$. Hence, for any sheaf $\widetilde{\mathcal{M}}$ coming from an algebraic representation $\mathcal{M}$ of $M_P/\mathbb{Q}$ we have

$$H^\bullet(\underline{\mathcal{S}}^{M_P}, \widetilde{\mathcal{M}}) = H^\bullet(\mathcal{S}^{M_P}, \widetilde{\mathcal{M}})^{\pi_0(K_\infty^{M_P})}.$$

This means that $H^\bullet(\underline{\mathcal{S}}^{M_P}, \widetilde{H^\bullet(\mathfrak{u}_P, \mathcal{M}_{\lambda,E})})$ is a module for $\pi_0(P(\mathbb{R})) \times P(\mathbb{A}_f)$ on which $\pi_0(K_\infty^{M_P})$ acts trivially and so is naturally a module for $\pi_0(G(\mathbb{R})) \times P(\mathbb{A}_f)$, which is then induced up to get a module for $\pi_0(G(\mathbb{R})) \times G(\mathbb{A}_f)$. See also Rem. 4.4.

4.2.2 Kostant's theorem on unipotent cohomology

The structure of the unipotent cohomology group $H^\bullet(\mathfrak{u}_P, \mathcal{M}_{\lambda,E})$ is well known by results of Kostant [46], and we briefly describe these results in our situation.

The calculation of the unipotent cohomology group is over the field E. Recall that we are dealing with the split group $G \times E = \prod_{\tau:F\to E} G_0^\tau$, where $G_0^\tau = G_0 \times_{F,\tau} E$.

Let $\boldsymbol{\Delta}_{G_0}$ stand for the set of roots of G_0 and $\boldsymbol{\Delta}^+_{G_0}$ the subset of positive roots (for choice of Borel subgroup being the upper triangular subgroup). Let $\boldsymbol{\Pi}_{G_0}$ be the set of simple roots. The notations $\boldsymbol{\Delta}_{G_0^\tau}$, $\boldsymbol{\Delta}^+_{G_0^\tau}$, and $\boldsymbol{\Pi}_{G_0^\tau}$ are clear. Let $P = R_{F/\mathbb{Q}}(P_0)$ be a parabolic subgroup of G as earlier, and we let $P_0^\tau := P_0 \times_\tau E$. The Weyl group factors as $W = \prod_{\tau:F\to E} W_0^\tau$, with each W_0^τ isomorphic to the permutation group $\mathfrak{S}_n$ on n-letters.

Let W^P be the set of Kostant representatives in the Weyl group W of G corresponding to the parabolic subgroup P which is defined as

$$W^P = \{w = (w^\tau) : w^\tau \in W_0^{\tau P_0^\tau}\},$$

where

$$W_0^{\tau P_0^\tau} := \{w^\tau \in W_0^\tau : (w^\tau)^{-1}\alpha > 0, \ \forall \alpha \in \boldsymbol{\Pi}_{M_{P_0^\tau}}\}.$$

(Here $\boldsymbol{\Pi}_{M_{P_0^\tau}} \subset \boldsymbol{\Pi}_{G_0^\tau}$ denotes the set of simple roots in the Levi quotient $M_{P_0^\tau}$ of P_0^τ.)

For $w \in W$, and in particular for $w \in W^P$, and for $\lambda \in X^*(T)$, by $w \cdot \lambda$ we mean the twisted action of w on the weight λ, i.e.,

$$w \cdot \lambda = (w^\tau \cdot \lambda^\tau)_{\tau:F\to E}, \qquad w^\tau \cdot \lambda^\tau = w^\tau(\lambda^\tau + \boldsymbol{\rho}_n) - \boldsymbol{\rho}_n.$$

Since $w \in W^P$, the weight $w \cdot \lambda$ is dominant and integral as a weight for $M_P \times E$ and we consider the associated irreducible finite-dimensional representation $\mathcal{M}_{w\cdot\lambda}$ of $M_P \times E$. Kostant's theorem asserts that as $M_P \times E$-modules, one has a multiplicity-free decomposition:

$$H^q(\mathfrak{u}_P, \mathcal{M}_{\lambda,E}) \simeq \bigoplus_{\substack{w\in W^P \\ l(w)=q}} \mathcal{M}_{w\cdot\lambda,E}. \tag{4.5}$$

The reader should bear in mind that the above result of Kostant can be parsed further over the set of embeddings $\tau : F \to E$. We have

$$\begin{aligned} H^q(\mathfrak{u}_P, \mathcal{M}_{\lambda,E}) &= \bigoplus_{\sum q_\tau = q} \bigotimes_{\tau:F\to E} H^{q_\tau}(\mathfrak{u}_{P_0^\tau}, \mathcal{M}_{\lambda^\tau,E}) \quad \text{(Künneth theorem)} \\ &= \bigoplus_{\sum q_\tau = q} \bigotimes_{\tau:F\to E} \bigoplus_{\substack{w^\tau \in W_0^{\tau} {}^{P_0^\tau} \\ l(w^\tau)=q_\tau}} \mathcal{M}_{w^\tau\cdot\lambda^\tau,E} \quad \text{(Kostant for each } \mathfrak{u}_{P_0^\tau}) \\ &= \bigoplus_{\sum q_\tau = q} \bigoplus_{\substack{w^\tau \in W_0^{\tau} {}^{P_0^\tau} \\ l(w^\tau)=q_\tau}} \bigotimes_{\tau:F\to E} \mathcal{M}_{w^\tau\cdot\lambda^\tau,E} \quad (\otimes \text{ commutes } \oplus) \\ &= \bigoplus_{\substack{w\in W^P \\ l(w)=q}} \mathcal{M}_{w\cdot\lambda,E} \quad (\text{since } w = (w^\tau) \text{ and } l(w) = \textstyle\sum_\tau l(w^\tau)). \end{aligned}$$

Applying (4.5) to the boundary cohomology as in Prop. 4.2, while using the description of the action of $\pi_0(P(\mathbb{R}))$ in Sect. 4.2.1, we get the following proposition.

Proposition 4.3. *The cohomology of $\partial_P \mathcal{S}^G$ is given by*

$$H^q(\partial_P \mathcal{S}^G, \widetilde{\mathcal{M}}_{\lambda,E}) = \bigoplus_{w\in W^P} {}^{\mathrm{a}}\mathrm{Ind}_{\pi_0(P(\mathbb{R}))\times P(\mathbb{A}_f)}^{\pi_0(G(\mathbb{R}))\times G(\mathbb{A}_f)} \left(H^{q-l(w)}(\mathcal{S}^{M_P}, \widetilde{\mathcal{M}}_{w\cdot\lambda,E})^{\pi_0(K_\infty^{M_P})} \right).$$

Remark 4.4. Suppose we have an irreducible $M_P(\mathbb{A}_f)$-module π_f such that for some $\lambda \in X^*(T)$ and some $w \in W^P$, we have $\pi_f \in \mathrm{Coh}(M_P, w\cdot\lambda)$. Let ε be a character of $\pi_0(M_P(\mathbb{R})) = \pi_0(P(\mathbb{R}))$. Then the induction to $\pi_0(G(\mathbb{R})) \times G(\mathbb{A}_f)$ of $\varepsilon \times \pi_f$ will contribute to $H^\bullet(\partial_P \mathcal{S}^G, \widetilde{\mathcal{M}}_{\lambda,E})$ if and only if ε is trivial on $\pi_0(K_\infty^{M_P})$. In other words, not all of the cohomology of the Levi contributes to boundary cohomology. This is a subtle point that will be relevant later in the monograph; see Sect. 5.3.5.

4.3 CUSPIDAL SPECTRUM VERSUS RESIDUAL SPECTRUM

We describe the contribution of the discrete but noncuspidal spectrum to cohomology. This means we formulate the consequences of the description of the discrete spectrum in Mœglin–Waldspurger for the square integrable cohomology, in a sense we make their results more explicit. We work at a transcendental level: our coefficient systems are $\mathbb{C}$-vector spaces.

We go back to the global situation. Take a parabolic (not necessarily maximal) subgroup $P \subset G$, and let M be its reductive quotient. Let $d(P)$ be the parabolic rank of P and let $\underline{\gamma}_P = \{\dots, \gamma_{P_i}, \dots\}$ be the array of dominant weight characters γ_{P_i} on the maximal parabolic subgroups $P_i \supset P$. We consider an isotypical σ_f in the cuspidal cohomology $H^\bullet_{\text{cusp}}(\mathcal{S}^M_{K^M_f}, \mathbb{D}_\mu \otimes \mathcal{M}_{w\cdot\lambda})$. Then we have the induced representation

$$ {}^{\mathrm{a}}\mathrm{Ind}_{P(\mathbb{A})}^{G(\mathbb{A})} \mathbb{D}_\mu \otimes V_{\sigma_f} \otimes |\underline{\gamma}|^{\underline{z}} \subset \mathcal{C}_\infty(P(\mathbb{Q})U(\mathbb{A})\backslash G(\mathbb{A})) $$

where $|\underline{\gamma}|^{\underline{z}} = \prod |\gamma_{iu}|^{z_i}$. We get the Eisenstein intertwining operator (Sect. 6.3.4)

$$ \mathrm{Eis}(\underline{z}, \mathbb{D}_\mu \otimes \sigma_f) \; : \; {}^{\mathrm{a}}\mathrm{Ind}_{P(\mathbb{A})}^{G(\mathbb{A})} \mathbb{D}_\mu \otimes V_{\sigma_f} \otimes |\underline{\gamma}|^{\underline{z}} \hookrightarrow \mathcal{C}_\infty(G(\mathbb{Q})\backslash G(\mathbb{A})). $$

We know that this series defines a holomorphic intertwining operator as long as if $\Re(z_i) \gg 0$ for all i, and it extends to a meromorphic function in the variable $\underline{z}$. We have to evaluate at $\underline{z} = 0$. In this section we are interested in the case that this operator is residual, which means it has simple poles along the lines $z_i = 0$. If $\mathrm{Eis}(\underline{z}, \mathbb{D}_\mu \otimes \sigma_f)$ is residual we can take the residue; i.e., we put

$$ \mathrm{Res}_{\underline{z}=0}\mathrm{Eis}(\mathbb{D}_\mu \otimes V_{\sigma_f}) \otimes |\underline{\gamma}|^{\underline{z}} \; = \; (\prod z_i)\mathrm{Eis}(\mathbb{D}_\mu \otimes V_{\sigma_f})|_{\underline{z}=0} $$

and then this defines an intertwining operator

$$ \begin{aligned} \mathrm{Res}_{\underline{z}=0}\mathrm{Eis}(\sigma \otimes \underline{z}) : {}^{\mathrm{a}}\mathrm{Ind}_{P(\mathbb{A})}^{G(\mathbb{A})} \mathbb{D}_\mu \otimes V_{\sigma_f} \\ \longrightarrow L^2_{\mathrm{disc}}(G(\mathbb{Q})\backslash G(\mathbb{A})/K_f, \omega^{-1}_{\mathcal{M}_\lambda}|_{S(\mathbb{R})^0}). \end{aligned} \tag{4.6} $$

In [52] it is proved that these residual classes only occur in a very specific situation, which we are going to describe. First of all the parabolic subgroup P is of a special type: We can write $N = uv$ and take the parabolic subgroup P_0 of G_0 containing the standard Borel subgroup, and with reductive quotient $M_0 = \mathrm{GL}_u \times \cdots \times \mathrm{GL}_u$, which is a v-fold product of copies of GL_u. Then $P = R_{F/\mathbb{Q}}(P_0)$ and $M = R_{F/\mathbb{Q}}(M_0)$. Using the Künneth formula we can write σ_f as $\sigma_f = \sigma_{1,f} \times \sigma_{2,f} \times \cdots \times \sigma_{v,f}$, where all the $\sigma_{i,f}$ occur in the cuspidal cohomology of GL_u; hence they may be compared. We say that σ_f is a *segment* if

$$ \sigma_{i+1,f} = \sigma_{i,f} \otimes |\det{}_u|_f. \tag{4.7} $$

It follows from the main result in [52] that σ_f provides a nonzero residual intertwining operator if and only if σ_f is a segment.

If σ_f is a segment and $H^\bullet_{\text{cusp}}(\mathcal{S}^M_{K^M_f}, \mathcal{M}_{w\cdot\lambda})(\sigma_f) \neq 0$ then this implies

$$\begin{aligned} w(\lambda+\rho)-\rho &= w\cdot\lambda = \mu \\ &= \mu_1^{(1)} - d_0\mathrm{det}_u^{(1)} + \mu_2^{(1)} - (d_0+1)\mathrm{det}_u^{(2)} + \dots \mu_v^{(1)} - (d_0+(v-1))\mathrm{det}_u^{(v)}, \end{aligned} \tag{4.8}$$

where all the $\mu_i^{(1)}$ are equal to $\mu^{(1)}$ and the τ component is given by

$$\mu^{(\tau,1)} = a_1^\tau \boldsymbol{\gamma}^M_{(i-1)u+1} + \cdots + a_{u-1}^\tau \boldsymbol{\gamma}^M_{iu-1} \text{ and } a_\nu^\tau = a_{u-\nu}^\tau.$$

We need some facts from the theory of induced Harish-Chandra modules (see [30]). Given μ we have to solve the system of equations (4.8) for the unknown highest weight λ and the Kostant represntative w. In [30] it is shown that we have a unique solution if and only if $a_\nu^\tau \geq v-1$ and then

$$\lambda = (a_1^\tau - v+1)\boldsymbol{\gamma}_v + \cdots + (a_{u-1}^\tau - v+1)\boldsymbol{\gamma}_{(u-1)v} + d\,\det.$$

Let w_P be the longest Kostant representative that sends all the roots in U_P into negative roots. Let $w' = w_P^{-1}w$ and $\mu' = w'(\lambda+\rho)-\rho$. Then we have the nonzero (local) intertwining operator

$$T^{(\mathrm{loc})}(\mathbb{D}_\mu) : {}^{\mathrm{a}}\mathrm{Ind}_{P(\mathbb{R})}^{G(\mathbb{R})}\mathbb{D}_\mu \longrightarrow {}^{\mathrm{a}}\mathrm{Ind}_{P(\mathbb{R})}^{G(\mathbb{R})}\mathbb{D}_{\mu'}.$$

The image of this operator is a unitary irreducible submodule J_{σ_∞}. We can identify this irreducible module with a module of type $A_{\mathfrak{q}}(\lambda)$ in [69]. The above highest weight λ is a character on the parabolic subgroup $P^\vee = R_{F/\mathbb{Q}}(P_0^\vee/F)$, and where the reductive quotient of $P_0^\vee$ is $M_0^\vee = \mathrm{GL}_v \times \mathrm{GL}_v \times \cdots \times \mathrm{GL}_v$. This parabolic subgroup can be interpreted as a Θ stable parabolic subgroup in the sense of [69] and hence together with λ provides a Harish-Chandra module $A_{\mathfrak{q}}(\lambda)$ that has nontrivial cohomology $H^\bullet(\mathfrak{g}, K_\infty, A_{\mathfrak{q}}(\lambda) \otimes \mathcal{M}_{\lambda,\mathbb{C}}) \neq 0$.

Theorem 4.5. *We get a diagram*

$$\begin{array}{ccc} {}^{\mathrm{a}}\mathrm{Ind}_{P(\mathbb{R})}^{G(\mathbb{R})}\mathbb{D}_\mu & \xrightarrow{T^{(\mathrm{loc})}(D_\mu)} & A_{\mathfrak{q}}(\lambda) \\ & & \downarrow \\ & & {}^{\mathrm{a}}\mathrm{Ind}_{P(\mathbb{R})}^{G(\mathbb{R})}\mathbb{D}_{\mu'} \end{array}$$

and we have the following:

1. *The horizontal arrow is surjective.*
2. *The vertical arrow is injective.*

3. The map induced by the vertical arrow in cohomology

$$H^q(\mathfrak{g}, K_\infty; A_\mathfrak{q}(\lambda) \otimes \mathcal{M}_\lambda) \longrightarrow H^q(\mathfrak{g}, K_\infty; {}^{\mathrm{a}}\mathrm{Ind}_{P(\mathbb{R})}^{G(\mathbb{R})}\mathbb{D}_{\mu'} \otimes \mathcal{M}_\lambda)$$

is an isomorphism in the lowest degree of nonzero cohomology; this lowest degree is

$$q_{min} = v\left[\frac{u^2}{4}\right] + \frac{N(u-1)(v-1)}{4}.$$

For a proof, see [30] and also [22].

The image of the intertwining operator

$$\mathrm{Res}_{\underline{z}=0}\mathrm{Eis}(\sigma \otimes \underline{z}) : {}^{\mathrm{a}}\mathrm{Ind}_{P(\mathbb{A})}^{G(\mathbb{A})}\mathbb{D}_\mu \otimes V_{\sigma_f} \to L^2_{\mathrm{disc}}(G(\mathbb{Q})\backslash G(\mathbb{A})/K_f, \omega^{-1}_{\mathcal{M}_\lambda}|_{S(\mathbb{R})^0})$$

is a nonzero submodule $J_{\sigma_\infty} \otimes J_{\sigma_f} \subset L^2_{\mathrm{disc}}(G(\mathbb{Q})\backslash G(\mathbb{A})/K_f, \omega^{-1}_{\mathcal{M}_\lambda}|_{S(\mathbb{R})^0})$. If we compose the intertwining operator with the constant Fourier coefficient $\mathcal{F}^P$ (see Sect. 6.3.5) then we get extension of our above diagram:

$$\begin{array}{ccc} {}^{\mathrm{a}}\mathrm{Ind}_{P(\mathbb{A})}^{G(\mathbb{A})}\mathbb{D}_\mu \otimes V_{\sigma_f} & \xrightarrow{T^{(\mathrm{loc})}(\sigma)} & J_{\sigma_\infty} \otimes J_{\sigma_f} \\ & & \downarrow \\ & & {}^{\mathrm{a}}\mathrm{Ind}_{P(\mathbb{A})}^{G(\mathbb{A})}\mathbb{D}_{\mu'} \otimes V_{\sigma'_f}. \end{array} \tag{4.9}$$

Since we know that

$$H^\bullet(\mathfrak{g}, K_\infty, {}^{\mathrm{a}}\mathrm{Ind}_{P(\mathbb{A})}^{G(\mathbb{A})}\mathbb{D}_{\mu'} \otimes \mathcal{M}_{\lambda,\mathbb{C}}) \otimes V_{\sigma'_f} \hookrightarrow H^\bullet(\partial_P\mathcal{S}^G_{K_f}, \mathcal{M}_{\lambda,\mathbb{C}}),$$

we conclude the following.

Corollary 4.6. *For the degree $q = q_{min}$ the homomorphism*

$$H^q(\mathfrak{g}, K_\infty, J_{\sigma_\infty} \otimes \mathcal{M}_{\lambda,\mathbb{C}}) \otimes J_{\sigma_f} \to H^q_{(2)}(\mathcal{S}^G_{K_f}, \mathcal{M}_{\lambda,\mathbb{C}})$$

is nonzero.

These results have also been proved by Grobner [22]. Observe that if $v \neq 1$, then $q_{min} < b_n^F$ and hence we have the following theorem.

Theorem 4.7. *(a) An isotypical subspace $H^\bullet_{(2)}(\mathcal{S}^G_{K_f}, \mathcal{M}_\lambda)(\pi_f)$ in the square integrable cohomology is not cuspidal if and only if there is a $q < b_n^F$ such that*

$$H^q_{(2)}(\mathcal{S}^G_{K_f}, \mathcal{M}_\lambda)(\pi_f) \longrightarrow H^q(\partial\mathcal{S}^G_{K_f}, \mathcal{M}_\lambda)$$

is nonzero.

(b) Equivalently,

$$\pi_f \text{ is not cuspidal} \iff H^\bullet_!(\mathcal{S}^G_{K_f}, \mathcal{M}_\lambda)(\pi_f) \neq H^\bullet_{(2)}(\mathcal{S}^G_{K_f}, \mathcal{M}_\lambda)(\pi_f).$$

In other words, a noncuspidal π_f cannot be "inner" in all degrees.

The above result, which we announced earlier in (3.12), can be sharpened as in the following theorem. During the induction argument in the proof we will use Thm. 4.7 for the reductive quotients M of the parabolic subgroups.

Theorem 4.8. *The isomorphism type π_f is cuspidal if and only if π_f does not occur as an isomorphism type in the Jordan–Hölder filtration of $H^\bullet(\partial\mathcal{S}^G_{K_f}, \widetilde{\mathcal{M}}_\lambda)$.*

Proof. If π_f does not occur in the Jordan–Hölder filtration of $H^\bullet(\partial\mathcal{S}^G_{K_f}, \widetilde{\mathcal{M}}_\lambda)$ then it is inner and by the previous theorem it is cuspidal. If π_f is cuspidal, then we restrict it to the central subalgebra $\mathcal{H}^{G,\mathsf{S}}$ (see (3.3.2)); we know that π_f is determined by this restriction π_f^{S}. If π_f occurs in the Jordan–Hölder filtration of $H^\bullet(\partial\mathcal{S}^G_{K_f}, \widetilde{\mathcal{M}}_\lambda)$ then an irreducible, hence one-dimensional, $V_{\pi_f^{\mathsf{S}}}$ occurs as a submodule in $H^\bullet(\partial\mathcal{S}^G_{K_f}, \widetilde{\mathcal{M}}_\lambda)(\pi_f^{\mathsf{S}})$. It follows from the spectral sequence considered in Sect. 4.1 that it occurs—for a suitable choice of q, M_P, w—as a submodule in a module induced from $H^{q-l(w)}(\mathcal{S}^{M_P}_{K_f^M}, \widetilde{\mathcal{M}}_{w\cdot\lambda,E})^{\pi_0(K_\infty^{M_P})}(\sigma_f^{\mathsf{S}})$.

We claim that σ_f cannot be cuspidal: since π_f is cuspidal, it cannot be a submodule of the finite part of a representation parabolically induced from a cuspidal representation σ_f of a proper parabolic subgroup; such an induced representation cannot be almost everywhere equivalent to a cuspidal representation by the well-known classification theorem of Jacquet and Shalika [39, Thm. 4.4].

But if σ_f is not cuspidal then for the cohomology of M_P, the restriction map

$$H^{\bullet-l(w)}(\mathcal{S}^{M_P}_{K_f^M}, \widetilde{\mathcal{M}}_{w\cdot\lambda,E})^{\pi_0(K_\infty^{M_P})}(\sigma_f) \to H^{\bullet-l(w)}(\partial\mathcal{S}^{M_P}_{K_f^M}, \widetilde{\mathcal{M}}_{w\cdot\lambda,E})^{\pi_0(K_\infty^{M_P})}$$

is nontrivial, by an obvious generalization of Thm. 4.7(b) to the case where G is a product $\mathrm{GL}_{n_1} \times \cdots \times \mathrm{GL}_{n_r}$. Hence, we get for a strictly smaller parabolic subgroup $Q \subset P$ and a suitable q', w' a nontrivial

$$H^{q'-l(w')}(\mathcal{S}^{M_Q}_{K_f^{M_Q}}, \widetilde{\mathcal{M}}_{w'\cdot w\cdot\lambda,E})^{\pi_0(K_\infty^{M_P})}(\tau_f^{\mathsf{S}})$$

such that π_f occurs in the module induced from τ_f. Again the theorem of Jacquet and Shalika implies that τ_f is not cuspidal. This cannot go on forever, and at some point τ_f must become cuspidal, contradicting the assumption that the cuspidal π_f occurs in the Jordan–Hölder filtration of $H^\bullet(\partial\mathcal{S}^G_{K_f}, \widetilde{\mathcal{M}}_\lambda)$. □

Chapter Five

The Strongly Inner Spectrum and Applications

5.1 THE STRONGLY INNER SPECTRUM: DEFINITION AND PROPERTIES

5.1.1 Identifying the cuspidal spectrum within the inner cohomology

We return to the arithmetic context; i.e., our coefficient systems are vector spaces over E. Consider the isotypical decomposition

$$H^\bullet_!(\mathcal{S}^G_{K_f}, \mathcal{M}_{\lambda,E}) = \bigoplus_{\pi_f \in \mathrm{Coh}_!(G,\lambda,K_f)} H^\bullet_!(\mathcal{S}^G_{K_f}, \mathcal{M}_{\lambda,E})(\pi_f).$$

For an embedding $\iota : E \to \mathbb{C}$, Thm. 4.7 and Thm. 4.8 say that the isomorphism type ${}^\iota\pi_f$ is cuspidal if and only if for all values of q we have the equality $H^q_!(\mathcal{S}^G_{K_f}, \widetilde{\mathcal{M}}_{{}^\iota\lambda})({}^\iota\pi_f) = H^q(\mathcal{S}^G_{K_f}, \mathcal{M}_{{}^\iota\lambda})({}^\iota\pi_f)$. This motivates us to define

$$\begin{aligned} &\mathrm{Coh}_{!!}(G,\lambda,K_f) \\ &\quad = \{\pi_f \in \mathrm{Coh}_!(G,\lambda,K_f) \mid H^\bullet_!(\mathcal{S}^G_{K_f}, \widetilde{\mathcal{M}}_{\lambda,E})(\pi_f) = H^\bullet(\mathcal{S}^G_{K_f}, \widetilde{\mathcal{M}}_{\lambda,E})(\pi_f)\}; \end{aligned}$$

i.e., every occurrence of π_f (as a module for the commutative algebra $\mathcal{H}^{G,\mathrm{S}}$) as a subquotient of full cohomology already occurs as a subquotient of inner cohomology; we then define the *strongly inner cohomology* as

$$H^\bullet_{!!}(\mathcal{S}^G_{K_f}, \widetilde{\mathcal{M}}_{\lambda,E}) := \bigoplus_{\pi_f \in \mathrm{Coh}_{!!}(G,\lambda,K_f)} H^\bullet(\mathcal{S}^G_{K_f}, \widetilde{\mathcal{M}}_{\lambda,E})(\pi_f). \tag{5.1}$$

Then we get an arithmetic characterization of the cuspidal cohomology

$$H^\bullet_{!!}(\mathcal{S}^G_{K_f}, \widetilde{\mathcal{M}}_{\lambda,E}) \otimes_{E,\iota} \mathbb{C} = H^\bullet_{\mathrm{cusp}}(\mathcal{S}^G_{K_f}, \widetilde{\mathcal{M}}_{{}^\iota\lambda,\mathbb{C}}). \tag{5.2}$$

5.1.2 Cuspidality of π_f in terms of estimates on the Satake parameters

Let $\pi_f \in \mathrm{Coh}_!(G, \lambda, K_f)$. Let S be a finite set of places including all the archimedean places such that for any $\mathfrak{p} \notin \mathsf{S}$, we have $K_\mathfrak{p} = \mathrm{GL}_n(\mathcal{O}_\mathfrak{p})$. Consider the local factor $\pi_\mathfrak{p}$ at $\mathfrak{p}$, which is an algebra homomorphism $\pi_\mathfrak{p} : \mathcal{H}_{K_\mathfrak{p}} \to E$ of the local Hecke algebra $\mathcal{H}_{K_\mathfrak{p}} = \mathcal{C}_c^\infty(\mathrm{GL}_n(F_\mathfrak{p})/\!/\mathrm{GL}_n(\mathcal{O}_\mathfrak{p}))$. Let $\varpi_\mathfrak{p}$ be a uniformizer at $\mathfrak{p}$ and $q_\mathfrak{p}$ the cardinality of the residue field $\mathcal{O}_\mathfrak{p}/\mathfrak{p}\mathcal{O}_\mathfrak{p}$. The Satake isomorphism says that $\pi_\mathfrak{p}$ is obtained from a homomorphism $\vartheta_\mathfrak{p} : T(F_\mathfrak{p}) \to \bar{E}^\times$, which is unique modulo the action of the Weyl group. Here $\bar{E}$ is an algebraic closure of E. For $1 \leq i \leq n$, we have the cocharacter $\eta_i : \mathbb{G}_m \to T_0$ given by the diagonal matrix $\eta_i(t) = \mathrm{diag}(1, 1, \ldots, t, \ldots, 1)$, where the t is in the i-th place. Then $\vartheta_\mathfrak{p}$ is determined by the values $\vartheta_{i,\mathfrak{p}} = \vartheta_\mathfrak{p}(\eta_i(\varpi_\mathfrak{p}))$. The *Satake parameter* of $\pi_\mathfrak{p}$ is the diagonal matrix $\mathrm{diag}(\vartheta_{1,\mathfrak{p}}, \ldots, \vartheta_{n,\mathfrak{p}})$ modulo conjugation by the Weyl group. The parameters $\vartheta_{i,\mathfrak{p}}$, for a fixed $\mathfrak{p}$, lie in some finite algebraic extension $\tilde{E}/E$ of E contained in $\bar{E}$. Cuspidality of π_f can be recognized by the size of the Satake parameters.

Proposition 5.1. *Let $\pi_f \in \mathrm{Coh}_!(G, \lambda, K_f)$ and $\iota : E \to \mathbb{C}$. The following are equivalent:*

1. *${}^\iota\pi_f$ is cuspidal.*
2. *For all $\mathfrak{p} \notin S$, any $\bar{\iota} : \bar{E} \to \mathbb{C}$ extending $\iota : E \to \mathbb{C}$, and any i we have*

$$q_\mathfrak{p}^{d-\frac{1}{2}} \ < \ |\bar{\iota}(\vartheta_{i,\mathfrak{p}})| \ < \ q_\mathfrak{p}^{d+\frac{1}{2}}.$$

Proof. (1) $\Longrightarrow$ (2) follows from the well-known estimates, due to Jacquet and Shalika [38, Cor. 2.5], for the Satake parameters of a *unitary* cuspidal automorphic representation of GL_n over a number field; note that ${}^\iota\pi_f \otimes |\ |^d$ would be the finite part of a unitary cuspidal representation. For (2) $\Longrightarrow$ (1), we prove the contrapositive statement as follows: Since $\pi_f \in \mathrm{Coh}_!$, we know that ${}^\iota\pi_f \in \mathrm{Coh}_{(2)}$; hence from the discussion in Sect. 3.2.3, ${}^\iota\pi_f \otimes |\ |^d$ is the finite part of an automorphic representation appearing in the discrete spectrum. Suppose ${}^\iota\pi_f$ is not cuspidal then by Mœglin and Waldspurger [52], there exists a, b with $n = ab$ and $b > 1$, a unitary cuspidal representation σ of $R_{F/\mathbb{Q}}(\mathrm{GL}_a/F)$ such that ${}^\iota\pi_f \otimes |\ |^d$ is the finite part of the unique irreducible quotient of the representation of $G(\mathbb{A})$ parabolically induced from $\sigma[(b-1)/2] \times \sigma[(b-3)/2] \times \cdots \times \sigma[-(b-1)/2]$. (Here, $\sigma[s] := \sigma \otimes |\det|^s$.) It's now an easy exercise to show that inequalities in (2) are violated. □

5.1.3 Cuspidal cohomology for GL_n

Consider $H^\bullet_?(\mathcal{S}^G_{K_f}, \widetilde{\mathcal{M}}_\lambda)$, where $? = \{c, !, \text{empty}, \partial\}$. Define $\mathrm{Coh}(H^\bullet_?(\mathcal{S}^G_{K_f}, \widetilde{\mathcal{M}}_\lambda))$ to be the set of isomorphism classes of absolutely irreducible $\mathcal{H}^G_{K_f}$-modules that occur in a Jordan–Hölder series; this definition makes sense even if the module is not semisimple. For an isomorphism class π_f let $\pi_f^{(S)}$ be the restriction to the central subalgebra $\mathcal{H}_{K_f}^{(S)}$. The various criteria for π_f to be cuspidal are collected together in the following theorem.

Theorem 5.2. *Let $\lambda \in X_0^*(T)$ be a pure weight and $\pi_f \in \mathrm{Coh}(H_!^\bullet(\mathcal{S}_{K_f}^G, \mathcal{M}_{\lambda,E})) = \mathrm{Coh}_!(G, \lambda, K_f)$. The following are equivalent:*

1. *There exists $\iota : E \to \mathbb{C}$ such that ${}^\iota\pi_f$ is cuspidal.*
2. *For all $\iota : E \to \mathbb{C}$, ${}^\iota\pi_f$ is cuspidal.*
3. *The isomorphism type $\pi_f^{(S)}$ does not occur in $\mathrm{Coh}(H^\bullet(\partial\mathcal{S}_{K_f}^G, \widetilde{\mathcal{M}}_{\lambda,E}))$.*
4. *We have $H^q(\mathcal{S}_{K_f}^G, \widetilde{\mathcal{M}}_{\lambda,E})(\pi_f) \neq (0)$ if and only if $b_n^F \leq q \leq \tilde{t}_n^F$.*
5. *There exists an $\iota : E \to \mathbb{C}$ such that $(\mathbb{D}_{{}^\iota\lambda} \otimes \varepsilon_\infty({}^\iota\pi_f)) \otimes {}^\iota\pi_f$ is automorphic; see (3.11).*
6. *For $\iota : E \to \mathbb{C}$ and any $\bar{\iota} : \bar{E} \to \mathbb{C}$ extending ι, the Satake parameters satisfy the estimates*
$$q_{\mathfrak{p}}^{d-\frac{1}{2}} \;<\; |\bar{\iota}(\vartheta_{i,\mathfrak{p}})| \;<\; q_{\mathfrak{p}}^{d+\frac{1}{2}}.$$

The above theorem has already been proved by all that was discussed before stating it; however, for the reader's benefit let's add a series of comments concerning the proof of the above theorem:

- (1) $\Longleftrightarrow$ (3): see Thm. 4.7 and Thm. 4.8.
- (1) $\Longrightarrow$ (5): see first paragraph of Sect. 3.1.4 and Prop. 3.3.
- (5) $\Longrightarrow$ (4): standard computation using Delorme's lemma; see Prop. 3.4.
- (4) $\Longrightarrow$ (1): see Thm. 4.7.
- (1) $\Longleftrightarrow$ (2): follows from (1) $\Longleftrightarrow$ (4).
  (This equivalence is a theorem of Clozel [14].)
- (2) $\Longleftrightarrow$ (6): see Prop. 5.1.

5.1.4 An arithmetic definition of Eisenstein cohomology

Let V be a finite-dimensional E-vector space that is also a module for a commutative $\mathbb{Q}$-algebra $\mathcal{A}$. Let us assume that all absolutely irreducible subquotients are one-dimensional (i.e., all eigenvalues lie in E). Define $\mathrm{Spec}_{\mathcal{A}}(V)$ to be the set of isomorphism classes of absolutely irreducible $\mathcal{A}$-modules over E that appear as irreducible subquotients of V. Let W be an $\mathcal{A}$-stable E-subspace of V such that
$$\mathrm{Spec}_{\mathcal{A}}(W) \cap \mathrm{Spec}_{\mathcal{A}}(V/W) = \emptyset.$$

Then there is an $\mathcal{A}$-equivariant projection $\pi_W : V \to W$; i.e., we have a splitting $V \simeq W \oplus V/W$ of $\mathcal{A}$-modules. We say that the summands have *disjoint sup-*

port with respect to $\mathcal{A}$, or simply that they have disjoint support with $\mathcal{A}$ being understood from the context at hand.

We apply this to the global cohomology and for $\mathcal{A}$ we take the central subalgebra $\mathcal{H}_{K_f}^{(S)}$ acting on $H^\bullet(\mathcal{S}^G_{K_f}, \widetilde{\mathcal{M}}_{\lambda,E})$. Then Thm. 4.8 gives that strongly inner cohomology splits off within global cohomology; i.e., we have a canonical decomposition with disjoint supports:

$$H^\bullet(\mathcal{S}^G_{K_f}, \widetilde{\mathcal{M}}_{\lambda,E}) \;=\; H^\bullet_{!!}(\mathcal{S}^G_{K_f}, \widetilde{\mathcal{M}}_{\lambda,E}) \oplus H^\bullet_{\mathrm{Eis}}(\mathcal{S}^G_{K_f}, \widetilde{\mathcal{M}}_{\lambda,E}), \tag{5.3}$$

which defines the Eisenstein cohomology denoted $H^\bullet_{\mathrm{Eis}}(\mathcal{S}^G_{K_f}, \widetilde{\mathcal{M}}_{\lambda,E})$. Such a decomposition, and a finer one at that, is also given by Franke and Schwermer [19]; but their decomposition is given only at a transcendental level.

5.2 DEFINITION OF THE RELATIVE PERIODS

In this section we will consider the group $G = R_{F/\mathbb{Q}}(\mathrm{GL}_n/F)$ only when n is an even positive integer. The purpose of this section then is to define and analyze certain relative periods $\Omega^\varepsilon({}^\iota\pi_f) \in \mathbb{C}^\times$, where $\lambda = (\lambda^\tau)_{\tau:F\to E}$ is a pure weight, $\pi_f \in \mathrm{Coh}_{!!}(G, K_f, \lambda)$ for some level structure K_f, $\iota : E \to \mathbb{C}$, and $\varepsilon = (\varepsilon_v)_{v\in \mathsf{S}_\infty}$ is a character of $\pi_0(G(\mathbb{R}))$.

5.2.1 The arithmetic identification

Consider inner cohomology in the bottom degree b_n^F, and take the field E large enough so that we have the decomposition in (3.9). Fix an E-linear isomorphism of $\mathcal{H}^G_{K_f}$-modules

$$T^\varepsilon_{\mathrm{arith}}(\lambda, \pi_f) \;:\; H^{b_n^F}_!(\mathcal{S}^G_{K_f}, \widetilde{\mathcal{M}}_{\lambda,E})(\pi_f \times \varepsilon) \;\longrightarrow\; H^{b_n^F}_!(\mathcal{S}^G_{K_f}, \widetilde{\mathcal{M}}_{\lambda,E})(\pi_f \times -\varepsilon),$$

which we call an arithmetic identification between these absolutely irreducible modules. The choice of $T^\varepsilon_{\mathrm{arith}}(\lambda, \pi_f)$ is well defined up to $E(\pi_f)^\times$-multiples. We may and will ask that these isomorphisms be compatibly chosen; i.e., for any $\iota : E \to E'$ we ask for the commutativity of

$$\begin{array}{ccc} H^{b_n^F}_!(\mathcal{S}^G_{K_f}, \widetilde{\mathcal{M}}_{\lambda,E})(\pi_f \times \varepsilon) & \xrightarrow{\;T^\varepsilon_{\mathrm{arith}}(\lambda,\pi_f)\;} & H^{b_n^F}_!(\mathcal{S}^G_{K_f}, \widetilde{\mathcal{M}}_{\lambda,E})(\pi_f \times -\varepsilon) \\ \Big\downarrow \iota^\bullet & & \Big\downarrow \iota^\bullet \\ H^{b_n^F}_!(\mathcal{S}^G_{K_f}, \widetilde{\mathcal{M}}_{{}^\iota\lambda,E'})({}^\iota\pi_f \times \varepsilon) & \xrightarrow{\;T^\varepsilon_{\mathrm{arith}}({}^\iota\lambda,{}^\iota\pi_f)\;} & H^{b_n^F}_!(\mathcal{S}^G_{K_f}, \widetilde{\mathcal{M}}_{{}^\iota\lambda,E'}))({}^\iota\pi_f \times -\varepsilon). \end{array} \tag{5.4}$$

We also want $T^\varepsilon_{\mathrm{arith}}(\lambda, \pi_f)^{-1} = T^{-\varepsilon}_{\mathrm{arith}}(\lambda, \pi_f)$.

5.2.2 The transcendental identification

We start from $\pi_f \in \mathrm{Coh}_{!!}(G, K_f, \lambda)$ and our aim is to construct—using the transcendental description of the cohomology—a family of isomorphisms:

$$T^{\varepsilon}_{\mathrm{trans}}({}^\iota\lambda, {}^\iota\pi_f) \;:\; H^{b_n^F}_{\mathrm{cusp}}(\mathcal{S}^G_{K_f}, \mathcal{M}_{{}^\iota\lambda,\mathbb{C}})({}^\iota\pi_f \times \varepsilon) \;\longrightarrow\; H^{b_n^F}_{\mathrm{cusp}}(\mathcal{S}^G_{K_f}, \mathcal{M}_{{}^\iota\lambda,\mathbb{C}})({}^\iota\pi_f \times -\varepsilon).$$

Fix an $\varepsilon_0 : \pi_0(G(\mathbb{R})) \to \{\pm 1\}$ and put $V_{\pi_f} = H^{b_n^F}_!(\mathcal{S}^G_{K_f}, \widetilde{\mathcal{M}}_{\lambda,E})(\pi_f \times \varepsilon_0)$. We do this simply because we want to realize the isomorphism type by an explicitly given module. It follows from Thm. 5.2 that for any $\iota : E \to \mathbb{C}$ there is a unique irreducible admissible $(\mathfrak{g}, K_\infty)$-module $\mathbb{D}_{{}^\iota\lambda}$ so that there is a nonzero intertwining operator

$$\Phi \in \mathrm{Hom}_{(\mathfrak{g},K_\infty)\otimes\mathcal{H}^G_{K_f}}\left(\mathbb{D}_{{}^\iota\lambda} \otimes V_{\pi_f} \otimes_{E,\iota} E'\right), \mathcal{C}^\infty_{\mathrm{cusp}}(G(\mathbb{Q})\backslash G(\mathbb{A})/K_f, \omega_\infty^{-1})).$$

(See Sect. 3.3.1.) Appealing to the multiplicity one result for cusp forms for GL_n/F we see that Φ is unique up to a scalar. This induces a map still called Φ on the relative Lie algebra complexes:

$$\begin{aligned}&\mathrm{Hom}_{K_\infty}(\Lambda^\bullet(\mathfrak{g}/\mathfrak{k}), \mathbb{D}_{{}^\iota\lambda} \otimes \mathcal{M}_{{}^\iota\lambda,\mathbb{C}} \otimes V_{\pi_f} \otimes_{E,\iota} E')\\ &\qquad \to \mathrm{Hom}_{K_\infty}(\Lambda^\bullet(\mathfrak{g}/\mathfrak{k}), \mathcal{C}^\infty_{\mathrm{cusp}}\left(G(\mathbb{Q})\backslash G(\mathbb{A})/K_f, \omega_\infty^{-1}\right) \otimes \mathcal{M}_{{}^\iota\lambda,\mathbb{C}})\end{aligned}$$

and hence a map

$$\Phi^\bullet : H^\bullet(\mathfrak{g}, K_\infty, \mathbb{D}_{{}^\iota\lambda} \otimes \mathcal{M}_{{}^\iota\lambda,\mathbb{C}}) \otimes V_{\pi_f} \otimes_{E,\iota} \mathbb{C} \;\to\; H^\bullet_!(\mathcal{S}^G_{K_f}, \mathcal{M}_{{}^\iota\lambda,\mathbb{C}}).$$

This map is an injection and in degree b_n^F it yields an isomorphism:

$$H^{b_n^F}(\mathfrak{g}, K_\infty, \mathbb{D}_{{}^\iota\lambda} \otimes \mathcal{M}_{{}^\iota\lambda,\mathbb{C}})) \otimes V_{\pi_f} \otimes_{E,\iota} \mathbb{C} \xrightarrow{\sim} H^{b_n^F}_!(\mathcal{S}^G_{K_f}, \mathcal{M}_{{}^\iota\lambda,\mathbb{C}})({}^\iota\pi_f). \tag{5.5}$$

We decompose the factor on the left into eigenspaces:

$$H^{b_n^F}(\mathfrak{g}, K_\infty, \mathbb{D}_{{}^\iota\lambda} \otimes \mathcal{M}_{{}^\iota\lambda,\mathbb{C}}) = \bigoplus_{\varepsilon:\pi_0(G(\mathbb{R}))\to\{\pm1\}} H^{b_n^F}(\mathfrak{g}, K_\infty, \mathbb{D}_{{}^\iota\lambda} \otimes \mathcal{M}_{{}^\iota\lambda,\mathbb{C}})(\varepsilon).$$

The spaces $H^{b_n^F}(\mathfrak{g}, K_\infty, \mathbb{D}_{{}^\iota\lambda} \otimes \mathcal{M}_{{}^\iota\lambda,\mathbb{C}})(\varepsilon)$ are one-dimensional. Our next step will be the construction of an *entangled* system of basis elements $\{w^\varepsilon({}^\iota\lambda)\}_\varepsilon$ for these one-dimensional spaces, and this system will be well defined up to a scalar in $\mathbb{C}^\times$.

THE CONSTRUCTION OF AN ENTANGLED SYSTEM $\{w^\varepsilon({}^\iota\lambda)\}_\varepsilon$.

We will now discuss the choice of basis $w^\varepsilon({}^\iota\lambda)$. Let ${}^\iota\lambda = ({}^\iota\lambda^\tau)_{\tau:F\to E} = (\lambda^\nu)_{\nu:F\to\mathbb{C}}$; here $\nu = \iota \circ \tau$. Since n is even, the restriction to $\mathrm{GL}_n(\mathbb{R})^0$ of $\mathbb{D}_{\lambda^\nu}$ breaks up as

$\mathbb{D}_{\lambda^\nu} = \mathbb{D}^+_{\lambda^\nu} \oplus \mathbb{D}^-_{\lambda^\nu}$, and the nontrivial element of $\pi_0(\mathrm{GL}_n(\mathbb{R}))$ switches these two summands. Implicit in the proof of Prop. 3.4 is the fact that cohomology in the bottom degree $H^{b_n}(\mathfrak{gl}_n, \mathrm{SO}(n)\mathbb{R}^\times_+; \mathbb{D}^+_{\lambda^\nu} \otimes \mathcal{M}_{\lambda^\nu})$ is one-dimensional, and we fix a basis for this space, say $w^+(\lambda^\nu)$; i.e.,

$$H^{b_n}(\mathfrak{gl}_n, \mathrm{SO}(n)\mathbb{R}^\times_+; \mathbb{D}^+_{\lambda^\nu} \otimes \mathcal{M}_{\lambda^\nu}) \;=\; \mathbb{C}\, w^+(\lambda^\nu).$$

Apply the Künneth theorem, to get that the element

$$w^{++}({}^\iota\lambda) \;:=\; \otimes_\nu\, w^+(\lambda^\nu)$$

generates the one-dimensional space $H^{b_n^F}(\mathfrak{g}, K^0_\infty; \mathbb{D}^{++}_{{}^\iota\lambda} \otimes \mathcal{M}_{{}^\iota\lambda})$, where $\mathbb{D}^{++}_{{}^\iota\lambda} = \otimes_\nu \mathbb{D}^+_{\lambda^\nu}$. For any character ε of $\pi_0(G_n(\mathbb{R}))$, let's define

$$w^\varepsilon({}^\iota\lambda) \;=\; \sum_{a \in \pi_0(G_n(\mathbb{R}))} \varepsilon(a) a \cdot w^{++}({}^\iota\lambda). \tag{5.6}$$

Then, clearly,

$$H^{b_n^F}\left(\mathfrak{g}, K^0_\infty; \mathbb{D}_{{}^\iota\lambda} \otimes \mathcal{M}_{{}^\iota\lambda,\mathbb{C}}\right)(\varepsilon) \;=\; \mathbb{C}\, w^\varepsilon({}^\iota\lambda).$$

If we change $w^{++}({}^\iota\lambda)$ by a scalar μ, then each $w^\varepsilon({}^\iota\lambda)$ is multiplied by the same scalar, and hence the entangled system $\{w^\varepsilon({}^\iota\lambda)\}_\varepsilon$ is well defined up to scalars.

Starting from this entangled system we get isomorphisms

$$\Psi(w^\varepsilon({}^\iota\lambda)) : V_{\pi_f} \otimes_{E,\iota} \mathbb{C} \xrightarrow{\sim} H_!^{b_n^F}(\mathcal{S}^G_{K_f}, \mathcal{M}_{{}^\iota\lambda,\mathbb{C}})({}^\iota\pi_f \times \varepsilon),$$

which are given by the composition $v \mapsto w^\varepsilon({}^\iota\lambda) \otimes v \mapsto \Phi^\bullet(w^\varepsilon({}^\iota\lambda) \otimes v)$. Then we define

$$T^\varepsilon_{\mathrm{trans}}({}^\iota\lambda, {}^\iota\pi_f, \underline{w}) \;:=\; i^{rn/2}\Psi(w^{-\varepsilon}({}^\iota\lambda)) \circ \Psi(w^\varepsilon({}^\iota\lambda))^{-1}. \tag{5.7}$$

It's clear that these operators $T^\varepsilon_{\mathrm{trans}}$ do not depend on the choices for Φ, $\{w^\varepsilon({}^\iota\lambda)\}$, or $w^+(\lambda^\nu)$. (The scaling factor of $i^{rn/2}$ ensures that the map induced by the standard intertwining operator in $(\mathfrak{g}, K^0_\infty)$-cohomology is defined over $\mathbb{Q}$; this will be relevant later in Chap. 8.)

We may orient our thoughts a bit differently for the transcendental identification. Consider the linear map of one-dimensional spaces:

$$T^\varepsilon({}^\iota\lambda) : \mathbb{C}w^\varepsilon({}^\iota\lambda) \;\longrightarrow\; \mathbb{C}w^{-\varepsilon}({}^\iota\lambda) \tag{5.8}$$

defined by

$$T^\varepsilon({}^\iota\lambda)(w^\varepsilon({}^\iota\lambda)) \;=\; i^{rn/2} w^{-\varepsilon}({}^\iota\lambda). \tag{5.9}$$

It is clear that $T^\varepsilon({}^\iota\lambda)$ is independent of the choice of basis $w^{++}({}^\iota\lambda)$ because, if we change $w^{++}({}^\iota\lambda)$ by a nonzero scalar, then from (5.6), the same scalar appears in both $w^{\pm\varepsilon}({}^\iota\lambda)$. Then the transcendental identification may be parsed as

$$\begin{array}{ccccc} & & V_{\pi_f} \otimes_{E,\iota} \mathbb{C}w^{\varepsilon}({}^\iota\lambda) & \longrightarrow & H_!^{b_n^F}(\mathcal{S}^G_{K_f}, \mathcal{M}_{{}^\iota\lambda,\mathbb{C}})({}^\iota\pi_f \times \varepsilon) \\ & \nearrow & & & \\ V_{\pi_f} \otimes_{E,\iota} \mathbb{C} & & \Big\downarrow {\scriptstyle 1\otimes T^{\varepsilon}({}^\iota\lambda)} & & \Big\downarrow {\scriptstyle T^{\varepsilon}_{\mathrm{trans}}({}^\iota\lambda, {}^\iota\pi_f)} \\ & \searrow & & & \\ & & V_{\pi_f} \otimes_{E,\iota} \mathbb{C}w^{-\varepsilon}({}^\iota\lambda) & \longrightarrow & H_!^{b_n^F}(\mathcal{S}^G_{K_f}, \mathcal{M}_{{}^\iota\lambda,\mathbb{C}})({}^\iota\pi_f \times -\varepsilon), \end{array}$$

and we may deduce, as before, that $T^{\varepsilon}_{\mathrm{trans}}({}^\iota\lambda, {}^\iota\pi_f)$ is independent of any choice of basis elements.

5.2.3 The relative periods

Definition 5.3. *There exists $\Omega^{\varepsilon}({}^\iota\lambda, {}^\iota\pi_f) \in \mathbb{C}^\times$ such that*

$$\Omega^{\varepsilon}({}^\iota\lambda, {}^\iota\pi_f)\, T^{\varepsilon}_{\mathrm{trans}}({}^\iota\lambda, {}^\iota\pi_f) \;=\; T^{\varepsilon}_{\mathrm{arith}}(\lambda, \pi_f) \otimes_{E,\iota} \mathbf{1}.$$

If we change the choice of $T^{\varepsilon}_{\mathrm{arith}}(\lambda, \pi_f)$ to $\alpha\, T^{\varepsilon}_{\mathrm{arith}}(\lambda, \pi_f)$ for an $\alpha \in E(\pi_f)^\times$ then $\Omega^{\varepsilon}({}^\iota\lambda, {}^\iota\pi_f)$ changes to $\iota(\alpha)\, \Omega^{\varepsilon}({}^\iota\lambda, {}^\iota\pi_f)$; i.e., we have an array of nonzero complex numbers

$$\{\dots, \Omega^{\varepsilon}({}^\iota\lambda, {}^\iota\pi_f), \dots\}_{\iota : E \to \mathbb{C}}$$

well defined up to multiplication by $E(\pi_f)^\times$.

In other words, we have defined a period $\Omega^{\varepsilon}(\lambda, \pi_f) \in (E \otimes \mathbb{C})^\times / E(\pi_f)^\times$. Sometimes, we suppress the λ in the notation, and will just write $\Omega^{\varepsilon}(\pi_f)$.

The reader is referred to Raghuram [54] to see how the relative periods $\Omega^{\varepsilon}({}^\iota\pi_f)$ are related to some other periods attached to ${}^\iota\pi_f$ that have played an important role in certain recent results on the special values of L-functions. To get these relations we have to replace the above choice of a "model space" V_{π_f} by its Whittaker realization $\mathcal{W}_{\pi_f}$, which is also an E-vector space. If $\mathcal{W}_{\mathbb{D}_{{}^\iota\lambda}}$ is the Whittaker model of $\mathbb{D}_{{}^\iota\lambda}$ then we have the canonical inclusion

$$\mathcal{W}_{\mathbb{D}_{{}^\iota\lambda}} \otimes \mathcal{W}_{\pi_f} \subset \mathcal{C}^{\infty}_{\mathrm{cusp}}(G(\mathbb{Q})\backslash G(\mathbb{A})/K_f, \omega_\infty^{-1})).$$

and therefore we get equalities

$$\begin{aligned} &H^{b_n^F}(\mathfrak{g}, K_\infty, \mathcal{W}_{\mathbb{D}_{{}^\iota\lambda}} \otimes \mathcal{M}_{{}^\iota\lambda,\mathbb{C}})(\varepsilon) \otimes (\mathcal{W}_{\pi_f} \otimes_{E,\iota} \mathbb{C}) \\ &\qquad\qquad = H^{b_n^F}(\mathcal{S}^G_{K_f}, \mathcal{M}_{\lambda,E})(\pi_f \times \epsilon) \otimes_{E,\iota} \otimes \mathbb{C}. \qquad (5.10) \end{aligned}$$

The comparison of the two E-structures on both sides of this equality yields the aforementioned periods $c({}^\iota\pi_f,\varepsilon)$. They are related to our relative periods as:

$$\Omega^\varepsilon({}^\iota\lambda,{}^\iota\pi_f) \;=\; \frac{c({}^\iota\pi_f,\varepsilon)}{c({}^\iota\pi_f,-\varepsilon)},$$

with equality to be construed as elements of $(E\otimes\mathbb{C})^\times/E^\times$.

5.2.4 Period relations under Tate twists

The periods $\Omega^\varepsilon({}^\iota\pi_f)$ have a simple behavior under Tate twists. For the moment, let n be any positive integer. Let $\lambda=(\lambda^\tau)_{\tau:F\to E}$ be a pure weight, where, as before, $\lambda^\tau=\sum_{i=1}^{n-1}(a_i^\tau-1)\boldsymbol{\gamma}_i+d\cdot\boldsymbol{\delta}_n=(b_1^\tau,\ldots,b_n^\tau)$. For $m\in\mathbb{Z}$, define a pure weight $\lambda+m\boldsymbol{\delta}_n:=(\lambda^\tau+m\boldsymbol{\delta}_n)_{\tau:F\to E}$, where

$$\lambda^\tau+m\boldsymbol{\delta}_n \;=\; \sum_{i=1}^{n-1}(a_i^\tau-1)\boldsymbol{\gamma}_i+(d+m)\cdot\boldsymbol{\delta}_n.$$

Also, let $m\delta_n$ stand for the weight $(m\boldsymbol{\delta}_n)_{\tau:F\to E}$. We have the fundamental class

$$e_{\delta_n}\;\in\; H^0(\mathcal{S}^G_{K_f},\mathbb{Q}[\delta_n])$$

(see [30]) and the cup product by this class yields an isomorphism

$$T^\bullet_{\mathrm{Tate}}(m): H^\bullet_?(\mathcal{S}^G_{K_f},\mathcal{M}_{\lambda,E}) \;\xrightarrow{\cup e^m_{\delta_n}}\; H^\bullet_?(\mathcal{S}^G_{K_f},\mathcal{M}_{\lambda+m\delta_n,E})$$

called the m-th Tate twist. If $\pi_f\in\mathrm{Coh}_?(G,\lambda,K_f)$ with $?\in\{!,!!\}$ then it is clear that

$$T^\bullet_{\mathrm{Tate}}(m)(\pi_f\times\varepsilon) \;=\; \pi_f(-m)\times(-1)^m\varepsilon, \tag{5.11}$$

where $\pi_f(-m):=\pi_f\otimes|\;|^{-m}$ stands for $\underline{g}_f\mapsto|\det(\underline{g}_f)|^{-m}\pi_f(\underline{g}_f)$, and ε is a character of $\pi_0(G(\mathbb{R}))$.

For later use, we introduce the notation $T^\varepsilon_{\mathrm{Tate}}(\lambda,\pi_f,m)$ for the m-th Tate twist on the module $\pi_f\times\varepsilon$ appearing in $\mathrm{Coh}_{!!}(G,K_f,\lambda)$.

Proposition 5.4. *For any integer m, if $\pi_f\in\mathrm{Coh}_{!!}(G,K_f,\lambda)$ then $\pi_f(m)\in\mathrm{Coh}_{!!}(G,K_f,\lambda-m\boldsymbol{\delta}_n)$. Assume now that n is even; then, for any ε we have*

$$\Omega^\varepsilon({}^\iota\pi_f(m)) \;=\; \Omega^{(-1)^m\varepsilon}({}^\iota\pi_f).$$

Proof. Follows from Thm. 5.2, Def. 5.3, and (5.11). □

We will say that the isomorphism types π_f and $\pi_f(m)$ are *conformally equivalent* and we will denote by $\{\pi_f\}=\{\ldots,\pi_f(m),\ldots\}_{m\in\mathbb{Z}}$ the conformal equivalence class of isomorphism classes. We say that an isomorphism type $\pi_f \in \mathrm{Coh}_{!!}(G, K_f, \lambda)$ is realized if we have picked—by a certain rule—an irreducible submodule $H(\pi_f) \subset H^{\bullet}_{!!}(\mathcal{S}^G_{K_f}, \mathcal{M}_{\lambda,E})$; such a rule may be given by the choice of a character ε, i.e., we may choose $H(\pi_f) = H^{\bullet}_{!!}(\mathcal{S}^G_{K_f}, \mathcal{M}_{\lambda,E})(\pi_f \times \varepsilon)$.

5.2.5 Period relations under twisting by Dirichlet characters

Let us record the behavior of the relative periods under twisting π_f by a finite-order character of the idèle class group of F. Suppose $\chi : F^\times \backslash \mathbb{A}_F^\times \to E^\times$ is a character of finite order. We let $\varepsilon_\chi = (\chi_v(-1))_{v \in \mathsf{S}_\infty}$ be the signature of χ. We have the following proposition.

Proposition 5.5. *If $\pi_f \in \mathrm{Coh}_{!!}(G, K_f, \lambda)$ then $\pi_f \otimes \chi_f \in \mathrm{Coh}_{!!}(G, K_f, \lambda)$. Assume now that n is even; then, for any ε we have*

$$\Omega^{\varepsilon}({}^{\iota}\pi_f \otimes {}^{\iota}\chi_f) \;=\; \Omega^{\varepsilon\varepsilon_\chi}({}^{\iota}\pi_f).$$

Proof. We will only briefly sketch the proof and leave the details to the reader. The character χ gives a degree zero class $\theta_\chi = \chi_f \otimes \varepsilon_\chi$ for GL_1 with constant coefficients. Pulling this back via the determinant $\delta : \mathrm{GL}_n \to \mathrm{GL}_1$, we get a class $\delta^*\theta_\chi$ in $H^0(\mathcal{S}^G_{K_f}, E)$. We may cup the class for σ_f with $\delta^*\theta_\chi$, or, using an embedding $\iota : E \to \mathbb{C}$, we may wedge the differential for ${}^{\iota}\sigma_f$ with ${}^{\iota}\delta^*\theta_\chi = \delta^*\theta_{{}^{\iota}\chi}$, and the key point is that both of these have the same effect. Furthermore, if we have two characters χ and χ', both of these operations are associative, i.e., cupping by $\theta_{\chi\chi'}$ is the same as first cupping with θ_χ followed by $\theta_{\chi'}$, etc.; i.e., we can define T_{trans} and T_{arith} in a compatible manner for the family of twists $\pi_f \otimes \chi_f$. From here on it is a routine exercise to see the rest of the proof of the period relation. (The reader should compare with the proof of the main theorem of [56].) ☐

We may derive the following corollary to Prop. 5.4 and Prop. 5.5: Suppose n is even, and let π_f be as above, and χ be an algebraic Hecke character of F with coefficients in E, then $\chi = |\ |^m \chi^\circ$ for $m \in \mathbb{Z}$ and χ° is a character of finite order with values in E. Let $\varepsilon_\chi = (-1)^m \varepsilon_{\chi^\circ}$ be the signature of χ. Then

$$\Omega^{\varepsilon}({}^{\iota}\pi_f \otimes {}^{\iota}\chi_f) \;=\; \Omega^{\varepsilon\varepsilon_\chi}({}^{\iota}\pi_f).$$

5.3 THE STRONGLY INNER COHOMOLOGY OF THE BOUNDARY

5.3.1 The restriction to the maximal strata

We have the restriction from the cohomology of the boundary to the cohomology of the maximal boundary strata

$$H^\bullet(\partial\mathcal{S}^G_{K_f},\widetilde{\mathcal{M}}_{\lambda,E}) \longrightarrow \bigoplus_{P:\,\mathrm{maximal}} H^\bullet(\partial_P\mathcal{S}^G_{K_f},\widetilde{\mathcal{M}}_{\lambda,E}) \tag{5.12}$$

and for the cohomology of each boundary strata $\partial_P\mathcal{S}^G_{K_f}$, after passing to the limit over all K_f, from Prop. 4.3 we have

$$H^\bullet(\partial_P\mathcal{S}^G,\widetilde{\mathcal{M}}_{\lambda,E}) = \bigoplus_{w\in W^P} {}^{\mathrm{a}}\mathrm{Ind}_{\pi_0(P(\mathbb{R}))\times P(\mathbb{A}_f)}^{\pi_0(G(\mathbb{R}))\times G(\mathbb{A}_f)} H^{\bullet-l(w)}(\mathcal{S}^{M_P},\widetilde{\mathcal{M}}_{w\cdot\lambda,E}).$$

Our previous result (5.3) applied to the quotients M_P yields a decomposition

$$H^{\bullet-l(w)}(\mathcal{S}^{M_P},\widetilde{\mathcal{M}}_{w\cdot\lambda,E}) = H^{\bullet-l(w)}_{!!}(\mathcal{S}^{M_P},\widetilde{\mathcal{M}}_{w\cdot\lambda,E}) \oplus H^{\bullet-l(w)}_{\mathrm{Eis}}(\mathcal{S}^{M_P},\widetilde{\mathcal{M}}_{w\cdot\lambda,E})$$

and therefore we get homomorphisms between Hecke modules:

$$\begin{aligned} r_{\max}: H^\bullet(\partial\mathcal{S}^G,\widetilde{\mathcal{M}}_{\lambda,E}) \longrightarrow & \bigoplus_P \bigoplus_{w\in W^P} \\ & \Big({}^{\mathrm{a}}\mathrm{Ind}_{\pi_0(P(\mathbb{R}))\times P(\mathbb{A}_f)}^{\pi_0(G(\mathbb{R}))\times G(\mathbb{A}_f)} \Big(H^{\bullet-l(w)}_{!!}(\mathcal{S}^{M_P},\widetilde{\mathcal{M}}_{w\cdot\lambda,E})\Big) \oplus \\ & \quad {}^{\mathrm{a}}\mathrm{Ind}_{\pi_0(P(\mathbb{R}))\times P(\mathbb{A}_f)}^{\pi_0(G(\mathbb{R}))\times G(\mathbb{A}_f)} \Big(H^{\bullet-l(w)}_{\mathrm{Eis}}(\mathcal{S}^{M_P},\widetilde{\mathcal{M}}_{w\cdot\lambda,E})\Big)\Big). \end{aligned} \tag{5.13}$$

On the right-hand side we project to the first set of summands to get

$$\begin{aligned} r_{max,!}: & H^\bullet(\partial\mathcal{S}^G,\widetilde{\mathcal{M}}_{\lambda,E}) \\ & \longrightarrow \bigoplus_P \bigoplus_{w\in W^P} {}^{\mathrm{a}}\mathrm{Ind}_{\pi_0(P(\mathbb{R}))\times P(\mathbb{A}_f)}^{\pi_0(G(\mathbb{R}))\times G(\mathbb{A}_f)} H^{\bullet-l(w)}_{!!}(\mathcal{S}^{M_P},\widetilde{\mathcal{M}}_{w\cdot\lambda,E}). \end{aligned}$$

This homomorphism is now surjective because the right-hand side is in the kernel of all differentials in the spectral sequence. Take K_f-invariants and if S is the finite set of places outside of which K_f is the full maximal compact, then we restrict the action of the Hecke algebra $\mathcal{H}^G_{K_f}$ to the commutative subalgebra $\mathcal{H}^{G,\mathsf{S}} = \otimes_{p\notin\mathsf{S}}\mathcal{H}^G_{K_p}$. Then, as in the proof of Thm. 4.8, once again applying

Jacquet and Shalika [39, Thm. 4.4], we have

$$\mathrm{Spec}_{\mathcal{H}^{G,\mathrm{s}}}\left(\bigoplus_{P}\bigoplus_{w\in W^P}\left({}^{\mathrm{a}}\mathrm{Ind}_{\pi_0(P(\mathbb{R}))\times P(\mathbb{A}_f)}^{\pi_0(G(\mathbb{R}))\times G(\mathbb{A}_f)} H_{!!}^{\bullet-l(w)}(\mathcal{S}^{M_P},\widetilde{\mathcal{M}}_{w\cdot\lambda,E})\right)^{K_f}\right)$$
$$\cap\ \mathrm{Spec}_{\mathcal{H}^{G,\mathrm{s}}}\left(\ker(r_{max,!})^{K_f}\right) = \emptyset. \quad (5.14)$$

5.3.2 An interlude on Kostant's representatives

Let the notations be as in Sect. 4.2.2, but take P_0 to be a maximal parabolic subgroup of G_0. Let $\mathbf{\Pi}_{M_{P_0}} = \mathbf{\Pi}_{G_0} - \{\alpha_{P_0}\}$. Let w_{P_0} be the unique element of $W_0 = W_{G_0}$ such that $w_{P_0}(\mathbf{\Pi}_{M_{P_0}}) \subset \mathbf{\Pi}_{G_0}$ and $w_{P_0}(\alpha_{P_0}) < 0$; it is the longest Kostant representative for W^{P_0}. Let Q_0 be the parabolic subgroup associated to P_0; we have

(i) $w_{P_0}(\mathbf{\Pi}_{M_{P_0}}) = \mathbf{\Pi}_{M_{Q_0}}$;
(ii) $w_{P_0}(\mathbf{\Delta}_{U_{P_0}}) = -\mathbf{\Delta}_{U_{Q_0}}$.
(Here $\mathbf{\Delta}_{U_{P_0}}$ is the set of those positive roots whose root spaces are in U_{P_0}; similarly, $\mathbf{\Delta}_{U_{Q_0}}$.)

For (ii), observe that if $\alpha \in \mathbf{\Delta}_{U_{P_0}}$ then in the expression for α in terms of simple roots, the root α_{P_0} has to appear with a positive integral coefficient, and since $w_{P_0}(\alpha_{P_0}) < 0$, there is a negative coefficient in the expression for $w_{P_0}(\alpha)$; but all coefficients have the same sign and hence $w_{P_0}(\alpha) < 0$.

Lemma 5.6. *With notations as above, we have:*

1. *The map $w \mapsto w' := w_P w$ gives a bijection $W^P \to W^Q$. If $w = (w^\tau)_{\tau:F\to E}$, then by definition, $w_P w = (w_{P_0} w^\tau)_{\tau:F\to E}$.*
2. *This bijection has the property that $l(w^\tau) + l(w'^\tau) = \dim(U_{P_0^\tau})$. Hence $l(w) + l(w') = \dim(U_P)$.*

Proof. The proof is the same for every component indexed by τ. We will suppress τ from the notation and just work over the group G_0/F and its subgroups. Let $w_0 \in W_0$. (In our simplified notation, this w_0 is any of the w^τ for the original w in (1) above.) To prove the first statement of the lemma:

$$\begin{aligned} w_{P_0} w_0 \in W^{Q_0} &\iff w_0^{-1} w_{P_0}^{-1}(\beta) > 0,\ \forall \beta \in \mathbf{\Pi}_{M_{Q_0}} \quad (\text{put } \beta = w_{P_0}\alpha;\ \text{see (i)}) \\ &\iff w_0^{-1}(\alpha) > 0,\ \forall \alpha \in \mathbf{\Pi}_{M_P} \\ &\iff w_0 \in W^{P_0}. \end{aligned}$$

To prove $l(w_0) + l(w_0') = \dim(U_{P_0})$, partition $\mathbf{\Delta}_{U_{P_0}}$ as a disjoint union

$\mathbf{\Delta}^+_{U_{P_0},w_0^{-1}} \cup \mathbf{\Delta}^-_{U_{P_0},w_0^{-1}}$, where

$$\mathbf{\Delta}^+_{U_{P_0},w_0^{-1}} := \{\alpha \in \mathbf{\Delta}_{U_{P_0}} : w_0^{-1}(\alpha) > 0\}$$

and

$$\mathbf{\Delta}^-_{U_{P_0},w_0^{-1}} := \{\alpha \in \mathbf{\Delta}_{U_{P_0}} : w_0^{-1}(\alpha) < 0\}.$$

Since $w_0 \in W^{P_0}$, it follows that

$$\{\alpha \in \mathbf{\Delta}^+_{G_0} : w_0^{-1}\alpha < 0\} \;=\; \{\alpha \in \mathbf{\Delta}_{U_{P_0}} : w_0^{-1}\alpha < 0\} \;=\; \mathbf{\Delta}^-_{U_{P_0},w_0^{-1}}.$$

Hence, $l(w_0) = l(w_0^{-1}) = |\mathbf{\Delta}^-_{U_P,w_0^{-1}}|$. Next, observe that the map $\alpha \mapsto -w_{P_0}\alpha$ gives a bijection $\mathbf{\Delta}^+_{U_{P_0},w_0^{-1}} \to \mathbf{\Delta}^-_{U_{Q_0},w_0'^{-1}}$ since

$$w_0'^{-1}(-w_{P_0}(\alpha)) = -w_0'^{-1}w_{P_0}(\alpha) = -w_0^{-1}\alpha < 0.$$

We have

$$\begin{aligned} \dim(U_{P_0}) = |\mathbf{\Delta}_{U_{P_0}}| &= |\mathbf{\Delta}^-_{U_{P_0},w_0^{-1}}| + |\mathbf{\Delta}^+_{U_{P_0},w_0^{-1}}| \\ &= |\mathbf{\Delta}^-_{U_{P_0},w_0^{-1}}| + |\mathbf{\Delta}^-_{U_{Q_0},w_0'^{-1}}| \\ &= l(w_0^{-1}) + l(w_0'^{-1}) = l(w_0) + l(w_0'). \end{aligned}$$

□

Similarly, we have the following self-bijection of W^P:

Lemma 5.7. *Let the notations be as in Lem. 5.6. Let w_G be the element of longest length in the Weyl group W_G of G, and similarly, let w_{M_P} be the element of longest length in the Weyl group W_{M_P} of the Levi quotient M_P of P. Then:*

1. *The map $w \mapsto w^\vee := w_{M_P}\, w\, w_G$ gives a bijection $W^P \to W^P$.*
2. *This bijection has the property that $l(w) + l(w^\vee) = \dim(U_P)$.*

Proof. Similar to the proof of Lem. 5.6. □

If $\mathcal{M}_\lambda$ is a highest weight module and $\mathcal{M}_{\lambda^\vee}$ is its dual, then we have a nondegenerate pairing

$$H^q(\mathfrak{u}_P, \mathcal{M}_{\lambda,E}) \;\times\; H^{d_U-q}(\mathfrak{u}_P, \mathcal{M}_{\lambda^\vee,E}) \;\longrightarrow\; E. \tag{5.15}$$

If we decompose the two cohomology modules according to (4.5) then the pairing becomes a direct sum over $w \in W^P$ of nondegenerate pairings

$$\mathcal{M}_{w\cdot\lambda,E} \;\times\; \mathcal{M}_{w^\vee\cdot\lambda^\vee,E} \;\longrightarrow\; E. \tag{5.16}$$

5.3.3 Further decomposition into isotypical pieces

We define

$$H^{\bullet}_{!!}(\partial\mathcal{S}^G, \widetilde{\mathcal{M}}_{\lambda,E}) = \bigoplus_{P:\,\text{maximal}} \bigoplus_{w\in W^P} {}^{\mathrm{a}}\mathrm{Ind}_{\pi_0(P(\mathbb{R}))\times P(\mathbb{A}_f)}^{\pi_0(G(\mathbb{R}))\times G(\mathbb{A}_f)} \left(H^{\bullet-l(w)}_{!!}(\mathcal{S}^{M_P}, \widetilde{\mathcal{M}}_{w\cdot\lambda,E})\right),$$

where each term decomposes further into isotypic components:

$$H^{\bullet-l(w)}_{!!}(\mathcal{S}^{M_P}, \widetilde{\mathcal{M}}_{w\cdot\lambda,E}) = \bigoplus_{\underline{\sigma}_f\in \mathrm{Coh}_{!!}(M_P, w\cdot\lambda)} H^{\bullet-l(w)}_{!!}(\mathcal{S}^{M_P}, \widetilde{\mathcal{M}}_{w\cdot\lambda,E})(\underline{\sigma}_f).$$

Given two maximal parabolic subgroups P and Q, with reductive quotients M_P and M_Q, respectively, we have to understand under what conditions we have nontrivial intertwining operators between the induced modules

$$\begin{aligned} &{}^{\mathrm{a}}\mathrm{Ind}_{\pi_0(P(\mathbb{R}))\times P(\mathbb{A}_f)}^{\pi_0(G(\mathbb{R}))\times G(\mathbb{A}_f)} H^{\bullet-l(w)}_{!!}(\mathcal{S}^{M_P}, \widetilde{\mathcal{M}}_{w\cdot\lambda,E})(\underline{\sigma}_f) \quad \text{and} \\ &\qquad {}^{\mathrm{a}}\mathrm{Ind}_{\pi_0(Q(\mathbb{R}))\times Q(\mathbb{A}_f)}^{\pi_0(G(\mathbb{R}))\times G(\mathbb{A}_f)} H^{\bullet-l(w_1)}_{!!}(\mathcal{S}^{M_Q}, \widetilde{\mathcal{M}}_{w_1\cdot\lambda,E})(\underline{\sigma}_{f,1}), \end{aligned} \tag{5.17}$$

where $w\in W^P, w_1\in W^Q$, $\underline{\sigma}_f \in \mathrm{Coh}_{!!}(M_P, w\cdot\lambda)$, and $\underline{\sigma}_{f,1}\in \mathrm{Coh}_{!!}(M_Q, w_1\cdot\lambda)$. It is again the theorem of Jacquet–Shalika that tells us that this will almost never be the case unless we are in a special situation which is summarized in the following proposition.

Proposition 5.8. *We have a nontrivial intertwining between two such modules as in (5.17) if and only if*

1. *$P=Q$ or P and Q are associate, and furthermore, if this condition is fulfilled, then*
2. *under the bijection $w\mapsto w'$ from $W^P\to W^Q$ (Lem. 5.6) which gives a bijection $\underline{\sigma}_f \to \underline{\sigma}_f^*$ between $\mathrm{Coh}_{!!}(M_P, w\cdot\lambda)$ and $\mathrm{Coh}_{!!}(M_Q, w'\cdot\lambda)$, we have $w_1 = w'$ and $\underline{\sigma}_{f,1} = \underline{\sigma}_f^*$.*

Therefore, strongly inner cohomology of the boundary has an isotypical decomposition

$$\begin{aligned} H^{\bullet}_{!!}(\partial\mathcal{S}^G, \widetilde{\mathcal{M}}_{\lambda,E}) = &\bigoplus_{P:\,\text{maximal}} \bigoplus_{w\in W^P} \bigoplus_{\underline{\sigma}_f\in\mathrm{Coh}_{!!}(M_P,w\cdot\lambda)} \\ &\left[{}^{\mathrm{a}}\mathrm{Ind}_{\pi_0(P(\mathbb{R}))\times P(\mathbb{A}_f)}^{\pi_0(G(\mathbb{R}))\times G(\mathbb{A}_f)} \left[H^{\bullet-l(w)}_{!!}(\mathcal{S}^{M_P}, \widetilde{\mathcal{M}}_{w\cdot\lambda,E})(\underline{\sigma}_f) \oplus \right.\right. \\ &\qquad \left.\left. {}^{\mathrm{a}}\mathrm{Ind}_{\pi_0(Q(\mathbb{R}))\times Q(\mathbb{A}_f)}^{\pi_0(G(\mathbb{R}))\times G(\mathbb{A}_f)} H^{\bullet-l(w')}_{!!}(\mathcal{S}^{M_Q}, \widetilde{\mathcal{M}}_{w'\cdot\lambda,E})(\underline{\sigma}_f^*)\right]\right]. \end{aligned} \tag{5.18}$$

Here isotypical means that two different summands of the form $[\dots]$ have disjoint supports (see Sect. 5.1.4).

Observe that in the above decomposition the cohomology in the two summands is possibly living in different degrees because of the shifts by $l(w)$ and $l(w')$. This motivates the following definition.

Definition 5.9. *An* $w = (w^\tau)_{\tau:F\to E} \in W$ *is called a* <u>*balanced*</u> *Kostant representative for* P *if* $w \in W^P$ *and*

$$l(w^\tau) \;=\; \dim(U_{P_0})/2, \quad \forall \tau : F \to E.$$

It follows from Lem. 5.6 that if w is a balanced Kostant representative for P then w' is also balanced as a Kostant representative for Q. An obvious necessary condition for the existence of balanced elements in W^P is that $\dim(U_{P_0}) = nn'$ is even. In this monograph we are concerned only with the contribution coming from such balanced Kostant representatives.

5.3.4 Interlude on induced representations

Let $\underline{\sigma}_f \in \mathrm{Coh}_{!!}(M_P, w\cdot\lambda)$ be as earlier. Write $M_P = G_n \times G_{n'}$, $\underline{\sigma}_f = \sigma_f \otimes \sigma'_f$ with $\sigma_f \in \mathrm{Coh}_{!!}(G_n, \mu)$ and $\sigma'_f \in \mathrm{Coh}_{!!}(G_{n'}, \mu')$ with pure weights μ and μ' such that $w\cdot\lambda = \mu + \mu'$. We will henceforth assume that μ and μ' are such that there is a balanced $w \in W^P$ for which $w^{-1}\cdot(\mu+\mu')$ is a dominant weight. This condition on the weights μ and μ' has many interesting and crucial consequences that are captured by the *combinatorial lemma*; see Sect. 7.2.3.

Consider now the associate parabolic subgroup Q with reductive quotient $M_Q = G_{n'} \times G_n$. The Kostant representative $w' \in W^Q$ given as in Lem. 5.6 is also balanced, and furthermore it is easy to see that

$$w'\cdot\lambda \;=\; w'\cdot(w^{-1}\cdot(\mu+\mu')) \;=\; (\mu' - n\boldsymbol{\delta}_{n'}) + (\mu + n'\boldsymbol{\delta}_n).$$

If $\sigma_f \in \mathrm{Coh}_{!!}(G_n, \mu)$ then $\sigma_f(-n') \in \mathrm{Coh}_{!!}(G_n, \mu + n'\boldsymbol{\delta}_n)$. Similarly, $\sigma'_f \in \mathrm{Coh}_{!!}(G_{n'}, \mu')$ implies $\sigma'_f(n) \in \mathrm{Coh}_{!!}(G_{n'}, \mu' - n\boldsymbol{\delta}_{n'})$. Here the module $\underline{\sigma}^*_f$ is nothing but $\sigma'_f(n) \times \sigma_f(-n')$. Consider the corresponding algebraically induced representations appearing as in (5.18):

$$ {}^{\mathrm{a}}\mathrm{Ind}_{P(\mathbb{A}_f)}^{G(\mathbb{A}_f)}\left(\sigma_f \otimes \sigma'_f\right) \quad \text{and} \quad {}^{\mathrm{a}}\mathrm{Ind}_{Q(\mathbb{A}_f)}^{G(\mathbb{A}_f)}\left(\sigma'_f(n) \otimes \sigma_f(-n')\right). \tag{5.19}$$

By Jacquet and Shalika [39, (4.3)], they are equivalent almost everywhere. Take any open-compact subgroup K_f, and an associated finite set of finite places S such that K_f is unramified outside S, and consider the K_f-invariants as an $\mathcal{H}^{G,\mathsf{S}}$-module. There is only one isomorphism type of simple $\mathcal{H}^{G,\mathsf{S}}$-module in the K_f-invariants of the induced representations in (5.19); denote this particular isomorphism type as $I^{\mathsf{S}}(\sigma_f, \sigma'_f)$.

Taking contragredients we have:

$$ {}^{\mathrm{a}}\mathrm{Ind}_{P(\mathbb{A}_f)}^{G(\mathbb{A}_f)} \left(\sigma_f \otimes \sigma_f'\right)^{\mathsf{v}} \stackrel{\sim}{\longrightarrow} {}^{\mathrm{a}}\mathrm{Ind}_{P(\mathbb{A}_f)}^{G(\mathbb{A}_f)} \left(\sigma_f{}^{\mathsf{v}}(n') \otimes (\sigma_f')^{\mathsf{v}}(-n)\right), \tag{5.20} $$

with $\sigma_f^{\mathsf{v}}(n') \in \mathrm{Coh}_{!!}(G_n, \mu^{\mathsf{v}} - n'\boldsymbol{\delta}_n)$ and $\sigma_f'^{\mathsf{v}}(-n) \in \mathrm{Coh}_{!!}(G_{n'}, \mu'^{\mathsf{v}} + n\boldsymbol{\delta}_{n'})$. As a notational artifice, given weights $\mu \in X^*(T_n)$ and $\mu' \in X^*(T_{n'})$, we will denote the weight $\mu + \mu' \in X^*(T_N)$ also as

$$ \mu + \mu' = \begin{bmatrix} \mu & \\ & \mu' \end{bmatrix}. $$

Note that the inducing data $\sigma_f^{\mathsf{v}}(n') \otimes \sigma_f'^{\mathsf{v}}(-n)$ for the contragredient representation considered earlier has cohomology with respect to the weight

$$ \begin{bmatrix} \mu^{\mathsf{v}} - n' & \\ & \mu'^{\mathsf{v}} + n \end{bmatrix}. $$

Lemma 5.10. *Let $\mu \in X_0^*(T_n)$ be a pure weight, and similarly, $\mu' \in X_0^*(T_n)$. Assume that there exists a balanced Kostant representative $w \in W^P$ such that $\lambda := w^{-1} \cdot (\mu + \mu')$ is dominant. Let $w^{\mathsf{v}} \in W^P$ be the Kostant representative associated to w given by Lem. 5.7. Then w^{v} is also balanced and furthermore*

$$ w^{\mathsf{v}} \cdot \lambda^{\mathsf{v}} = \begin{bmatrix} \mu^{\mathsf{v}} - n' & \\ & \mu'^{\mathsf{v}} + n \end{bmatrix}. $$

Proof. That w^{v} is balanced follows from Lem. 5.7. The rest of the proof may be parsed over the embeddings $\tau : F \to E$ and for each τ it is the same calculation, and so we just suppress τ from notation. Next, as a notational artifice, let w_N be the element of longest length in the Weyl group of $G_0 := \mathrm{GL}_N/F$. As a permutation matrix, we have $w_N(i,j) = \delta_{i,N-j+1}$. For $\tau : F \to E$, let w_N^τ be the same element but now thought of as the element of longest length in $G_0 \times_{F,\tau} E$. Then, $w_G = (w_N^\tau)_{\tau:F\to E}$. Similarly, representing M_{P_0} as block diagonal matrices permits us to think of w_{M_P} as an array indexed by τ with each entry being the block diagonal matrix: $w_{M_{P_0}} = \left(\begin{smallmatrix} w_n & \\ & w_{n'} \end{smallmatrix}\right)$, where $w_n(i,j) = \delta_{i,n-j+1}$, etc. Note that $-w_N\lambda = \lambda^{\mathsf{v}}$, $-w_n\mu = \mu^{\mathsf{v}}$ and $-w_{n'}\mu' = \mu'^{\mathsf{v}}$. We have:

$$\begin{aligned}
w^{\vee}\cdot\lambda^{\vee} &= \left(\begin{pmatrix} w_n & \\ & w_{n'}\end{pmatrix} w w_N\right)\cdot(-w_N\lambda)\\
&= \left(\begin{pmatrix} w_n & \\ & w_{n'}\end{pmatrix} w\right)\cdot(-\lambda-2\boldsymbol{\rho}_N)\\
&= \begin{pmatrix} w_n & \\ & w_{n'}\end{pmatrix}\cdot\left(\begin{bmatrix} -\mu & \\ & -\mu'\end{bmatrix}-2\boldsymbol{\rho}_N\right)\\
&= \begin{bmatrix} \mu^{\vee} & \\ & \mu'^{\vee}\end{bmatrix}-\begin{pmatrix} w_n & \\ & w_{n'}\end{pmatrix}\boldsymbol{\rho}_N-\boldsymbol{\rho}_N\\
&= \begin{bmatrix} \mu^{\vee}-n' & \\ & \mu'^{\vee}+n\end{bmatrix}.
\end{aligned}$$

□

Again, there is only one isomorphism type of simple $\mathcal{H}^{G,\mathsf{S}}$-module in the K_f-invariants of the induced representation in (5.20).

5.3.5 The character of the component group II

The assumption of wanting a balanced Kostant representative necessitates nn' to be even. Without loss of generality, we take n to be even and let n' be any positive integer. Given $\sigma_f \in \mathrm{Coh}_{!!}(G_n,\mu)$ and $\sigma'_f \in \mathrm{Coh}_{!!}(G_{n'},\mu')$ we begin by taking a character ε' of $\pi_0(G_{n'}(\mathbb{R}))$ as

$$\varepsilon' = \begin{cases} \varepsilon(\sigma'_f) & \text{if } n' \text{ is odd, and} \\ \text{any character of } \pi_0(G_{n'}(\mathbb{R})) & \text{if } n' \text{ is even,} \end{cases} \tag{5.21}$$

where $\varepsilon(\sigma'_f)$ is as in (3.10). Note that $\varepsilon(\sigma'_f(n)) = \varepsilon(\sigma'_f)$ since n is an even integer. Now take any character ε of $\pi_0(G_n(\mathbb{R}))$. From the short exact sequence of the group of connected components in (4.4) it follows that

$$\text{the character } \varepsilon\times\varepsilon' \text{ of } \pi_0(P(\mathbb{R})) \text{ is trivial on } \pi_0(K_\infty^{M_P}) \iff \varepsilon=\varepsilon'.$$

Hence $(\varepsilon'\times\sigma_f)\times(\varepsilon'\times\sigma'_f) \hookrightarrow H^{\bullet}_{!!}(\mathcal{S}^{M_P},\widetilde{\mathcal{M}}_{\mu+\mu',E})^{\pi_0(K_\infty^{M_P})}$. The character $\varepsilon'\times\varepsilon'$ of $\pi_0(P(\mathbb{R}))$ maps to ε' on $\pi_0(G(\mathbb{R}))$. From Sect. 4.2.1 it follows that

$${}^{\mathrm{a}}\mathrm{Ind}_{\pi_0(P(\mathbb{R}))\times P(\mathbb{A}_f)}^{\pi_0(G(\mathbb{R}))\times G(\mathbb{A}_f)}\left((\varepsilon'\otimes\sigma_f)\otimes(\varepsilon'\otimes\sigma'_f)\right) = \varepsilon'\otimes{}^{\mathrm{a}}\mathrm{Ind}_{P(\mathbb{A}_f)}^{G(\mathbb{A}_f)}\left(\sigma_f\otimes\sigma'_f\right).$$

Proposition 5.11. *We assume that N is odd. Then the Hecke module $\underline{\sigma}_f = \sigma_f\otimes\sigma'_f$ occurs with multiplicity at most one in $H_{!!}^{b_n^F+b_{n'}^F}(\mathcal{S}^{M_P},\widetilde{\mathcal{M}}_{w\cdot\lambda,E})^{\pi_0(K_\infty^{M_P})}$.*

Proof. Assume n is even and n' is odd. Take pure weights μ and μ', and write $\mu + \mu' =: \underline{\mu}$. We abbreviate

$$H^b_P(\underline{\sigma}_f) \ := \ H^{b^F_n + b^F_{n'}}_{!!}(\mathcal{S}^{M_P}, \widetilde{\mathcal{M}}_{\underline{\mu},E})^{\pi_0(K^{M_P}_\infty)}(\underline{\sigma}_f). \tag{5.22}$$

Then we get a decomposition into irreducible Hecke modules

$$H^{b^F_n + b^F_{n'}}_{!!}(\mathcal{S}^{M_P}, \widetilde{\mathcal{M}}_{\underline{\mu},E})(\underline{\sigma}_f) = \bigoplus_{\varepsilon} H^{b^F_n}_{!!}(\mathcal{S}^{G_n}, \mathcal{M}_\mu)(\sigma_f \times \varepsilon) \otimes H^{b^F_{n'}}_{!!}(\mathcal{S}^{G_{n'}}, \mathcal{M}_{\mu'})(\sigma'_f)$$

and the submodule $H^b_P(\underline{\sigma}_f)$ is the (irreducible) summand with $\varepsilon = \varepsilon(\sigma'_f)$. (See Sect. 3.3.3.) Hence we see that $H^b_P(\underline{\sigma}_f)$ realizes $\underline{\sigma}_f$. □

5.3.6 Interlude on arithmetic identifications

Following Sect. 5.2.4, we call two isomorphism types $\underline{\sigma}_f$ and $\underline{\sigma_1}_f$ to be conformally equivalent if there is a $\delta = d\delta_n + d'\delta_{n'}$ such that $\underline{\sigma}_f \otimes |\delta_f|^{-1} = \underline{\sigma_1}_f$. For any such δ we have the canonical generator $e_\delta \in H^0(\mathcal{S}^{M_P}, \mathbb{Q}[\delta])$, and the cup product yields an isomorphism

$$T^\bullet_{\text{Tate}}(\delta) \ = \ \cup e_\delta : \ H^{b^F_n + b^F_{n'}}_{!!}(\mathcal{S}^{M_P}, \widetilde{\mathcal{M}}_{\underline{\mu},E})(\underline{\sigma}_f) \xrightarrow{\sim} H^{b^F_n + b^F_{n'}}_{!!}(\mathcal{S}^{M_P}, \widetilde{\mathcal{M}}_{\underline{\mu}+\delta,E})(\underline{\sigma}_f \otimes |\delta_f|^{-1}).$$

We may use these identifications and form the direct limit

$$H^{b^F_n + b^F_{n'}}_{!!}(\mathcal{S}^{M_P}, \widetilde{\mathcal{M}}_{\underline{\mu}+,E})(\{\underline{\sigma}_f\}) \ := \ \varinjlim_{\delta} H^{b^F_n + b^F_{n'}}_{!!}(\mathcal{S}^{M_P}, \widetilde{\mathcal{M}}_{\underline{\mu}+\delta,E})(\underline{\sigma}_f \otimes |\delta_f|^{-1}\}).$$

This is a vector space that has a distinguished isomorphism to any member of the family

$$\{H^{b^F_n + b^F_{n'}}_{!!}(\mathcal{S}^{M_P}, \widetilde{\mathcal{M}}_{\underline{\mu}+\delta,E})(\underline{\sigma}_f \otimes |\delta_f|^{-1})\}_\delta.$$

Now we notice that $T^\bullet_{\text{Tate}}(\delta)$ gives an isomorphism from

$$H^{b^F_n}_P(\mathcal{S}^{G_n}, \mathcal{M}_\mu)(\sigma_f \times \epsilon) \otimes H^{b^F_{n'}}_P(\mathcal{S}^{G_{n'}}, \mathcal{M}_{\mu'})(\sigma'_f)$$

to

$$H^{b^F_n}_P(\mathcal{S}^{G_n}, \mathcal{M}_{\mu+d\delta_n})(\sigma_f \otimes |\delta_{n,f}|^{-d} \times (-1)^d \epsilon) \otimes H^{b^F_{n'}}_P(\mathcal{S}^{G_{n'}}, \mathcal{M}_{\mu'+d'\delta_{n'}})(\sigma'_f \otimes |\delta_{n',f}|^{-d'}).$$

Since $\epsilon(\sigma'_f \otimes |\delta_{n',f}|^{-d'}) = \epsilon(\sigma'_f)(-1)^{d'}$ we get

$$T^{\bullet}_{\text{Tate}}(\delta) : H^b_P(\underline{\sigma}_f) \xrightarrow{\sim} H^b_P(\underline{\sigma}_f \otimes |\delta_f|^{-1}) \iff d(\delta) = d - d' \equiv 0 \mod 2. \quad (5.23)$$

This suggests that we divide the family of weights $\{\underline{\mu}+\delta\}$ into two classes $\{+,-\}$ according to the value of $d(\delta)$ mod 2. We can define the two limits

$$H^{b+}_P(\{\underline{\sigma}_f\}) := \varinjlim_{\underline{\mu}+\delta:d(\delta)\equiv 0(2)} H^b_P(\underline{\sigma}_f \otimes |\delta_f|^{-1}),$$

$$H^{b-}_P(\{\underline{\sigma}_f\}) := \varinjlim_{\underline{\mu}+\delta:d(\delta)\equiv 1(2)} H^b_P(\underline{\sigma}_f \otimes |\delta_f|^{-1}).$$

Finally, we can construct an isomorphism

$$T_{\text{arith}}(\{\underline{\sigma}_f\}, \pm) : H^{b+}_P(\{\underline{\sigma}_f\}) \xrightarrow{\sim} H^{b-}_P(\{\underline{\sigma}_f\}). \quad (5.24)$$

To do so we go back to our $\underline{\sigma}_f = \sigma_f \otimes \sigma'_f$ and the description

$$H^{b+}_P(\{\underline{\sigma}_f\}) = H^{b^F_n}_P(\mathcal{S}^{G_n}, \mathcal{M}_\mu)(\sigma_f \times \epsilon(\sigma'_f)) \otimes H^{b^F_{n'}}_P(\mathcal{S}^{G_{n'}}, \mathcal{M}_{\mu'})(\sigma'_f).$$

In Sect. 5.2.1 we have chosen an isomorphism between irreducible Hecke modules

$$T^{\varepsilon}_{\text{arith}}(\mu, \sigma_f) : H^{b^F_n}_P(\mathcal{S}^{G_n}, \mathcal{M}_\mu)(\sigma_f \times \epsilon(\sigma'_f)) \xrightarrow{\sim} H^{b^F_n}_P(\mathcal{S}^{G_n}, \mathcal{M}_\mu)(\sigma_f \times -\epsilon(\sigma'_f)), \quad (5.25)$$

which is unique up to an element of $E^\times$. We take $\delta_1 = \delta_{n'}$ and then

$$T^{\varepsilon}_{\text{arith}}(\lambda, \sigma_f) \otimes T^{\bullet}_{\text{Tate}}(\delta_1) : H^b_P(\underline{\sigma}_f) \xrightarrow{\sim} H^b_P(\underline{\sigma}_f \otimes |\delta_{1,f}|^{-1}), \quad (5.26)$$

and this induces our searched for $T_{\text{arith}}(\{\underline{\sigma}_f\}, \pm)$.

If we are in the case that n, n' are both even then

$$\bigoplus_{\varepsilon} H^{b^F_n}_{!!}(\mathcal{S}^{G_n}, \mathcal{M}_\mu)(\sigma_f \times \varepsilon) \otimes H^{b^F_{n'}}_{!!}(\mathcal{S}^{G_{n'}}, \mathcal{M}_{\mu'})(\sigma'_f \times \varepsilon)$$
$$= H^b_P(\underline{\sigma}_f) = \bigoplus_{\varepsilon} H^b_P(\underline{\sigma}_f \times \varepsilon).$$

It will be important that $T^{\bullet}_{\text{Tate}}(\delta_Q)$ maps $H^b_P(\underline{\sigma}'_f \times \varepsilon)$ to $H^b_P(\underline{\sigma}'_f \otimes |\delta_Q|_f \times \varepsilon)$.

5.3.7 A summary of our notation with some simplifications

The ambient group will be $G = R_{F/\mathbb{Q}}(\mathrm{GL}_N/F)$. Let n and n' be positive integers so that $N = n+n'$. Let $G_0 = \mathrm{GL}_N/F$ and let P_0 be the standard (i.e., containing the standard Borel subgroup) maximal parabolic subgroup of G_0 of type (n, n'),

U_{P_0} its unipotent radical, and $M_{P_0} = P_0/U_{P_0} = \mathrm{GL}_n \times \mathrm{GL}_{n'}/F$ is the Levi quotient. Let Q_0 be the standard maximal parabolic subgroup of G_0 of type (n', n), and U_{Q_0} and M_{Q_0} are similarly defined. Let G, P, U_P, M_P, Q, U_Q, and M_Q be the restriction of scalars from F to $\mathbb{Q}$ of G_0, P_0, U_{P_0}, M_{P_0}, Q_0, U_{Q_0}, and M_{Q_0}, respectively. If $n = n'$ then $Q_0 = P_0$, $Q = P$, etc.; we will call this the self-associate case. We identify M_P and M_Q as in the introduction.

Let $\mu \in X_0^*(T_n)$ be a pure weight, and similarly, $\mu' \in X_0^*(T_{n'})$. Assume there exists a balanced Kostant representative $w \in W^P$ such that $\lambda := w^{-1} \cdot (\mu + \mu')$ is dominant. So nn' is even, and without loss of generality, we take n to be even, and n' may be even or odd. (See Sect. 7.4.1 for the case nn' is odd.) Let $w^{\vee} \in W^P$ correspond to w as in Lem. 5.7. Let w' (resp., $w^{\vee\prime}$) be in W^Q corresponding to w (resp., $w^{\vee}$) via Lem. 5.6. All the elements $w, w', w^{\vee}$, and $w^{\vee\prime}$ are balanced; i.e., their lengths are $dim(U_P)/2 = dim(U_Q)/2$. In the self-associate case ($n = n'$, $P = Q$), we note that $w' \neq w$ and $w^{\vee} \neq w^{\vee\prime}$. Write $\mu = \mu^{(1)} + \mu_{\mathrm{ab}}$ with $\mu_{\mathrm{ab}} = d\delta_n$, and similarly, $\mu = \mu'^{(1)} + \mu'_{\mathrm{ab}}$ with $\mu'_{\mathrm{ab}} = d'\delta_{n'}$. Write $\underline{\mu} = \mu + \mu'$, and put

$$a(\underline{\mu}) = a(\mu, \mu') = d - d'.$$

If $d(\underline{\mu}) = (nd + n'd')/N$ then note that $\underline{\mu} = \mu^{(1)} + \mu'^{(1)} + a(\underline{\mu})\gamma_P + d(\underline{\mu})\delta_N$ (see Sect. 6.3.1 for γ_P); the quantity $a(\underline{\mu})$ plays an important role for much that follows; however, $d(\underline{\mu})$ is uninteresting.

Let $\sigma_f \in \mathrm{Coh}_{!!}(G_n, \mu)$ and $\sigma'_f \in \mathrm{Coh}_{!!}(G_{n'}, \mu')$. Take a large enough finite set of finite places S containing all the places where either σ_f or σ'_f is ramified. Take $K_f = \prod_p K_p$ an open-compact subgroup of $G(\mathbb{A}_f)$ such that $w_P K_f w_P^{-1} = K_f$ and $K_{\mathfrak{p}} = \mathrm{GL}_N(\mathcal{O}_{\mathfrak{p}})$ for and $\mathfrak{p}|p$ with $p \notin \mathsf{S}$. By $\mathcal{H}^{G,\mathsf{S}}$ we mean the product of local spherical Hecke algebras outside of S.

Take ε' as in (5.21), and for brevity, we also denote $\tilde{\varepsilon}'$ for the character $\varepsilon' \otimes \varepsilon'$ of $\pi_0(M_P(\mathbb{R}))$ or of $\pi_0(M_Q(\mathbb{R}))$. In the bottom degree, we observe

$$b_n^F + b_{n'}^F + \tfrac{1}{2}\dim(U_P) = b_N^F,$$

and write $I_b^{\mathsf{S}}(\sigma_f, \sigma'_f, \varepsilon')_{P,w}$ for

$$^{\mathrm{a}}\mathrm{Ind}_{\pi_0(P(\mathbb{R}))\times P(\mathbb{A}_f)}^{\pi_0(G(\mathbb{R}))\times G(\mathbb{A}_f)} \left(H_{!!}^{b_n^F + b_{n'}^F}(\mathcal{S}^{M_P}, \widetilde{\mathcal{M}}_{w\cdot\lambda,E})^{\pi_0(K_\infty^{M_P})}(\tilde{\varepsilon}' \otimes (\sigma_f \otimes \sigma'_f)) \right)^{K_f}.$$

Note that the inducing data is $H_P^b(\underline{\sigma}_f)$ as in (5.22) when n' is odd; however, if n' is even, we have a choice of ε'; nevertheless, the reader should bear in mind that the explicit model for $\varepsilon' \otimes \sigma_f$ or $\varepsilon' \otimes \sigma'_f$ is as it appears (with multiplicity one) in strongly inner cohomology in the bottom degree. Similarly, $I_b^{\mathsf{S}}(\sigma'_f(n), \sigma_f(-n'), \varepsilon')_{Q,w'}$ denotes

$$^{\mathrm{a}}\mathrm{Ind}_{\pi_0(Q(\mathbb{R}))\times Q(\mathbb{A}_f)}^{\pi_0(G(\mathbb{R}))\times G(\mathbb{A}_f)} \left(H_{!!}^{b_n^F + b_{n'}^F}(\mathcal{S}^{M_Q}, \widetilde{\mathcal{M}}_{w'\cdot\lambda,E})^{\pi_0(K_\infty^{M_Q})}(\tilde{\varepsilon}' \otimes (\sigma'_f(n) \otimes \sigma_f(-n'))) \right)^{K_f}.$$

In the top degree we work with the contragredient modules. The module $\sigma_f^{\vee}(n') \otimes \sigma_f'^{\vee}(-n)$ is strongly inner for M_P with respect to the weight $w^{\vee} \cdot \lambda^{\vee}$. Likewise, $\sigma_f^{\vee\prime} \otimes \sigma_f^{\vee}$ is strongly inner for M_Q for the weight $w^{\vee\prime} \cdot \lambda^{\vee}$. We also have

$$\tilde{t}_n^F + \tilde{t}_{n'}^F + \tfrac{1}{2}\dim(U_P) = \tilde{t}_N^F - 1.$$

In this case $I_t^{\mathsf{S}}(\sigma_f^{\vee}(n'), \sigma_f'^{\vee}(-n), \varepsilon')_{P,w^{\vee}}$ denotes

$${}^{\mathrm{a}}\mathrm{Ind}_{\pi_0(P(\mathbb{R}))\times P(\mathbb{A}_f)}^{\pi_0(G(\mathbb{R}))\times G(\mathbb{A}_f)} \left(H_{!!}^{\tilde{t}_n^F + \tilde{t}_{n'}^F}(\mathcal{S}^{M_P}, \widetilde{\mathcal{M}}_{w^{\vee}\cdot\lambda^{\vee},E})^{\pi_0(K_\infty^{M_P})}(\tilde{\varepsilon}' \otimes (\sigma_f^{\vee}(n') \otimes \sigma_f'^{\vee}(-n)))\right)^{K_f},$$

and similarly $I_t^{\mathsf{S}}(\sigma_f'^{\vee}, \sigma_f^{\vee}, \varepsilon')_{Q,w^{\vee\prime}}$.

5.3.8 A strong form of the Manin–Drinfeld principle

Our main theorem on boundary cohomology as a Hecke module is the following theorem, which is the culmination of the entire discussion in this section.

Theorem 5.12. *Let the notations be as in Sect. 5.3.7. Then:*

1. *The $\pi_0(G(\mathbb{R}))\times\mathcal{H}^{G,\mathsf{S}}$-modules $I_b^{\mathsf{S}}(\sigma_f, \sigma_f', \varepsilon')_{P,w}$ and $I_b^{\mathsf{S}}(\sigma_f'(n), \sigma_f(-n'), \varepsilon')_{Q,w'}$ and similarly, the modules $I_t^{\mathsf{S}}(\sigma_f^{\vee}(n'), \sigma_f'^{\vee}(-n), \varepsilon')_{P,w^{\vee}}$ and $I_t^{\mathsf{S}}(\sigma_f'^{\vee}, \sigma_f^{\vee}, \varepsilon')_{Q,w^{\vee\prime}}$ are finite-dimensional E-vector spaces, all of which have the same dimension, say, denoted k.*
2. *The sum*
$$I_b^{\mathsf{S}}(\sigma_f, \sigma_f', \varepsilon')_{P,w} \ \oplus\ I_b^{\mathsf{S}}(\sigma_f'(n), \sigma_f(-n'), \varepsilon')_{Q,w'}$$
is a $2\mathsf{k}$-dimensional E-vector space that is isotypic in $H^{b_N^F}(\partial\mathcal{S}^G, \widetilde{\mathcal{M}}_{\lambda,E})^{K_f}$. (When $P = Q$, note that $w' \neq w$.) Furthermore, there is a $\pi_0(G(\mathbb{R}))\times\mathcal{H}^{G,\mathsf{S}}$-equivariant projection:
$$\mathfrak{R}^b_{\sigma_f,\sigma_f',\varepsilon'} : H^{b_N^F}(\partial\mathcal{S}^G, \widetilde{\mathcal{M}}_{\lambda,E})^{K_f} \longrightarrow I_b^{\mathsf{S}}(\sigma_f, \sigma_f', \varepsilon')_{P,w} \ \oplus\ I_b^{\mathsf{S}}(\sigma_f'(n), \sigma_f(-n'), \varepsilon')_{Q,w'}.$$
3. *The sum*
$$I_t^{\mathsf{S}}(\sigma_f^{\vee}(n'), \sigma_f'^{\vee}(-n), \varepsilon')_{P,w^{\vee}} \ \oplus\ I_t^{\mathsf{S}}(\sigma_f'^{\vee}, \sigma_f^{\vee}, \varepsilon')_{Q,w^{\vee\prime}}$$
is a $2\mathsf{k}$-dimensional E-vector space that is isotypic in $H^{\tilde{t}_N^F-1}(\partial\mathcal{S}^G, \widetilde{\mathcal{M}}_{\lambda^{\vee},E})^{K_f}$. (When $P = Q$, note that $w^{\vee\prime} \neq w^{\vee}$.) Furthermore, there is a $\pi_0(G(\mathbb{R}))\times\mathcal{H}^{G,\mathsf{S}}$-equivariant projection:
$$\mathfrak{R}^t_{\sigma_f,\sigma_f',\varepsilon'} : H^{\tilde{t}_N^F-1}(\partial\mathcal{S}^G, \widetilde{\mathcal{M}}_{\lambda^{\vee},E})^{K_f} \longrightarrow I_t^{\mathsf{S}}(\sigma_f^{\vee}(n'), \sigma_f'^{\vee}(-n), \varepsilon')_{P,w^{\vee}} \ \oplus\ I_t^{\mathsf{S}}(\sigma_f'^{\vee}, \sigma_f^{\vee}, \varepsilon')_{Q,w^{\vee\prime}}.$$

Chapter Six

Eisenstein Cohomology

Recall from Sect. 2.3.4 the long exact sequence of $\pi_0(G(\mathbb{R})) \times \mathcal{H}^G_{K_f}$-modules

$$\cdots \longrightarrow H^\bullet_c(\mathcal{S}^G_{K_f}, \widetilde{\mathcal{M}}_{\lambda,E}) \xrightarrow{\mathfrak{i}^\bullet} H^\bullet(\bar{\mathcal{S}}^G_{K_f}, \widetilde{\mathcal{M}}_{\lambda,E}) \xrightarrow{\mathfrak{r}^\bullet} H^\bullet(\partial\mathcal{S}^G_{K_f}, \widetilde{\mathcal{M}}_{\lambda,E}) \xrightarrow{\mathfrak{d}^\bullet} \cdots$$

Eisenstein cohomology is defined as:

$$H^\bullet_{\rm Eis}(\partial\mathcal{S}^G_{K_f}, \widetilde{\mathcal{M}}_{\lambda,E}) \ := \ \mathrm{Image}\left(H^\bullet(\bar{\mathcal{S}}^G_{K_f}, \widetilde{\mathcal{M}}_{\lambda,E}) \xrightarrow{\mathfrak{r}^\bullet} H^\bullet(\partial\mathcal{S}^G_{K_f}, \widetilde{\mathcal{M}}_{\lambda,E})\right). \quad (6.1)$$

We may pass to the limit over all K_f and also look at $\pi_0(G(\mathbb{R}))\times G(\mathbb{A}_f)$-modules

$$H^\bullet_{\rm Eis}(\partial\mathcal{S}^G, \widetilde{\mathcal{M}}_{\lambda,E}) \ := \ \mathrm{Image}\left(H^\bullet(\bar{\mathcal{S}}^G, \widetilde{\mathcal{M}}_{\lambda,E}) \xrightarrow{\mathfrak{r}^\bullet} H^\bullet(\partial\mathcal{S}^G, \widetilde{\mathcal{M}}_{\lambda,E})\right). \quad (6.2)$$

The reader should note the subtle difference between $H^\bullet_{\rm Eis}(\partial\mathcal{S}^G, \widetilde{\mathcal{M}}_{\lambda,E})$ and $H^\bullet_{\rm Eis}(\mathcal{S}^G, \widetilde{\mathcal{M}}_{\lambda,E})$. The latter, we recall, is a complement to strongly inner cohomology in global cohomology. Since strongly inner cohomology does not intertwine with boundary cohomology, we get a surjective map:

$$H^\bullet_{\rm Eis}(\mathcal{S}^G, \widetilde{\mathcal{M}}_{\lambda,E}) \longrightarrow\!\!\!\!\!\rightarrow H^\bullet_{\rm Eis}(\partial\mathcal{S}^G, \widetilde{\mathcal{M}}_{\lambda,E}).$$

6.1 POINCARÉ DUALITY AND MAXIMAL ISOTROPIC SUBSPACE OF BOUNDARY COHOMOLOGY

6.1.1 Poincaré duality

Let the notations be as in Sect. 5.3.7. Let $\mathsf{d} := \mathsf{d}^F_N = \dim(\mathcal{S}^G_{K_f})$. We have the following Poincaré duality pairing for sheaf cohomology on $\mathcal{S}^G_{K_f}$:

$$H^\bullet(\mathcal{S}^G_{K_f}, \widetilde{\mathcal{M}}_{\lambda,E}) \ \times \ H^{\mathsf{d}-\bullet}_c(\mathcal{S}^G_{K_f}, \widetilde{\mathcal{M}}_{\lambda^\vee,E}) \ \longrightarrow \ E. \quad (6.3)$$

The boundary $\partial\mathcal{S}^G_{K_f}$ is a compact manifold with corners with $\dim(\partial\mathcal{S}^G_{K_f}) = \mathsf{d}-1$. We also have Poincaré duality on $\partial\mathcal{S}^G_{K_f}$:

$$H^\bullet(\partial\mathcal{S}^G_{K_f}, \widetilde{\mathcal{M}}_{\lambda,E}) \times H^{\mathsf{d}-1-\bullet}(\partial\mathcal{S}^G_{K_f}, \widetilde{\mathcal{M}}_{\lambda^{\mathsf{v}},E}) \longrightarrow E. \tag{6.4}$$

We will denote either of the two dualities simply by (,).

6.1.2 Compatibility of duality isomorphisms with the connecting homomorphism

Consider the following diagram:

$$\begin{array}{ccccccc} H^\bullet(\mathcal{S}^G_{K_f}, \widetilde{\mathcal{M}}_{\lambda,E}) & \times & H_c^{\mathsf{d}-\bullet}(\mathcal{S}^G_{K_f}, \widetilde{\mathcal{M}}_{\lambda^{\mathsf{v}},E}) & \longrightarrow & E \\ \downarrow \mathfrak{r}^* & & \uparrow \mathfrak{d}^* & & \\ H^\bullet(\partial\mathcal{S}^G_{K_f}, \widetilde{\mathcal{M}}_{\lambda,E}) & \times & H^{\mathsf{d}-1-\bullet}(\partial\mathcal{S}^G_{K_f}, \widetilde{\mathcal{M}}_{\lambda^{\mathsf{v}},E}) & \longrightarrow & E \end{array} \tag{6.5}$$

where the horizontal arrows are the Poincaré duality pairings (6.3) and (6.4), and the vertical arrows $\mathfrak{r}^*$ and $\mathfrak{d}^*$ are as in the long exact sequence. For any class $\xi \in H^\bullet(\mathcal{S}^G_{K_f}, \widetilde{\mathcal{M}}_{\lambda,E})$ and any $\varsigma \in H^{\mathsf{d}-1-\bullet}(\partial\mathcal{S}^G_{K_f}, \widetilde{\mathcal{M}}_{\lambda^{\mathsf{v}},E})$ we have

$$(\mathfrak{r}^*(\xi), \varsigma) = (\xi, \mathfrak{d}^*(\varsigma)). \tag{6.6}$$

6.1.3 Maximal isotropic subspaces

The following proposition asserts that Eisenstein cohomology is a maximal isotropic subspace of boundary cohomology under Poincaré duality.

Proposition 6.1. *Under the duality pairing (6.4) for boundary cohomology, we have:*

$$H^\bullet_{\mathrm{Eis}}(\partial\mathcal{S}^G_{K_f}, \widetilde{\mathcal{M}}_{\lambda,E}) = H^{\mathsf{d}-1-\bullet}_{\mathrm{Eis}}(\partial\mathcal{S}^G_{K_f}, \widetilde{\mathcal{M}}_{\lambda^{\mathsf{v}},E})^\perp.$$

Proof. This is an exercise in using (6.6). Let $\mathfrak{r}^*(\xi) \in H^\bullet_{\mathrm{Eis}}(\partial\mathcal{S}^G_{K_f}, \widetilde{\mathcal{M}}_{\lambda,E})$. Then

$$(\mathfrak{r}^*(\xi), \mathfrak{r}^*(\varsigma')) = (\xi, \mathfrak{d}^*\mathfrak{r}^*(\varsigma')) = (\xi, 0) = 0, \quad \forall r^*(\varsigma') \in H^{\mathsf{d}-1-\bullet}_{\mathrm{Eis}}(\partial\mathcal{S}^G_{K_f}, \widetilde{\mathcal{M}}_{\lambda^{\mathsf{v}},E}).$$

Hence the left-hand side is contained in the right-hand side. For the reverse inclusion, suppose $\xi' \in H^\bullet(\partial\mathcal{S}^G_{K_f}, \widetilde{\mathcal{M}}_{\lambda,E})$ is orthogonal to $H^{\mathsf{d}-1-\bullet}_{\mathrm{Eis}}(\partial\mathcal{S}^G_{K_f}, \widetilde{\mathcal{M}}_{\lambda^{\mathsf{v}},E})$, then

$$0 = (\xi', \mathfrak{r}^*(\varsigma')) = (\mathfrak{d}^*(\xi'), \varsigma'), \quad \forall \varsigma' \in H^{\mathsf{d}-1-\bullet}(\mathcal{S}^G_{K_f}, \widetilde{\mathcal{M}}_{\lambda^{\mathsf{v}},E}).$$

Nondegeneracy of the duality pairing (6.3) in degree $\mathsf{d} - 1 - \bullet$ implies $\xi' \in$

$\mathrm{K}er(\mathfrak{d}^*) = \mathrm{Im}(\mathfrak{r}^*)$. Hence, $\xi' \in H^{\bullet}_{\mathrm{Eis}}(\partial \mathcal{S}^G_{K_f}, \widetilde{\mathcal{M}}_{\lambda,E})$. □

6.2 THE MAIN RESULT ON RANK-ONE EISENSTEIN COHOMOLOGY

Notations are as in Thm. 5.12. Consider the following maps starting from global cohomology $H^{b_N^F}(\mathcal{S}^G, \widetilde{\mathcal{M}}_{\lambda,E})^{K_f}$ and ending with an isotypic component in boundary cohomology:

$$\begin{array}{c} H^{b_N^F}(\mathcal{S}^G, \widetilde{\mathcal{M}}_{\lambda,E})^{K_f} \\ \Big\downarrow \mathfrak{r}^* \\ H^{b_N^F}(\partial\mathcal{S}^G, \widetilde{\mathcal{M}}_{\lambda,E})^{K_f} \\ \Big\downarrow \mathfrak{R}^b_{\sigma_f,\sigma'_f,\varepsilon'} \\ I^{\mathsf{S}}_b(\sigma_f,\sigma'_f,\varepsilon')_{P,w} \,\oplus\, I^{\mathsf{S}}_b(\sigma'_f(n),\sigma_f(-n'),\varepsilon')_{Q,w'} \end{array} \tag{6.7}$$

Recall, from Thm. 5.12, that $I^{\mathsf{S}}_b(\sigma_f,\sigma'_f,\varepsilon')_{P,w} \oplus I^{\mathsf{S}}_b(\sigma'_f(n),\sigma_f(-n'),\varepsilon')_{Q,w'}$ is a E-vector space of dimension $2\mathsf{k}$. In the self-associate case just change the Q to P. Our main result on Eisenstein cohomology (see Thm. 6.2) says that the image of $H^{b_N^F}_{\mathrm{Eis}}(\partial\mathcal{S}^G, \widetilde{\mathcal{M}}_{\lambda,E})^{K_f} = \mathrm{Im}(\mathfrak{r}^*)$ under $\mathfrak{R}^b_{\sigma_f,\sigma'_f,\varepsilon'}$ is a middle-dimensional (i.e., k-dimensional) subspace of this $2\mathsf{k}$-dimensional space. (It helps to have a mental picture of when $\mathsf{k}=1$, i.e., of a line in an ambient two-dimensional space; we will see later that the *slope* of this line contains arithmetic information about L-values.) The proof of this main result also needs the analogue of (6.7) in the top degree. We now state and prove this main result on Eisenstein cohomology.

6.2.1 The image of Eisenstein cohomology under $\mathfrak{R}^{\bullet}_{\sigma_f,\sigma'_f,\varepsilon'}$

Theorem 6.2. *Let the notations be as in 5.3.7. Furthermore, for brevity, let*

$$\begin{aligned} \mathfrak{I}^b(\sigma_f,\sigma'_f,\varepsilon') &:= \mathfrak{R}^b_{\sigma_f,\sigma'_f,\varepsilon'}(H^{b_N^F}_{\mathrm{Eis}}(\partial\mathcal{S}^G, \widetilde{\mathcal{M}}_{\lambda,E})^{K_f}), \\ \mathfrak{I}^t(\sigma_f,\sigma'_f,\varepsilon')^{\mathsf{v}} &:= \mathfrak{R}^t_{\sigma_f,\sigma'_f,\varepsilon'}(H^{\tilde{t}_N^F-1}_{\mathrm{Eis}}(\partial\mathcal{S}^G, \widetilde{\mathcal{M}}_{\lambda^{\mathsf{v}},E})^{K_f}). \end{aligned}$$

1. In the non-self-associate cases ($n \neq n'$) we have:

(a) $\mathfrak{I}^b(\sigma_f, \sigma'_f, \varepsilon')$ *is a* k*-dimensional* E*-subspace of*

$$I_b^{\mathsf{S}}(\sigma_f, \sigma'_f, \varepsilon')_{P,w} \oplus I_b^{\mathsf{S}}(\sigma'_f(n), \sigma_f(-n'), \varepsilon')_{Q,w'}.$$

(b) $\mathfrak{I}^t(\sigma_f, \sigma'_f, \varepsilon')^{\vee}$ *is a* k*-dimensional* E*-subspace of*

$$I_t^{\mathsf{S}}(\sigma_f^{\vee}(n'), \sigma_f'^{\vee}(-n), \varepsilon')_{P,w^{\vee}} \oplus I_t^{\mathsf{S}}(\sigma_f'^{\vee}, \sigma_f^{\vee}, \varepsilon')_{Q,w^{\vee\prime}}.$$

2. In the self-associate case ($n = n'$) *the same assertions hold by putting* $Q = P$.

6.2.2 Proof of Thm. 6.2

The proof of (2) is almost identical to the proof of (1), and so we give the details only for (1). The proof of (1) involves two steps:

(i) The first step is to show that both $\mathfrak{I}^b(\sigma_f, \sigma'_f, \varepsilon')$ and $\mathfrak{I}^t(\sigma_f, \sigma'_f, \varepsilon')^{\vee}$ are at least k-dimensional; this is achieved by going to a transcendental level and appealing to Langlands's constant term theorem and producing enough cohomology classes in the image.

(ii) The second step, after invoking properties of the Poincaré duality pairing, is to show that both $\mathfrak{I}^b(\sigma_f, \sigma'_f, \varepsilon')$ and $\mathfrak{I}^t(\sigma_f, \sigma'_f, \varepsilon')^{\vee}$ have dimension exactly k.

We take up these two arguments in the Sects. 6.2.2.1 and 6.2.2.2.

6.2.2.1 The cohomological meaning of the constant term theorem of Langlands

Take an embedding $\iota : E \to \mathbb{C}$ and pass to a transcendental level via ι. To show that $\mathfrak{I}^b(\sigma_f, \sigma'_f, \varepsilon')$ or $\mathfrak{I}^t(\sigma_f, \sigma'_f, \varepsilon')^{\vee}$ has dimension at least k as an E-vector space, it suffices to show that their base change to $\mathbb{C}$ via ι has dimension at least k as a $\mathbb{C}$-vector space; i.e., we would like to show:

$$\dim_{\mathbb{C}}\left(\mathfrak{R}^b_{{}^\iota\sigma_f, {}^\iota\sigma'_f, \varepsilon'}(H^{b_N^F}_{\mathrm{Eis}}(\partial\mathcal{S}^G, \widetilde{\mathcal{M}}_{{}^\iota\lambda})^{K_f})\right) \geq \mathsf{k} \quad \text{and}$$
$$\dim_{\mathbb{C}}\left(\mathfrak{R}^t_{{}^\iota\sigma_f, {}^\iota\sigma'_f, \varepsilon'}(H^{\tilde{t}_N^F}_{\mathrm{Eis}}(\partial\mathcal{S}^G, \widetilde{\mathcal{M}}_{{}^\iota\lambda})^{K_f})\right) \geq \mathsf{k}. \quad (6.8)$$

In Sect. 6.3 we briefly introduce the L-functions at hand, recall the celebrated theorem of Langlands on the constant term of an Eisenstein series, and then we will come back to the proof of this part in Sect. 6.3.7.

6.2.2.2 Application of Poincaré duality

The proofs of (1)(a) and (1)(b), assuming that we have proved (6.8), are an exercise involving properties of Poincaré duality especially that it is nondegenerate, Hecke-equivariant, and that the Eisenstein part is maximal isotropic.

The bare-bones linear algebra looks like: Suppose we have decompositions $V = V_P \oplus V_Q$ and $W = W_P \oplus W_Q$, where V_P, V_Q, W_P, and W_Q are all k-dimensional vector spaces over E; suppose also that we have a nondegenerate pairing $(\ ,\ ) : V \times W \to E$ such that $(V_P, W_Q) = (V_Q, W_P) = 0$ and the pairing is nondegenerate on $V_P \times W_P$ and $V_Q \times W_Q$; furthermore, suppose we are given subspaces $\mathfrak{I} \subset V$ and $\mathfrak{J} \subset W$ such that $\dim_E(\mathfrak{I}) \geq \mathsf{k}$, $\dim_E(\mathfrak{J}) \geq \mathsf{k}$, and $(\mathfrak{I}, \mathfrak{J}) = 0$. Then it is easy to see that $\dim_E(\mathfrak{I}) = \mathsf{k} = \dim_E(\mathfrak{J})$.

This concludes the proof of Thm. 6.2 under the assumption that we have proved (6.8).

6.3 A THEOREM OF LANGLANDS: THE CONSTANT TERM OF AN EISENSTEIN SERIES

We will need some details from the Langlands–Shahidi method in our context. The reader is referred to Kim [45] and Shahidi [65] for details, proofs, and further references.

6.3.1 Notations for certain characters attached to P

Let the notations be as in Sect. 5.3.7. In particular, $P_0 = M_{P_0} U_{P_0}$ is the standard (n, n') parabolic subgroup of GL_N/F, and $P = R_{F/\mathbb{Q}}(P_0)$, etc. We write

$$M_{P_0} = \left\{ m = \mathrm{diag}(h, h') = \begin{pmatrix} h & 0 \\ 0 & h' \end{pmatrix} \ : \ h \in \mathrm{GL}_n,\ h' \in \mathrm{GL}_{n'} \right\},$$

and let

$$A_{P_0} := Z_{M_{P_0}} = \left\{ a = \begin{pmatrix} t 1_n & 0 \\ 0 & t' 1_{n'} \end{pmatrix} \ : \ t, t' \in \mathrm{GL}_1 \right\}.$$

Fix an identification $X^*(A_{P_0}) = \mathbb{Z}^2$, by letting $(k, k') \in \mathbb{Z}^2$ correspond to the character that sends a to $t^k t'^{k'}$. We have

$$X^*(A_P \times E) = \bigoplus_{\tau : F \to E} X^*(A_{P_0} \times_\tau E) = \bigoplus_{\tau : F \to E} \mathbb{Z}^2.$$

Similarly, fix $X^*(M_{P_0}) = \mathbb{Z}^2$, by letting $(k, k') \in \mathbb{Z}^2$ correspond to the character that sends $\mathrm{diag}(h, h')$ to $\det(h)^k \det(h')^{k'}$. Restriction from M_{P_0} to A_{P_0} gives an inclusion $X^*(M_{P_0}) \hookrightarrow X^*(A_{P_0})$ that is given by $(k, k') \mapsto (nk, n'k')$. Clearly, $X^*(M_{P_0}) \otimes \mathbb{Q} = X(A_{P_0}) \otimes \mathbb{Q}$, which fixes an identification $X^*(M_{P_0}) \otimes \mathbb{Q} = \mathbb{Q}^2$ via $X(A_{P_0}) \otimes \mathbb{Q} = \mathbb{Q}^2$. Similarly, $X^*(M_{P_0}) \otimes \mathbb{R} = X^*(A_{P_0}) \otimes \mathbb{R}$ and fix $X^*(M_{P_0}) \otimes \mathbb{R} = \mathbb{R}^2$. This fixes $X^*(M_P) \otimes \mathbb{R} = \oplus_{\tau : F \to E} \mathbb{R}^2$.

Let $\boldsymbol{\rho}_{P_0}$ be half the sum of positive roots whose root spaces appear in U_{P_0}.

The restriction of $\boldsymbol{\rho}_{P_0}$ to A_{P_0} is in $X^*(A_{P_0}) \otimes \mathbb{Q} \hookrightarrow \mathfrak{a}^*_{P_0} := X(A_{P_0}) \otimes \mathbb{R} = X^*(M_{P_0}) \otimes \mathbb{R}$ and under the above identification of the latter with $\mathbb{R}^2$, one has $\boldsymbol{\rho}_{P_0} = (n'/2, -n/2)$, or using the notations of 2.2.4, we have

$$\boldsymbol{\rho}_{P_0} = \frac{1}{2} \sum_{\substack{1 \leq i \leq n \\ n+1 \leq j \leq N}} \mathbf{e}_i - \mathbf{e}_j = \frac{n'}{2}(\mathbf{e}_1 + \cdots + \mathbf{e}_n) - \frac{n}{2}(\mathbf{e}_{n+1} + \cdots + \mathbf{e}_{n+n'}).$$

And $\rho_P = (\boldsymbol{\rho}_{P_0^\tau})_{\tau:F\to E}$, with each $\boldsymbol{\rho}_{P_0^\tau}$ given as earlier.

Let $\boldsymbol{\alpha}_{P_0} = \boldsymbol{\alpha}_n = \mathbf{e}_n - \mathbf{e}_{n+1}$ be the unique simple root of G_0 that is not among the roots of M_{P_0}. Consider the corresponding fundamental weight $\boldsymbol{\gamma}_{P_0} = \boldsymbol{\gamma}_n = \langle \boldsymbol{\rho}_{P_0}, \boldsymbol{\alpha}_{P_0}\rangle^{-1} \boldsymbol{\rho}_{P_0}$. Identify $X^*(T_{N,0}) \otimes \mathbb{R}$ with $\mathbb{R}^N$ using the $\mathbf{e}_i$'s, and let $(\ ,\ )$ be the usual euclidean inner product on $\mathbb{R}^N$. It is easy to see that $\langle \boldsymbol{\rho}_{P_0}, \boldsymbol{\alpha}_{P_0}\rangle = \frac{2(\boldsymbol{\rho}_{P_0}, \boldsymbol{\alpha}_{P_0})}{(\boldsymbol{\alpha}_{P_0}, \boldsymbol{\alpha}_{P_0})} = N/2$, and hence

$$\boldsymbol{\gamma}_n = \boldsymbol{\gamma}_{P_0} = \frac{2}{N} \boldsymbol{\rho}_{P_0}.$$

We also have $\alpha_P = (\boldsymbol{\alpha}_{P_0^\tau})_{\tau:F\to E}$ and $\gamma_P = (\boldsymbol{\gamma}_{P_0^\tau})_{\tau:F\to E}$, with $N\,\boldsymbol{\gamma}_{P_0^\tau} = 2\,\boldsymbol{\rho}_{P_0^\tau}$.

Let $|\boldsymbol{\delta}_{P_0}|$ be the modular character of $M_{P_0}(k)$, where k is any local field (such as $k = \mathbb{R}$ or $k = F_v$ any p-adic completion of F), which is defined as $|\boldsymbol{\delta}_{P_0}|(m) = |\det(\mathrm{Ad}_{\mathfrak{u}_P}(m))|$ for $m \in M_{P_0}(k)$. If $m = \mathrm{diag}(h, h')$ with $h \in \mathrm{GL}_n(k)$ and $h' \in \mathrm{GL}_{n'}(k)$ then

$$|\boldsymbol{\delta}_{P_0}|(\mathrm{diag}(h, h')) = |\det(h)|^{n'} |\det(h')|^{-n}.$$

Note that $|2\boldsymbol{\rho}_{P_0}|(\cdot) = |\boldsymbol{\delta}_{P_0}|(\cdot)$. Also, $|\delta_P|$, the modular character of $M_P(k)$, for $k = \mathbb{R}$ or $k = \mathbb{Q}_p$, is defined via the various completions of F over that place; for example, $|\delta_P|$ on $M_P(\mathbb{Q}_p)$ is $\prod_{\mathfrak{p}|p} |\boldsymbol{\delta}_{P_0}|_{\mathfrak{p}}$, where by $|\boldsymbol{\delta}_{P_0}|_{\mathfrak{p}}$ we mean the character as above on $M_{P_0}(F_{\mathfrak{p}})$. For any character $\gamma : M_0 \to \mathbb{G}_m$ and $m \in M_0(F_{\mathfrak{p}})$ we can write $\gamma(m) = \varpi_{\mathfrak{p}}^{ord_{\mathfrak{p}}(\gamma(m))} \times$ unit, and then

$$|\gamma|(m) = q_{\mathfrak{p}}^{-ord_{\mathfrak{p}}(\gamma(m))}.$$

6.3.2 Induced representations

Let σ (resp., σ') be a cuspidal automorphic representation of $G_n(\mathbb{A})$ (resp., $G_{n'}(\mathbb{A})$). The relation with our previous arithmetic notation is that given $\sigma_f \in \mathrm{Coh}_{!!}(G_n, \mu)$ and given $\iota : E \to \mathbb{C}$, think of ${}^\iota\sigma_f$ to be the finite part of a cuspidal automorphic representation ${}^\iota\sigma$, etc. The ι is fixed, and we suppress it until otherwise mentioned. Consider the induced representation $I_P^G(s, \sigma \otimes \sigma')$

consisting of all smooth functions $f : G(\mathbb{A}) \to V_\sigma \otimes V_{\sigma'}$ such that

$$f(mug) = |\delta_P|(m)^{\frac{1}{2}} |\delta_P|(m)^{\frac{s}{N}} (\sigma \otimes \sigma')(m) f(g) \tag{6.9}$$

for all $m \in M_P(\mathbb{A})$, $u \in U_P(\mathbb{A})$, and $g \in G(\mathbb{A})$, where V_σ (resp., $V_{\sigma'}$) is the subspace inside the space of cusp forms on $G_n(\mathbb{A})$ (resp., $G_{n'}(\mathbb{A})$) realizing the representation σ (resp., σ'). In other words,

$$I_P^G(s, \sigma \otimes \sigma') = \mathrm{Ind}_{P(\mathbb{A})}^{G(\mathbb{A})}((\sigma \otimes |\ |^{\frac{n'}{N}s}) \otimes (\sigma' \otimes |\ |^{\frac{-n}{N}s})),$$

where Ind_P^G denotes the normalized parabolic induction. In terms of algebraic or un-normalized induction, we have

$$I_P^G(s, \sigma \otimes \sigma') = {}^{\mathrm{a}}\mathrm{Ind}_{P(\mathbb{A})}^{G(\mathbb{A})}((\sigma \otimes |\ |^{\frac{n'}{N}s + \frac{n'}{2}}) \otimes (\sigma' \otimes |\ |^{\frac{-n}{N}s - \frac{n}{2}})). \tag{6.10}$$

6.3.3 Standard intertwining operators

There is an element $w_P \in W_G$, the Weyl group of G, which is uniquely determined by the property $w_P(\mathbf{\Pi}_G - \{\boldsymbol{\alpha}_P\}) \subset \mathbf{\Pi}_G$ and $w_P(\boldsymbol{\alpha}_P) < 0$. This element looks like $w_P = (w_{P_0}^\tau)_{\tau : F \to E}$, where for each τ, as a permutation matrix in GL_N, we have

$$w_{P_0}^\tau = \begin{bmatrix} & 1_n \\ 1_{n'} & \end{bmatrix}.$$

The parabolic subgroup Q, which is associate to P, corresponds to $w_P(\mathbf{\Pi}_G - \{\boldsymbol{\alpha}_P\})$. Since ${w_{P_0}^\tau}^{-1} \mathrm{diag}(h, h') w_{P_0}^\tau = \mathrm{diag}(h', h)$ for all $\mathrm{diag}(h, h') \in M_{P_0^\tau}$, we get $w_P(\sigma \otimes \sigma') = \sigma' \otimes \sigma$ as a representation of $M_Q(\mathbb{A})$. The global standard intertwining operator:

$$T_{\mathrm{st}}^{PQ}(s, \sigma \otimes \sigma') : I_P^G(s, \sigma \otimes \sigma') \longrightarrow I_Q^G(-s, \sigma' \otimes \sigma)$$

is given by the integral

$$(T_{\mathrm{st}}^{PQ}(s, \sigma \otimes \sigma') f)(g) = \int_{U_Q(\mathbb{A})} f(w_{P_0}^{-1} u g)\, du. \tag{6.11}$$

Often, we will abbreviate $T_{\mathrm{st}}^{PQ}(s, \sigma \otimes \sigma')$ as $T_{\mathrm{st}}(s, \sigma \otimes \sigma')$. The global standard intertwining operator factorizes as a product of local standard intertwining operators: $T_{\mathrm{st}}(s, \sigma \otimes \sigma') = \otimes_v T_{\mathrm{st}}(s, \sigma_v \otimes \sigma'_v)$, where the local operator

$$T_{\mathrm{st}}(s, \sigma_v \otimes \sigma'_v) : I_P^G(s, \sigma_v \otimes \sigma'_v) \longrightarrow I_Q^G(-s, \sigma'_v \otimes \sigma_v) \tag{6.12}$$

is given by a similar local integral.

6.3.4 Eisenstein series

Let $f \in I_P^G(s, \sigma \times \sigma')$; for $\underline{g} \in G(\mathbb{A})$ the value $f(\underline{g})$ is a cusp form on $M_P(\mathbb{A})$. By the defining equivariance property of f, the complex number $f(\underline{g})(\underline{m})$ determines and is determined by $f(\underline{mg})(\underline{1})$ for any $\underline{m} \in M_P(\mathbb{A})$. Henceforth, we identify $f \in I_P^G(s, \sigma \times \sigma')$ with the complex valued function $\underline{g} \mapsto f(\underline{g})(\underline{1})$; i.e., we have embedded:

$$I_P^G(s, \sigma \times \sigma') \hookrightarrow \mathcal{C}^\infty\left(U_P(\mathbb{A})M_P(\mathbb{Q})\backslash G(\mathbb{A}), \omega_\infty^{-1}\right) \subset \mathcal{C}^\infty\left(P(\mathbb{Q})\backslash G(\mathbb{A}), \omega_\infty^{-1}\right),$$

where ω_∞^{-1} is a simplified notation for the central character of $\sigma \otimes \sigma'$ restricted to $S(\mathbb{R})^\circ$. If $\sigma_f \in \mathrm{Coh}(G_n, \mu)$, $\sigma'_f \in \mathrm{Coh}(G_{n'}, \mu')$, and $\iota : E \to \mathbb{C}$ then ω_∞ is the product of the central characters $\omega_{\mathcal{M}_{\iota_\mu}} \omega_{\mathcal{M}_{\iota_{\mu'}}}$ restricted to $S(\mathbb{R})$.

Given $f \in I_P^G(s, \sigma \times \sigma')$, thought of as a function on $P(\mathbb{Q})\backslash G(\mathbb{A})$, define the corresponding Eisenstein series $\mathrm{Eis}_P(s, f) \in \mathcal{C}^\infty\left(G(\mathbb{Q})\backslash G(\mathbb{A}), \omega_\infty^{-1}\right)$ by averaging over $P(\mathbb{Q})\backslash G(\mathbb{Q})$:

$$\mathrm{Eis}_P(s, f)(\underline{g}) := \sum_{\gamma \in P(\mathbb{Q})\backslash G(\mathbb{Q})} f(\gamma \underline{g}); \tag{6.13}$$

this is convergent if $\Re(s) \gg 0$ and has meromorphic continuation into the entire complex plane. This provides an intertwining operator

$$\mathrm{Eis}_P(s, \sigma \times \sigma') : I_P^G(s, \sigma \times \sigma') \hookrightarrow \mathcal{C}^\infty\left(G(\mathbb{Q})\backslash G(\mathbb{A}), \omega_\infty^{-1}\right)$$

and we write $\mathrm{Eis}_P(s, \sigma \times \sigma')(f) = \mathrm{Eis}_P(s, f)$. If we want to construct a map in cohomology we need to evaluate at $s = -N/2$, so we ask whether the Eisenstein series is holomorphic at $s = -N/2$.

A few words about the point of evaluation being $s = -N/2$. Recall the induced representations that appear in boundary cohomology as in Sect. 5.3.7, and noting how the complex variable s appears in an induced representation as in (6.10), we see

$$^{\mathrm{a}}\mathrm{Ind}_{P(\mathbb{A})}^{G(\mathbb{A})}(\sigma \otimes \sigma') = I_P^G(s, \sigma \otimes \sigma')|_{s=-N/2}.$$

Similarly, we have

$$^{\mathrm{a}}\mathrm{Ind}_{Q(\mathbb{A})}^{G(\mathbb{A})}(\sigma'(n) \otimes \sigma(-n')) = I_Q^G(s, \sigma' \otimes \sigma)|_{s=N/2}.$$

It might be helpful to bear in mind the following picture:

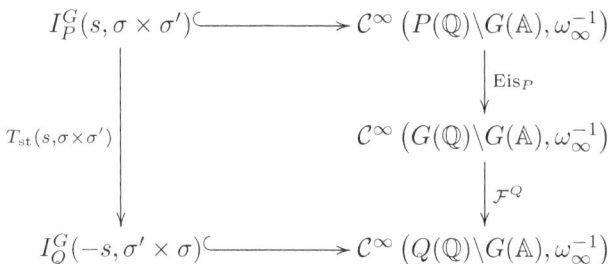

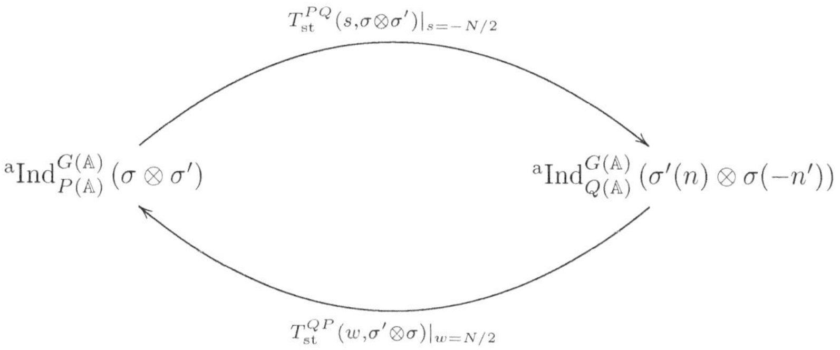

6.3.5 Constant term of an Eisenstein series

For holomorphy of the Eisenstein series at $s = -N/2$, it is well known that we have to show that the constant term of the Eisenstein series is holomorphic at $s = -N/2$. Toward this, we recall the results in [52] and [35, Chap. IV, Sect. 5]. For the parabolic subgroup Q, recall the constant term map denoted $\mathcal{F}^Q : \mathcal{C}^\infty(G(\mathbb{Q})\backslash G(\mathbb{A}), \omega_\infty^{-1}) \to \mathcal{C}^\infty(M_Q(\mathbb{Q})U_Q(\mathbb{A})\backslash G(\mathbb{A}), \omega_\infty^{-1})$, and given by

$$\mathcal{F}^Q(\phi)(\underline{g}) = \int\limits_{U_Q(\mathbb{Q})\backslash U_Q(\mathbb{A})} \phi(\underline{u}\,\underline{g})\,d\underline{u}.$$

Consider the following diagram of maps:

$$\begin{array}{ccc}
I_P^G(s,\sigma\times\sigma') & \hookrightarrow & \mathcal{C}^\infty\left(P(\mathbb{Q})\backslash G(\mathbb{A}),\omega_\infty^{-1}\right) \\
 & & \downarrow{\scriptstyle \mathrm{Eis}_P} \\
\downarrow{\scriptstyle T_{\mathrm{st}}(s,\sigma\times\sigma')} & & \mathcal{C}^\infty\left(G(\mathbb{Q})\backslash G(\mathbb{A}),\omega_\infty^{-1}\right) \\
 & & \downarrow{\scriptstyle \mathcal{F}^Q} \\
I_Q^G(-s,\sigma'\times\sigma) & \hookrightarrow & \mathcal{C}^\infty\left(Q(\mathbb{Q})\backslash G(\mathbb{A}),\omega_\infty^{-1}\right)
\end{array}$$

Theorem 6.3 (Langlands). *Let $f \in I_P^G(s,\sigma\times\sigma')$.*

1. *In the non-self-associate cases ($n \neq n'$), we have*

   (a) $\mathcal{F}^P \circ \mathrm{Eis}_P(s,f) = f$.

(b) $\mathcal{F}^Q \circ \mathrm{Eis}_P(s,f) \;=\; T_{\rm st}(s,\sigma\times\sigma')(f).$

2. In the self-associate cases ($n=n'$ and $P=Q$), we have

$$\mathcal{F}^P \circ \mathrm{Eis}_P(s,f) \;=\; f + T_{\rm st}(s,\sigma\times\sigma')(f).$$

6.3.6 Holomorphy of the Eisenstein series at the point of evaluation

Given pure weights μ and μ' (recall the notations of Sect. 5.3.7), we say that $\underline{\mu} = \mu + \mu'$ is *on the right of the unitary axis* to mean that $-a(\underline{\mu}) \geq N/2$; i.e., $-N/2 - d + d' \geq 0$ (see Rem. 7.2).

Theorem 6.4. *Let σ and σ' be cuspidal automorphic representations as above. Suppose μ and μ' are pure weights, such that σ (resp., σ') has cohomology with respect to μ (resp., μ'). Suppose nn' is even. Assume that $\underline{\mu}$ is on the right of the unitary axis. Then $\mathrm{Eis}_P(s,f)$ is holomorphic at $s=-N/2$, unless we are in the exceptional case $n=n'$ and $\sigma' = \sigma\otimes|\;|^{-n-1}$.*

The poles on the right of the unitary axis are simple and contribute to the residual spectrum [35] and then the assertion follows from the description of the residual spectrum [52]. We shall see later that, under the conditions imposed by the combinatorial lemma, the exceptional case does not occur: if σ' is a Tate twist of σ then $\mu^{(1)} = \mu'^{(1)}$ and the cuspidal width $\ell(\mu,\mu')=0$; however, under the combinatorial lemma $\ell(\mu,\mu') \geq 2$. See also Sect. 7.2.2.3.

6.3.7 Conclusion of proof of Thm. 6.2

We now prove the claim made in Sect. 6.2.2.1. The notations are now as in the statement of Thm. 6.2, and we would like to show

$$\dim_E(\mathfrak{I}^b(\sigma_f,\sigma'_f,\varepsilon')) \geq \mathsf{k} \quad\text{and}\quad \dim_E(\mathfrak{I}^t(\sigma_f,\sigma'_f,\varepsilon')^{\vee}) \geq \mathsf{k}.$$

The argument for the top degree is the same as that for the bottom degree, so we give the details of the proof only in the bottom degree. We assume that $\underline{\mu}$ is on the right of the unitary axis. Take an embedding $\iota : E \to \mathbb{C}$ and pass to a transcendental level via ι. To show $\dim_E(\mathfrak{I}^b(\sigma_f,\sigma'_f,\varepsilon')) \geq \mathsf{k}$, it suffices to show that its base change to $\mathbb{C}$ via ι has dimension at least k; i.e., it suffices to prove

$$\dim_{\mathbb{C}}(\mathfrak{I}^b({}^\iota\sigma_f,{}^\iota\sigma'_f,\varepsilon')) \geq \mathsf{k},$$

where, for brevity, we have

$$\mathfrak{I}^b({}^\iota\sigma_f,{}^\iota\sigma'_f,\varepsilon') \;:=\; \mathfrak{R}^b_{{}^\iota\sigma_f,{}^\iota\sigma'_f,\varepsilon'}(H^{b_N^F}_{\rm Eis}(\partial\mathcal{S}^G,\widetilde{\mathcal{M}}_{{}^\iota\lambda})^{K_f}).$$

Given ι, we know that ${}^\iota\sigma_f$ (resp., ${}^\iota\sigma'_f$) is the finite part of a cuspidal automorphic representation ${}^\iota\sigma$ (resp., ${}^\iota\sigma'$) of $G_n(\mathbb{A})$ (resp., $G_{n'}(\mathbb{A})$). We know that $\mathrm{Eis}_P(s,f)$ is holomorphic at $s=-N/2$. Then, consider the diagram in (6.17), where, for brevity, we denote ω_∞^{-1} for the inverse of the central character of $\mathcal{M}_{\iota\lambda}$ restricted to $S(\mathbb{R})^0$. Tensor the above diagram by $\widetilde{\mathcal{M}}_{\iota\lambda}$ and take cohomology; i.e., apply the functor $H^{b_N^F}(\mathfrak{g}, K_\infty^0; -)$. Then take the ε'-isotypic component for the action of $\pi_0(K_\infty)$ and also take K_f-invariants to get (6.18). Also the map $(\mathcal{F}^P \oplus \mathcal{F}^Q)^*$ is the same as the restriction $\mathfrak{r}^*$ to boundary cohomology followed by a projection afforded by the Manin–Drinfeld principle. It is clear now that the required image $\mathfrak{I}^b({}^\iota\sigma_f, {}^\iota\sigma'_f, \varepsilon')$ contains the image of $(\mathcal{F}^P \oplus \mathcal{F}^Q)^* \circ \mathrm{Eis}_P^*$; i.e., after applying Thm 6.3 we see that

$$\mathfrak{I}^b({}^\iota\sigma_f, {}^\iota\sigma'_f, \varepsilon') \supset \left\{\left(\xi\,,\, T_{\mathrm{st}}(-\tfrac{N}{2}, {}^\iota\sigma\otimes{}^\iota\sigma')^*\xi\right) \;:\; \xi \in I_b^{\mathsf{S}}({}^\iota\sigma_f, {}^\iota\sigma'_f, \varepsilon')_{P,{}^\iota w}\right\}. \quad (6.14)$$

Hence $\dim_{\mathbb{C}}(\mathfrak{I}^b({}^\iota\sigma_f, {}^\iota\sigma'_f, \varepsilon')) \geq \dim_{\mathbb{C}}(I_b^{\mathsf{S}}({}^\iota\sigma_f, {}^\iota\sigma'_f, \varepsilon')_{P,{}^\iota w}) = \mathsf{k}$.

Suppose we are on the left of the unitary axis; i.e., $-N/2+d-d' < 0$. Then we start from the parabolic subgroup Q and consider the situation

$$T_{\mathrm{st}}^{QP}(s, \sigma'\otimes\sigma) : I_Q^G(s, \sigma'\otimes\sigma) \;\to\; I_P^G(s, \sigma\otimes\sigma')$$

evaluated at $s = N/2$, for which we would be on the right of the unitary axis. In this case, we get

$$\mathfrak{I}^b({}^\iota\sigma_f, {}^\iota\sigma'_f, \varepsilon') \supset \left\{\left(T_{\mathrm{st}}(N/2, {}^\iota\sigma'\otimes{}^\iota\sigma)^*\tilde{\xi}\,,\, \tilde{\xi}\right) \;:\; \tilde{\xi} \in I_b^{\mathsf{S}}({}^\iota\sigma_f, {}^\iota\sigma'_f, \varepsilon')_{Q,{}^\iota w'}\right\}. \quad (6.15)$$

Hence, $\dim_{\mathbb{C}}(\mathfrak{I}^b({}^\iota\sigma_f, {}^\iota\sigma'_f, \varepsilon')) \geq \dim_{\mathbb{C}}(I_b^{\mathsf{S}}({}^\iota\sigma_f, {}^\iota\sigma'_f, \varepsilon')_{Q,{}^\iota w'}) = \mathsf{k}$. This concludes the proof of Thm. 6.2.

It follows from the proof of the above theorem that, if we are on the right of the unitary axis, then the k-dimensional subspace $\mathfrak{I}(\sigma_f, \sigma'_f, \varepsilon)$ consists of elements of the form $(\xi, T_{\mathrm{Eis}}(\sigma, \sigma', \varepsilon')(\xi))$ for an *Eisenstein* intertwining operator

$$T_{\mathrm{Eis}}(\sigma, \sigma', \varepsilon') \;:\; I_b^{\mathsf{S}}(\sigma_f, \sigma'_f, \varepsilon')_{P,w} \longrightarrow I_b^{\mathsf{S}}(\sigma'_f(n), \sigma_f(-n'), \varepsilon')_{Q,w'}, \quad (6.16)$$

that is defined over E. It will turn out that this linear map will contain in it arithmetic information about certain L-values giving us our main theorem; see Sect. 7.3.2.

$$
\begin{array}{ccccc}
 & & \mathcal{C}^\infty\left(G(\mathbb{Q})\backslash G(\mathbb{A}),\omega_\infty^{-1}\right) & & \\
 & \nearrow^{\mathrm{Eis}_P} & \downarrow{\scriptstyle \mathcal{F}^P \oplus \mathcal{F}^Q} & & \\
\mathcal{C}^\infty\left(P(\mathbb{Q})\backslash G(\mathbb{A}),\omega_\infty^{-1}\right) & \hookrightarrow & \mathcal{C}^\infty\left(P(\mathbb{Q})\backslash G(\mathbb{A}),\omega_\infty^{-1}\right) \oplus \mathcal{C}^\infty\left(Q(\mathbb{Q})\backslash G(\mathbb{A}),\omega_\infty^{-1}\right) & \hookleftarrow & \mathcal{C}^\infty\left(Q(\mathbb{Q})\backslash G(\mathbb{A}),\omega_\infty^{-1}\right) \\
\uparrow & & & & \uparrow \\
I_P^G(-N/2, {}^\iota\sigma \otimes {}^\iota\sigma') & & \xrightarrow{\quad T_{\mathrm{st}}(-N/2, {}^\iota\sigma \otimes {}^\iota\sigma') \quad} & & I_Q^G(N/2, {}^\iota\sigma' \otimes {}^\iota\sigma)
\end{array}
\tag{6.17}
$$

$$
\begin{array}{ccccc}
 & & H^{b_N^F}(\mathcal{S}^G, \widetilde{\mathcal{M}}_{\iota\lambda})^{K_f} & & \\
 & \nearrow^{\mathrm{Eis}_P^*} & \Big\downarrow{\scriptstyle (\mathcal{F}^P \oplus \mathcal{F}^Q)^*} & & \\
H_{!!}^{b_N^F}(\partial_P \mathcal{S}^G, \widetilde{\mathcal{M}}_{\iota\lambda})^{K_f} & \hookrightarrow & H_{!!}^{b_N^F}(\partial_P \mathcal{S}^G, \widetilde{\mathcal{M}}_{\iota\lambda})^{K_f} \oplus H_{!!}^{b_N^F}(\partial_Q \mathcal{S}^G, \widetilde{\mathcal{M}}_{\iota\lambda})^{K_f} & \hookleftarrow & H_{!!}^{b_N^F}(\partial_Q \mathcal{S}^G, \widetilde{\mathcal{M}}_{\iota\lambda})^{K_f} \\
\Big\uparrow & & & & \Big\uparrow \\
I_b^{\mathsf{S}}({}^\iota\sigma_f, {}^\iota\sigma'_f, \varepsilon')_{P,{}^\iota w} & & \xrightarrow{\;T_{\mathrm{st}}(-N/2, {}^\iota\sigma \otimes {}^\iota\sigma', \varepsilon')^*\;} & & I_b^{\mathsf{S}}({}^\iota\sigma'_f(n), {}^\iota\sigma_f(-n')))_{Q,{}^\iota w'}
\end{array}
\tag{6.18}
$$

Chapter Seven

L-Functions

7.1 MOTIVIC AND COHOMOLOGICAL L-FUNCTIONS

There is a well-known conjectural dictionary between cohomological cuspidal automorphic representations of GL_n and pure rank n motives. We briefly review this dictionary while recasting it in the context of strongly inner Hecke summands on the one hand and pure effective motives on the other. In the discussion that follows, we consider the motives only via their Betti, de Rham and $\mathfrak{l}$-adic realizations as in Deligne [15].

Let $\lambda \in X_0^*(T_n \times E)$ be a pure weight and $\sigma_f \in \mathrm{Coh}_{!!}(G_n, \lambda)$ be a strongly inner absolutely irreducible Hecke module. Recall that we have a Galois extension $E/\mathbb{Q}$ that contains a copy of F; furthermore, $\lambda = (\lambda^\tau)_{\tau:F\to E}$ and $E/\mathbb{Q}$ is taken large enough so that σ_f is defined over E. Then it is conjectured that we can attach a motive $\mathbf{M}(\sigma_f)$ that is defined over F, and has coefficients in $E(\sigma_f)$. There is also an effective motive $\mathbf{M}_{\mathrm{eff}}(\sigma_f)$ that is a Tate twist of $\mathbf{M}(\sigma_f)$. This Tate twist is chosen so that the eigenvalues of the Frobenius Φ_p^{-1} in any $\mathfrak{l}$-adic realization $\mathbf{M}_{\mathrm{eff}}(\sigma_f)_{\acute{e}t,\mathfrak{l}}$ are algebraic integers. The motivic L-function attached to $\mathbf{M}_{\mathrm{eff}}(\sigma_f)$ is the same as the cohomological L-function attached to σ_f.

7.1.1 Motivic L-functions

Fix $\iota : E \to \mathbb{C}$ then ${}^\iota\sigma := {}^\iota\sigma_f \otimes {}^\iota\sigma_\infty$ is a cohomological cuspidal automorphic representation. Identify the sets $\mathrm{Hom}(F,\mathbb{C}) = \mathrm{Hom}(F,\mathbb{R}) = \mathsf{S}_\infty$; say, $v \in \mathsf{S}_\infty$ corresponds to $\nu \in \mathrm{Hom}(F,\mathbb{C})$; we also write $v \mapsto \nu_v$ or $\nu \mapsto v_\nu$. As in Sect. 2.3.3, the map $\tau \mapsto \iota \circ \tau$ identifies $\mathrm{Hom}(F,E)$ with $\mathrm{Hom}(F,\mathbb{C})$, and the weight ${}^\iota\lambda \in X_0^*(T_n \times \mathbb{C})$ is written as ${}^\iota\lambda = ({}^\iota\lambda^\nu)_{\nu:F\to\mathbb{C}}$ where ${}^\iota\lambda^\nu = \lambda^{\iota^{-1}\circ\nu}$. The representation ${}^\iota\sigma_\infty$ factors as $\mathbb{D}_{{}^\iota\lambda} = \otimes_{v\in\mathsf{S}_\infty}\mathbb{D}_{{}^\iota\lambda^{\nu_v}}$ up to a signature character $\epsilon_\infty(\sigma_f)$ that is relevant when n is odd. For each ν, let $\varrho({}^\iota\lambda^\nu) := \varrho(\mathbb{D}_{{}^\iota\lambda^\nu})$ be the Langlands parameter of $\mathbb{D}_{{}^\iota\lambda^\nu}$, which is an n-dimensional semisimple representation of the Weil group $W_\mathbb{R}$ of $\mathbb{R}$, as given by the local Langlands correspondence for $\mathrm{GL}_n(F_{v_\nu}) = \mathrm{GL}_n(\mathbb{R})$; see Knapp [47]. The restriction to $\mathbb{C}^\times \simeq W_\mathbb{C} \subset W_\mathbb{R}$

up to a half-integral twist (see (7.2)) is given by

$$| \, |^{\frac{(1-n)}{2}} \varrho({}^\iota\lambda^\nu)|_{\mathbb{C}^\times} \;=\; \bigoplus_{j=1}^{n} (z \mapsto z^{p_j^\tau} \bar{z}^{q_j^\tau}), \quad p_j^\tau = p_j({}^\iota\lambda^\nu), q_j^\tau = q_j({}^\iota\lambda^\nu).$$

(Here $\tau = \iota^{-1} \circ \nu$.) When we restrict from $W_{\mathbb{R}}$ to $W_{\mathbb{C}}$ the signature $\epsilon_\infty(\sigma_f)$ disappears. One can read off the p_j^τ and q_j^τ from the cuspidal parameters; we have

$$p_j^\tau = \frac{\ell_j^\tau - 2d + 1 - n}{2}, \quad q_j = \frac{-\ell_j^\tau - 2d + 1 - n}{2}. \tag{7.1}$$

The parity condition in (3.5) gives that $p_j^\tau, q_j^\tau \in \mathbb{Z}$.

On the other hand, for $\iota : E \to \mathbb{C}$, we have a rank n motive ${}^\iota\mathbf{M} = \mathbf{M}({}^\iota\sigma)$ over F with coefficients in $\iota(E) \subset \mathbb{C}$. For $\nu : F \to \mathbb{C}$, the Betti realization $H_B(\nu, {}^\iota\mathbf{M})$ is an n-dimensional $\iota(E)$-vector space together with a Hodge decomposition:

$$H_B(\nu, {}^\iota\mathbf{M}) \otimes_{\iota(E)} \mathbb{C} \;=\; \bigoplus_{(p,q)\in\mathcal{H}od(\nu,{}^\iota\mathbf{M})} H^{p,q}(\nu, {}^\iota\mathbf{M}),$$

which is a representation of $W_{\mathbb{C}} = \mathbb{C}^\times$, where $z \in \mathbb{C}^\times$ acts on $H^{p,q}(\nu, {}^\iota\mathbf{M})$ via $z^{-p}\bar{z}^{-q}$. The set of Hodge pairs for ${}^\iota\mathbf{M}$ is

$$\mathcal{H}od({}^\iota\mathbf{M}) \;=\; \bigcup_{\nu:F\to\mathbb{C}} \mathcal{H}od(\nu, {}^\iota\mathbf{M}).$$

The motive $\mathbf{M}(\sigma_f)$ is pure; i.e., there exists an integer $\tilde{\mathbf{w}}$ such that for any ι if $(p, q) \in \mathcal{H}od({}^\iota\mathbf{M})$ then $p + q = \tilde{\mathbf{w}}$.

Under the conjectural correspondence $\sigma_f \longleftrightarrow \mathbf{M}(\sigma_f)$, the exponents in the Langlands parameter, up to a specific half-integral twist, and the Hodge numbers determine each other:

$$| \, |^{\frac{(1-n)}{2}} \varrho({}^\iota\lambda^\nu)|_{\mathbb{C}^\times} \;\simeq\; H_B(\nu, {}^\iota\mathbf{M}) \otimes_{\iota(E)} \mathbb{C}, \tag{7.2}$$

where the isomorphism is as representations of $W_{\mathbb{C}}$. Now (7.1) and (7.2) give the Hodge pairs:

$$\begin{aligned} \mathcal{H}od(\nu, {}^\iota\mathbf{M}) = \Bigg\{ & \left(\frac{\ell_1^\tau + 2d + n - 1}{2}, \frac{-\ell_1^\tau + 2d + n - 1}{2} \right), \\ & \left(\frac{\ell_2^\tau + 2d + n - 1}{2}, \frac{-\ell_2^\tau + 2d + n - 1}{2} \right), \\ & \qquad \vdots \\ & \left(\frac{\ell_n^\tau + 2d + n - 1}{2}, \frac{-\ell_n^\tau + 2d + n - 1}{2} \right) \Bigg\}. \end{aligned} \tag{7.3}$$

The purity weight $\tilde{\mathbf{w}}$ of $\mathbf{M}(\sigma_f)$ is given by $\tilde{\mathbf{w}} = 2d + n - 1$.

We will consider a suitable Tate twist of $\mathbf{M}(\sigma_f)$ to make it effective. (We say a motive $\mathbf{M}$ defined over F with coefficients in E is *effective* if for some $\iota : E \to \mathbb{C}$ and $\nu : F \to \mathbb{C}$ the Hodge pairs in $\mathcal{H}od(\nu, {}^{\iota}\mathbf{M})$ are of the form $\{(\mathbf{w}, 0), (\mathbf{w}-a_1, a_1), (\mathbf{w}-a_1-a_2, a_1+a_2), \dots, (0, \mathbf{w})\}$, where $\mathbf{w} = a_1 + \dots + a_{n-1}$ with $a_j \geq 1$.) For $m \in \mathbb{Z}$, let $\mathbf{M}(m) := \mathbf{M} \otimes \mathbb{Q}(m)$, where $\mathbb{Q}(m)$ is the Tate motive for m. If (p, q) is a Hodge pair for $\mathbf{M}$ then $(p-m, q-m)$ is a Hodge pair for $\mathbf{M}(m)$. Recall the definition of the motivic weight from Sect. 3.1.3: $\mathbf{w}_\lambda = \max\{\mathbf{w}_{\lambda^\tau} \mid \tau : F \to E\}$, where $\mathbf{w}_{\lambda^\tau} = a_1^\tau + \dots + a_{n-1}^\tau = \ell_1^\tau$. Define:

$$\mathbf{M}_{\mathrm{eff}}(\sigma_f) \;=\; \mathbf{M}(\sigma_f)\left(\frac{2d+n-1-\mathbf{w}_\lambda}{2}\right). \tag{7.4}$$

Suppose τ_0 is such that $\mathbf{w}_\lambda = \mathbf{w}_{\lambda^{\tau_0}}$. Take any ι and ν as above. The Hodge pairs in $\mathcal{H}od(\nu, {}^{\iota}\mathbf{M}_{\mathrm{eff}}(\sigma_f))$ are of the form

$$\left\{\left(\frac{\ell_1^\tau+\ell_1^{\tau_0}}{2}, \frac{-\ell_1^\tau+\ell_1^{\tau_0}}{2}\right), \left(\frac{\ell_2^\tau+\ell_1^{\tau_0}}{2}, \frac{-\ell_2^\tau+\ell_1^{\tau_0}}{2}\right), \dots, \left(\frac{\ell_n^\tau+\ell_1^{\tau_0}}{2}, \frac{-\ell_n^\tau+\ell_1^{\tau_0}}{2}\right)\right\}, \tag{7.5}$$

where $\tau = \iota^{-1} \circ \nu$. In particular, for a pair ι_0 and ν_0 such that $\iota_0^{-1} \circ \nu_0 = \tau_0$ we get

$$\begin{aligned}&\mathcal{H}od(\nu_0, {}^{\iota_0}\mathbf{M}_{\mathrm{eff}}(\sigma_f)) \\ &\quad= \{(\mathbf{w}_\lambda, 0), (\mathbf{w}_\lambda - a_1^{\tau_0}, a_1^{\tau_0}), (\mathbf{w} - a_1^{\tau_0} - a_2^{\tau_0}, a_1^{\tau_0} + a_2^{\tau_0}), \dots, (0, \mathbf{w}_\lambda)\}.\end{aligned}$$

Let's record a consequence of (7.2) for the behavior of the correspondences

$$\sigma_f \longleftrightarrow \mathbf{M}(\sigma_f) \quad \text{and} \quad \sigma_f \longleftrightarrow \mathbf{M}_{\mathrm{eff}}(\sigma_f)$$

under taking duals. Given $\sigma_f \in \mathrm{Coh}_{!!}(G_n, \lambda)$, the contragredient module $\sigma_f^\vee$ is in $\mathrm{Coh}_{!!}(G_n, \lambda^\vee)$, where $\lambda^\vee = -w_{G_n}(\lambda)$ is the dual weight. Further, the local Langlands correspondence commutes with taking contragredients. On the motivic side, given a motive $\mathbf{M}$ over F with coefficients in E, the dual motive $\mathbf{M}^\vee$ is also defined over F with coefficients in E whose realizations are the respective duals of the realizations of $\mathbf{M}$. In particular, if (p, q) is a Hodge pair for $\mathbf{M}$ then $(-p, -q)$ is a Hodge pair for $\mathbf{M}^\vee$. Putting these observations together with (7.2), we leave it to the reader to check that

$$\mathbf{M}(\sigma_f^\vee) \;=\; \mathbf{M}(\sigma_f)^\vee(1-n). \tag{7.6}$$

For effective motives, from (7.4) and (7.6), we get

$$\mathbf{M}_{\text{eff}}(\sigma_f^{\vee}) \;=\; \mathbf{M}_{\text{eff}}(\sigma_f)^{\vee}(-\mathbf{w}_{\lambda}). \tag{7.7}$$

(The reader who wishes to check this will need to use that $d(\lambda^{\vee}) = -d(\lambda)$ and $\mathbf{w}_{\lambda^{\vee}} = \mathbf{w}_{\lambda}$.)

Let $\mathbf{M}$ be a motive defined over F with coefficients in E. Its ℓ-adic realization $H_{\ell}(\mathbf{M})$ admits an action of the absolute Galois group of F, and for each $\iota : E \to \mathbb{C}$, we can take the associated L-function (see [15]). The coefficients appearing in the local Euler factors at all finite places initially take values in $E_{\mathfrak{l}}$ for $\mathfrak{l}|\ell$; the usual expectation of ℓ-independence make them take values in E, and after applying ι we get a $\mathbb{C}$-valued local L-function. Taking the Euler product over all finite places, we get the finite part of the L-function denoted $L_f(\iota, \mathbf{M}, s)$. (See Deligne [15, 2.2] for a precise definition.) For the L-factor at infinity, denoted $L_{\infty}(\iota, \mathbf{M}, s)$, see (7.18). The completed L-function, defined as $L(\iota, \mathbf{M}, s) := L_{\infty}(\iota, \mathbf{M}, s)L_f(\iota, \mathbf{M}, s)$, conjecturally extends to a meromorphic function to all of $\mathbb{C}$ and satisfies a functional equation of the form

$$L(\iota, \mathbf{M}, s) \;=\; \varepsilon(\iota, \mathbf{M}, s)L(\iota, \mathbf{M}^{\vee}, 1-s), \tag{7.8}$$

where the ε-factor on the right-hand side is a nonzero constant times an exponential function. All these comments will be applied to L-functions for $\mathbf{M}(\sigma_f)$ and $\mathbf{M}_{\text{eff}}(\sigma_f)$.

On the other hand, for $\iota : E \to \mathbb{C}$ we have the standard automorphic L-function of the cuspidal automorphic representation ${}^{\iota}\sigma$. This may be defined in terms of integral representations as, for example, in Jacquet [36], or may be defined as Langlands L-functions as in Sect. 7.2.1. It is known that one gets the same L-function irrespective of which definition is used. The product of local L-functions over all finite places is denoted $L_f^{\text{aut}}(s, {}^{\iota}\sigma) = L^{\text{aut}}(s, {}^{\iota}\sigma_f)$, and the product over all archimedean places giving the L-factor at infinity is denoted $L_{\infty}^{\text{aut}}(s, {}^{\iota}\sigma) = L^{\text{aut}}(s, {}^{\iota}\sigma_{\infty})$. The completed L-function is defined as $L^{\text{aut}}(s, {}^{\iota}\sigma) = L_{\infty}^{\text{aut}}(s, {}^{\iota}\sigma)L_f^{\text{aut}}(s, {}^{\iota}\sigma)$. For automorphic L-functions it is known that $L^{\text{aut}}(s, {}^{\iota}\sigma)$ admits an analytic continuation to an entire function (unless $n = 1$ and σ_f is the trivial representation whence the L-function is the Dedekind zeta function of F, which has a pole at $s = 1$), and satisfies a functional equation:

$$L^{\text{aut}}(s, {}^{\iota}\sigma) \;=\; \varepsilon(s, {}^{\iota}\sigma)L^{\text{aut}}(1-s, {}^{\iota}\sigma^{\vee}).$$

The key point in the expected comparison between the motivic L-functions and automorphic L-function is that there is an analogue at finite places of (7.2), i.e., for any rational prime p and a prime ideal $\mathfrak{p}$ of F above p, the Langlands parameter $\varrho({}^{\iota}\sigma_{\mathfrak{p}})$ (which is an n-dimensional semisimple representation of the Weil group $W_{F_{\mathfrak{p}}}$ of $F_{\mathfrak{p}}$) is related to the local Galois representation at $\mathfrak{p}$ on the

ℓ-adic realization $H_\ell({}^\iota\mathbf{M}(\sigma_f))$ via

$$| \, |^{\frac{(1-n)}{2}} \varrho({}^\iota\sigma_{\mathfrak{p}}) \;\simeq\; H_\ell({}^\iota\mathbf{M}(\sigma_f)). \tag{7.9}$$

Taking the corresponding Artin L-factors on both sides and then taking a product over all $\mathfrak{p}$ gives

$$L\left(\iota, \mathbf{M}(\sigma_f), s\right) \;=\; L^{\text{aut}}\left(s + \tfrac{1-n}{2}, {}^\iota\sigma_f\right). \tag{7.10}$$

Using (7.4) we get

$$L\left(\iota, \mathbf{M}_{\text{eff}}(\sigma_f), s\right) \;=\; L^{\text{aut}}\left(s + \tfrac{2d-\mathbf{w}}{2}, {}^\iota\sigma_f\right). \tag{7.11}$$

The shift $(2d-\mathbf{w})/2$ in the s-variable can be seen by working intrinsically in the context of the cohomology of arithmetic groups by considering a cohomological L-function attached to σ_f which we now discuss.

7.1.2 Cohomological L-functions

Let λ and $\sigma_f \in \text{Coh}_{!!}(G_n, \lambda)$ be as above. Let $E/\mathbb{Q}$ be a finite Galois extension that contains a copy of F; furthermore, $\lambda = (\lambda^\tau)_{\tau:F\to E}$ and $E/\mathbb{Q}$ is taken large enough so that σ_f is defined over E. Then the conjectured effective motive $\mathbf{M}_{\text{eff}}(\sigma_f)$ is defined over $E(\lambda)$ (see Sect. 2.2.1) and has coefficients in E. This motive gives rise to a compatible system of $\mathfrak{l}$-adic representation

$$\rho_{\mathfrak{l}}(\sigma_f) : \text{Gal}(\bar{\mathbb{Q}}/E(\lambda)) \to \text{GL}(\mathbf{M}(\sigma_f)_{et} \otimes E_{\mathfrak{l}}), \tag{7.12}$$

which is unramified outside S (say), and for any $v \notin \mathsf{S}$ the local Euler factor at v in (7.11) is of the form

$$\det(\text{Id} - \rho_{\mathfrak{l}}(\sigma_f)(\Phi_v^{-1})q_v^{-s}|\mathbf{M}(\sigma_f)_{et} \otimes E_{\mathfrak{l}}) \\ = \det(\text{Id} - q_v^{\frac{\mathbf{w}-2d}{2}} \vartheta_v(\sigma_v)q_v^{-s}) \in \mathcal{O}_E[q_v^{-s}], \tag{7.13}$$

where Φ_v is the Frobenius element at v and $\vartheta_v(\sigma_v)$ is the Satake parameter of σ_f at v considered as an element in the dual group $\text{GL}_n(\mathbb{C})$.

This needs some explanation, especially the integrality of the coefficients in the local factor. The prime v lies above a prime p in $\mathbb{Z}$; we assume in addition that p is unramified in the normal closure of F. Choose a prime v_1 in E that lies above v. Then the set $I_{F_v}^{E_{v_1}}$ of embeddings of $F_v/\mathbb{Q}_p$ into $E_{v_1}/\mathbb{Q}_p$ is a subset of $\{\tau : F \to E\}$. Of course we get a disjoint union $\cup_{v|p} I_{F_v}^{E_{v_1}} = \tau : F \to E$.

We look at the cocharacters $\chi_\nu : \mathbb{G}_m \to T_0$, which are given by

$$\chi_\nu := \sum_{i=1}^{\nu} \eta_i, \quad t \mapsto \mathrm{diag}(t, t, \dots t, 1, \dots, 1)$$

(see Sect. 5.1.2). We choose a uniformizing element ϖ_v in F_v and consider the double coset $\mathrm{GL}_n(\mathcal{O}_v)\chi_\nu(\varpi_v)\mathrm{GL}_n(\mathcal{O}_v)$ in $\mathrm{GL}_n(F_v)$. This double coset defines an endomorphism $\mathbf{c}h(\chi_\nu)$ of $H^\bullet(\mathcal{S}^G_{K_f}, \mathcal{M}_\lambda)$ by convolution. In [30] we describe the cohomology with integral coefficient systems, i.e., we define the cohomology groups $H^\bullet(\mathcal{S}^G_{K_f}, \widetilde{\mathcal{M}}_{\lambda,\mathcal{O}_E})$ and also define the action of the Hecke algebra on the integral cohomology. We put

$$< \chi_\nu, \lambda^{(1)}_v + \rho_v > := \sum_{\tau \in I^{E_{v1}}_{F_v}} < \chi_\nu, \lambda^{(1,\tau)} + \rho > .$$

Then a slight extension of the arguments in loc. cit. shows that we can define an endomorphism

$$p^{<\chi_\nu,\lambda^{(1)}_v+\rho_v>} q_v^{-d<\chi_\nu,\delta>} \mathbf{c}h(\chi_\nu) \ : \ H^\bullet(\mathcal{S}^G_{K_f}, \widetilde{\mathcal{M}}_{\lambda,\mathcal{O}_E}) \to H^\bullet(\mathcal{S}^G_{K_f}, \widetilde{\mathcal{M}}_{\lambda,\mathcal{O}_E}). \quad (7.14)$$

Such an endomorphism acts as a scalar on $H^\bullet(\mathcal{S}^G_{K_f}, \widetilde{\mathcal{M}}_{\lambda,\mathcal{O}_E})(\sigma_f)$, and this scalar is the ν-th elementary symmetric function evaluated at

$$\{\dots, p^{<\chi_1,\lambda^{(1)}_v+\rho>} q_v^{-d(\chi_1,\delta)} \vartheta_j, \dots\}_{j=1,\dots,n},$$

i.e., it is

$$A_\nu(\sigma_v) \ := \ S_\nu \left(\{\dots, p^{<\chi_1,\lambda^{(1)}_v+\rho>} q_v^{-d(\chi_1,\delta)} \vartheta_j, \dots\}_{j=1,\dots,n}\right) \in \mathcal{O}_E,$$

and this implies that the numbers $p^{<\chi_1,\lambda^{(1)}_v+\rho>} q_v^{-d(\chi_1,\delta)} \vartheta_j$ are algebraic integers. Furthermore, $p^{<\chi_\nu,\lambda^{(1)}_v+\rho>} q_v^{-d(\chi_\nu,\delta)} = q_v^{\frac{\mathrm{w}-2d}{2}}$, and hence we conclude that in fact the expressions in (7.13) are polynomials $1 + A_1(\sigma_v)q_v^{-s} + \cdots \in \mathcal{O}_E[q_v^{-s}]$. We also see that the eigenvalues of the inverse Frobenius $\rho_\mathfrak{l}(\sigma_f)(\Phi_v^{-1})$ are the numbers

$$\left\{\dots, q_v^{\frac{\mathrm{w}-2d}{2}} \vartheta_{v,j}(\sigma_v), \dots\right\}_{\{j=1,\dots,n\}}.$$

Let's note the following remarks:

(a) The exponent in $q_v^{\frac{\mathrm{w}-2d}{2}}$ is optimal in a certain sense. At a prime p that splits completely in the normal closure of F we find—by definition—a $v|p$ such that $< \chi_\nu, \lambda^{(1)}_v + \rho >= \frac{\mathrm{w}}{2}$. Since the exponent of p in (7.14) is chosen in such a way that the endomorphism is nonzero modulo ϖ_v, it becomes clear what it means that the exponent $\frac{\mathbf{w}-2d}{2}$ is optimal.

(b) The expression $\det(\mathrm{Id} - q_v^{\frac{\mathbf{w}-2d}{2}} \vartheta_v(\sigma_v) q_v^{-s}) \in \mathcal{O}_E[q_v^{-s}]$ is invariant under twisting; if we modify $d \to d' = d + k$, then the Satake parameter of the twisted representation $\sigma'_f = \sigma_f \otimes |\delta|^{-k}$ gets multiplied by $|\varpi_v|^{-k} = q_v^k$ and this cancels against the factor q_v^{-k} in $q_v^{-d'}$.

We now define the local Euler factor of the cohomological L-function at a prime $p \notin \mathsf{S}$ that is unramified in the normal closure of $F/\mathbb{Q}$ as

$$L_p^{\mathrm{Coh}}(\sigma_f, s) = \prod_{v|p} \frac{1}{\det\left(\mathrm{Id} - q_v^{\frac{\mathbf{w}-2d}{2}} \vartheta_v(\sigma_v) q_v^{-s}\right)}, \tag{7.15}$$

with the denominator being a polynomial in $\mathcal{O}_E[p^{-s}]$. For $\iota : E \to \mathbb{C}$, we will let $L_p^{\mathrm{Coh}}(\iota, \sigma_f, s)$ denote the effect of applying ι to coefficients of the polynomial. Given $\iota : E \to \mathbb{C}$, there is the cuspidal automorphic representation ${}^\iota\sigma = {}^\iota\sigma_\infty \otimes {}^\iota\sigma_f$. At the places $v \notin \mathsf{S}$ we have

$$L_v^{\mathrm{aut}}(s, {}^\iota\sigma_v) = \frac{1}{\det\left(\mathrm{Id} - \vartheta_v({}^\iota\sigma_v) q_v^{-s}\right)} = L_v^{\mathrm{Coh}}\left(\iota, \sigma_f, s + \frac{\mathbf{w}-2d}{2}\right), \tag{7.16}$$

where the last term is obtained by applying ι to the coefficients of the above polynomial in $\mathcal{O}_E[q_v^{-s}]$. With this as a cue, we define the local factors of the cohomological L-function for $v \in \mathsf{S}$ and for v archimedean via the same shift in the s-variable, and hence we can define the completed cohomological L-function as

$$L^{\mathrm{Coh}}(\iota, \sigma, s) = L^{\mathrm{aut}}\left(s - \frac{\mathbf{w}-2d}{2}, {}^\iota\sigma\right). \tag{7.17}$$

From (7.11) and (7.17) it is clear that

$$L^{\mathrm{Coh}}(\iota, \sigma, s) = L\left(\iota, \mathbf{M}_{\mathrm{eff}}(\sigma_f), s\right).$$

We should keep in mind that the cohomological L-function, as well as the motive $\mathbf{M}_{\mathrm{eff}}(\sigma_f)$, are attached to the conformal class of σ_f (see Remark (b) earlier).

7.1.3 Critical points of motivic L-functions

Suppose $\mathbf{M}$ is a pure regular rank n motive defined over F with coefficients in E. Let $\mathbf{w} = \mathbf{w}(\mathbf{M})$ be the purity weight of $\mathbf{M}$; i.e., for any Hodge type (p, q) of $\mathbf{M}$ we have $p + q = \mathbf{w}$. By regularity we mean that any nonzero Hodge number is 1. Fix $\iota : E \to \mathbb{C}$. For any $\nu : F \to \mathbb{C}$ suppose the Hodge types are $\mathcal{H}od(\nu, {}^\iota\mathbf{M}) = \{(p_1^\tau, q_1^\tau), \ldots, (p_n^\tau, q_n^\tau)\}$, where $\tau = \iota^{-1} \circ \nu : F \to E$. Further, we suppose that there is no middle Hodge type; i.e., for every τ and j, $p_j^\tau \neq \mathbf{w}/2$.

In this situation, the Γ-factor at the infinite place v_ν corresponding to ν is given by (Serre [61, (25)])

$$L_{v_\nu}(\iota, \mathbf{M}, s) = \prod_{p_j^\tau < \frac{\mathbf{w}}{2}} \Gamma_{\mathbb{C}}(s - p_j^\tau),$$

where $\Gamma_{\mathbb{C}}(s) = 2(2\pi)^{-s}\Gamma(s)$. Taking the product over all the archimedean places, we get

$$L_\infty(\iota, \mathbf{M}, s) = \prod_{\nu: F \to \mathbb{C}} \prod_{p_j^\tau < \frac{\mathbf{w}}{2}} \Gamma_{\mathbb{C}}(s - p_j^\tau) = \prod_{\tau: F \to E} \prod_{p_j^\tau < \frac{\mathbf{w}}{2}} \Gamma_{\mathbb{C}}(s - p_j^\tau). \quad (7.18)$$

We have the following relation between the Betti realizations:

$$H_B(\mathbf{M}^\vee(1)) = H_B(\mathbf{M}(\mathbf{w}+1)).$$

(The Betti realization is essentially self-dual.) Hence:

$$\begin{aligned} L_\infty(\iota, \mathbf{M}^\vee, 1-s) &= L_\infty(\iota, \mathbf{M}^\vee(1), -s) \\ &= L_\infty(\iota, \mathbf{M}(\mathbf{w}+1), -s) = L_\infty(\iota, \mathbf{M}, \mathbf{w}+1-s). \end{aligned}$$

The functional equation may be written as

$$L_\infty(\iota, \mathbf{M}, s) L_f(\iota, \mathbf{M}, s) = \varepsilon(\iota, \mathbf{M}, s) L_\infty(\iota, \mathbf{M}, \mathbf{w}+1-s) L_f(\iota, \mathbf{M}^\vee, 1-s).$$

An integer m is said to be *critical* for $L(\iota, \mathbf{M}, s)$ if the Γ-factors on either side of the functional equation are regular, i.e., do not have poles at $s = m$ (see Deligne [15]). This means that both $L_\infty(\iota, \mathbf{M}, s)$ and $L_\infty(\iota, \mathbf{M}, \mathbf{w}+1-s)$ are regular at $s = m$. Define

$$\begin{aligned} p_{\max} &:= \max\{p_j^\tau : p_j^\tau < \mathbf{w}/2\}, \text{ and} \\ & \qquad q_{\min} := \mathbf{w} - p_{\max} = \min\{q_j^\tau : q_j^\tau > \mathbf{w}/2\}. \end{aligned} \quad (7.19)$$

It is an easy exercise using (7.18) to see that

$$\begin{aligned} \text{The critical set for } L(\iota, \mathbf{M}, s) &= \{m \in \mathbb{Z} : p_{\max} < m \leq q_{\min}\} \\ &= \{p_{\max}+1, p_{\max}+2, \ldots, q_{\min}-1, q_{\min}\}. \end{aligned} \quad (7.20)$$

The critical set is centered around $(\mathbf{w}+1)/2$ and has $q_{\min} - p_{\max}$ integers. The assumption that there is no middle Hodge type assures us that the critical set is nonempty. It follows from (7.19) and (7.20) that the critical set for $L(\iota, \mathbf{M}, s)$ is independent of ι.

7.2 CRITICAL POINTS FOR L-FUNCTIONS AND THE COMBINATORIAL LEMMA

7.2.1 The Rankin–Selberg L-functions

(We refer the reader to Shahidi's book [65, 10.1] for further details and references.) Let σ (resp., σ') be a cuspidal automorphic representation of $G_n(\mathbb{A})$ (resp., $G_{n'}(\mathbb{A})$). For any place v the local L-function $L(s, \sigma_v \times \sigma_v'^{\mathsf{v}})$ is defined as a local Artin L-function via the local Langlands correspondence. For a finite unramified place v, if $\mathrm{diag}(\vartheta_{1,v}, \ldots, \vartheta_{n,v})$ (resp., $\mathrm{diag}(\vartheta'_{1,v}, \ldots, \vartheta'_{n',v})$) are the Satake parameters of σ_v (resp., σ'_v). Then the local L-function is defined as

$$L^{\mathrm{aut}}(s, \sigma_v \times \sigma_v'^{\mathsf{v}}) = \prod_{\substack{1 \le i \le n \\ 1 \le j \le n'}} (1 - \vartheta_{i,v}\vartheta'^{-1}_{j,v} q_v^{-s})^{-1}. \tag{7.21}$$

There are other ways of defining the local factor, for example, via the approach of zeta integrals and integral representations. It is a fact that we end up with the same local factor; see loc. cit. The global L-function is defined for $\Re(s) \gg 0$ as an Euler product: $L(s, \sigma \times \sigma'^{\mathsf{v}}) = \prod_v L(s, \sigma_v \times \sigma_v'^{\mathsf{v}})$. For any finite set of places S, the partial L-function is defined for $\Re(s) \gg 0$ as $L^{\mathsf{S}}(s, \sigma \times \sigma'^{\mathsf{v}}) = \prod_{v \notin \mathsf{S}} L(s, \sigma_v \times \sigma_v'^{\mathsf{v}})$. These Euler products, defined a priori only in a half-plane, admit a meromorphic continuation to all of $\mathbb{C}$. We need the following results (see [65, Thm. 10.1.1] and [39, Sect. 3]).

Theorem 7.1. *Let σ (resp., σ') be a cuspidal automorphic representation of $G_n(\mathbb{A})$ (resp., $G_{n'}(\mathbb{A})$).*

1. *Suppose $n \neq n'$. Then $L(s, \sigma \times \sigma'^{\mathsf{v}})$ extends to an entire function of s.*
2. *Suppose $n = n'$. Then $L(s, \sigma \times \sigma'^{\mathsf{v}})$ extends to an entire function of s, unless $\sigma \simeq \sigma'$, and in which case $L(s, \sigma \times \sigma^{\mathsf{v}})$ extends to a meromorphic function of s with only one pole, which is located at $s = 1$ and is a simple pole.*
3. *Functional equation: $L(s, \sigma \times \sigma'^{\mathsf{v}}) = \varepsilon(s, \sigma \times \sigma'^{\mathsf{v}})L(1-s, \sigma^{\mathsf{v}} \times \sigma')$. (The epsilon factor on the right-hand side is an exponential function.)*
4. *Suppose that σ and σ' are unitary. Then for any finite set of places S we have*

$$L^{\mathsf{S}}(s, \sigma \times \sigma'^{\mathsf{v}}) \neq 0, \quad \Re(s) \geq 1.$$

Now suppose we are in an arithmetic situation, and say $\sigma_f \in \mathrm{Coh}_{!!}(G_n, \mu)$ and $\sigma'_f \in \mathrm{Coh}_{!!}(G_{n'}, \mu')$, both defined over E. (Recall the notations in Sect. 5.3.7.) Take $\iota : E \to \mathbb{C}$, and define the cohomological (or motivic) Rankin–Selberg L-function by

$$L^{\mathrm{Coh}}(\iota, \sigma_f \times \sigma_f'^{\mathsf{v}}, s) = L(\iota, \mathbf{M}_{\mathrm{eff}}(\sigma_f) \otimes \mathbf{M}_{\mathrm{eff}}(\sigma_f'^{\mathsf{v}}), s). \tag{7.22}$$

This gives the following relation with the automorphic Rankin–Selberg L-function:

$$L\left(\iota, \mathbf{M}_{\text{eff}}(\sigma_f) \otimes \mathbf{M}_{\text{eff}}(\sigma_f'^{\mathsf{v}}), s\right) = L^{\text{Coh}}(\iota, \sigma_f \times \sigma_f'^{\mathsf{v}}, s)$$
$$= L\left(s - \tfrac{\mathbf{w}+\mathbf{w}'-2a(\underline{\mu})}{2}, {}^\iota\sigma_f \times {}^\iota\sigma_f'^{\mathsf{v}}\right). \quad (7.23)$$

The functional equation for the automorphic L-function implies a functional equation for the cohomological L-function:

$$L^{\text{Coh}}(\iota, \sigma_f \times \sigma_f'^{\mathsf{v}}, s) \approx L^{\text{Coh}}(\iota, \sigma_f^{\mathsf{v}} \times \sigma_f', 1 + \mathbf{w} + \mathbf{w}' - s), \quad (7.24)$$

where $\approx$ means equality up to a relevant epsilon-factor; the shift in the s-variable is compatible with (7.7).

Remark 7.2. Recall that the data (μ, μ') is said to be on the right of the unitary axis if $-a(\underline{\mu}) \geq N/2$ or that $-N/2 - d + d' \geq 0$. For $\iota : E \to \mathbb{C}$, the cuspidal representation ${}^\iota\sigma$ may be written as ${}^\iota\sigma = {}^\iota\sigma^\circ \otimes |\ |^{-d}$, where ${}^\iota\sigma^\circ$ is a unitary cuspidal representation, and similarly for ${}^\iota\sigma' = {}^\iota\sigma'^\circ \otimes |\ |^{-d'}$ with a unitary ${}^\iota\sigma'^\circ$. Hence, by (4) of the above theorem, for any S the L-value

$$L^{\mathsf{S}}(1 - \tfrac{N}{2}, \sigma \times \sigma'^{\mathsf{v}}) = L^{\mathsf{S}}\left(1 - \tfrac{N}{2} - d + d', \sigma^\circ \times \sigma'^{\circ\mathsf{v}}\right)$$

is nonzero. The data (μ, μ') being on the right of the unitary axis translates to the L-value being toward the right of the unitary line $\Re(s) = 1$. It is this L-value that appears in the denominator in Langlands's calculation of the standard intertwining operator, which in turn gives holomorphy of $\text{Eis}_P(s, f)$ at $s = -N/2$. See Thm. 7.10.

7.2.2 The critical set for Rankin–Selberg L-functions

The definition of a critical point as in Sect. 7.1.3 may be adapted as before for automorphic Rankin–Selberg L-functions. Given a cuspidal automorphic representation σ, we can write the representation at infinity as $\sigma_\infty = \otimes_{v \in \mathsf{S}_\infty} \sigma_v$; the tensor product is over all the archimedean places of F. The L-factor at infinity

$$L(s, \sigma_\infty \times \sigma_\infty'^{\mathsf{v}}) = \prod_{v \in \mathsf{S}_\infty} L(s, \sigma_v \times \sigma_v'^{\mathsf{v}})$$

is a product of Γ-factors times exponential functions.

Definition 7.3. *Let σ (resp., σ') be as above.*

1. *If $n \equiv n' \pmod 2$, then take $m \in \mathbb{Z}$.*
2. *If $n \not\equiv n' \pmod 2$, then take $m \in \frac{1}{2} + \mathbb{Z}$.*

We say that such an m is critical for the Rankin–Selberg L-function $L(s, \sigma \times \sigma'^{\mathsf{v}})$ if the local factors at infinity on both sides of the functional equation are regular (holomorphic) at $s = m$; i.e., $L(s, \sigma_\infty \times \sigma'^{\mathsf{v}}_\infty)$ and $L(1-s, \sigma^{\mathsf{v}}_\infty \times \sigma'_\infty)$ are finite at $s = m$.

We also have the notion of critical point for the cohomological or motivic Rankin–Selberg L-function—the definition is the same as in Sect. 7.1.3—where we have to take the shift in the variable s into account when we write the functional equation as in (7.24). On the motivic side a critical point is always an integer. We briefly present the calculation of the critical set of Rankin–Selberg L-functions; we may do this entirely in the automorphic setup or in the motivic setup; the result is the same up to the explicit shift in the s-variable.

7.2.2.1 The critical set for $L(\iota, \mathbf{M}_{\mathrm{eff}}(\sigma_f) \times \mathbf{M}_{\mathrm{eff}}(\sigma'^{\mathsf{v}}_f), s)$

Let $\sigma_f \in \mathrm{Coh}_{!!}(G_n, \mu)$ and $\sigma'_f \in \mathrm{Coh}_{!!}(G_{n'}, \mu')$, with $\mu = (\mu^\tau)_{\tau:F\to E}$ and $\mu' = (\mu'^\tau)_{\tau:F\to E}$. Then $\sigma'^{\mathsf{v}}_f \in \mathrm{Coh}_{!!}(G_{n'}, \mu'^{\mathsf{v}})$. Let the motivic weights be $\mathbf{w} = \mathbf{w}(\mu)$ and $\mathbf{w}' = \mathbf{w}(\mu') = \mathbf{w}(\mu'^{\mathsf{v}})$. Similarly, let the cuspidal parameters for μ be $\ell = (\ell^\tau)_{\tau:F\to E}$ with $\ell^\tau = (\ell^\tau_i)_{1\le i\le n}$. The cuspidal parameters for μ' are the same as the cuspidal parameters for dual weight μ'^{v} and these we denote by $\ell' = (\ell'^\tau)_{\tau:F\to E}$ with $\ell'^\tau = (\ell'^\tau_j)_{1\le j\le n'}$. For $\iota : E \to \mathbb{C}$ and $\nu : F \to \mathbb{C}$, using (7.5), the Hodge pairs of the tensor product motive are given by

$$\mathcal{H}od(\nu, {}^\iota\mathbf{M}_{\mathrm{eff}}(\sigma_f) \otimes {}^\iota\mathbf{M}_{\mathrm{eff}}(\sigma'^{\mathsf{v}}_f)) = \left\{ \left(\frac{\ell^\tau_i + \ell'^\tau_j + \mathbf{w} + \mathbf{w}'}{2}, \frac{-\ell^\tau_i - \ell'^\tau_j + \mathbf{w} + \mathbf{w}'}{2} \right)_{\substack{1\le i\le n \\ 1\le j\le n'}} \right\}. \tag{7.25}$$

We would like to see under what condition there is no middle Hodge type in such a tensor product motive. This begets the following definition.

Definition 7.4. *We say that $\sigma_f \in \mathrm{Coh}_{!!}(G_n, \mu)$ and $\sigma'_f \in \mathrm{Coh}_{!!}(G_{n'}, \mu')$ have disjoint cuspidal parameters if $\ell^\tau_i \neq \ell'^\tau_j$ for all $1 \le i \le n$, $1 \le j \le n'$ and $\tau : F \to E$. Also, we define the cuspidal width between σ_f and σ'_f to be*

$$\ell(\sigma_f, \sigma'_f) = \ell(\mu, \mu') = \min\{\, |\ell^\tau_i - \ell'^\tau_j| \;:\; 1 \le i \le n,\ 1 \le j \le n',\ \text{and } \tau : F \to E\}.$$

Note that the cuspidal parameters are disjoint if and only if $\ell(\sigma_f, \sigma'_f) > 0$. Also, the cuspidal width depends only on the semisimple parts of the weights, and furthermore, since the cuspidal parameters of a weight and its dual weight are equal, we deduce:

$$\ell(\mu, \mu') = \ell(\mu^{(1)}, \mu'^{(1)}) = \ell(\mu, \mu'^{\mathsf{v}}), \text{ etc.}$$

Proposition 7.5. *The tensor product motive* ${}^{\iota}\mathbf{M}_{\text{eff}}(\sigma_f) \otimes {}^{\iota}\mathbf{M}_{\text{eff}}(\sigma_f'^{\vee})$ *has no middle Hodge type if and only if* σ_f *and* σ_f' *have disjoint cuspidal parameters. Furthermore, cuspidal parameters being disjoint implies that* nn' *is even.*

Proof. The weight of the tensor product motive is $\mathbf{w} + \mathbf{w}'$; hence there is a middle Hodge type if and only if $\ell_i^\tau + \ell_j'^\tau = 0$ or that $\ell_j'^\tau = \ell_{n-i+1}^\tau$ for some i, j, τ. Suppose n and n' are both odd, then for each τ, we have $\ell_i^\tau = 0$ for $i = (n+1)/2$ and $\ell_j'^\tau = 0$ for $j = (n'+1)/2$; hence $\ell(\sigma_f, \sigma_f') = 0$. □

Proposition 7.6. *Suppose* $\sigma_f \in \text{Coh}_{!!}(G_n, \mu)$ *and* $\sigma_f' \in \text{Coh}_{!!}(G_{n'}, \mu')$ *have disjoint cuspidal parameters. Then the critical set for* $L(\iota, \mathbf{M}_{\text{eff}}(\sigma_f) \times \mathbf{M}_{\text{eff}}(\sigma_f'^{\vee}), s)$ *is the finite set:*

$$\left\{ m \in \mathbb{Z} \ : \ \frac{-\ell(\mu, \mu') + \mathbf{w} + \mathbf{w}'}{2} < m \leq \frac{\ell(\mu, \mu') + \mathbf{w} + \mathbf{w}'}{2} \right\}.$$

The critical set is centered around $(1 + \mathbf{w} + \mathbf{w}')/2$ *and the number of critical points is the cuspidal width* $\ell(\mu, \mu')$.

Let's recall the parity condition: $\ell(\mu, \mu') \equiv \mathbf{w} + \mathbf{w}' \pmod 2$, which tells us that the bounds for the critical set in the above proposition are both integers.

Proof. From (7.25) it follows that $p_{\max} = (-\ell(\mu, \mu') + \mathbf{w} + \mathbf{w}')/2$ and $q_{\min} = (\ell(\mu, \mu') + \mathbf{w} + \mathbf{w}')/2$. The rest is clear from the discussion in Sect. 7.1.3. □

7.2.2.2 *The critical set for* $L(s, \sigma \times \sigma'^{\vee})$

Suppose $\sigma_f \in \text{Coh}_{!!}(G_n, \mu)$ and $\sigma_f' \in \text{Coh}_{!!}(G_{n'}, \mu')$. Assume that n is even. For $\iota : E \to \mathbb{C}$, we know from Thm. 5.2 that the representations at infinity of ${}^{\iota}\sigma$ (resp., ${}^{\iota}\sigma'$) are given by $\mathbb{D}_{{}^{\iota}\mu}$ (resp., $\mathbb{D}_{{}^{\iota}\mu'}$ up to a signature character when n' is odd). For each archimedean place v and corresponding $\nu_v : F \to \mathbb{C}$ we know the local archimedean Langlands parameters $\varrho({}^{\iota}\sigma_v) = \varrho({}^{\iota}\mu^{\nu_v})$ (resp., $\varrho({}^{\iota}\sigma_v') = \varrho({}^{\iota}\mu'^{\nu_v})$). For notational brevity, for any integer k, let's denote by $I(k)$ the two-dimensional irreducible representation of $W_{\mathbb{R}}$ obtained by inducing the unitary character $z \mapsto (z/\bar{z})^{k/2}$, i.e., the character $re^{i\theta} \mapsto e^{ik\theta}$, of $\mathbb{C}^\times = W_{\mathbb{C}}$. Using the local Langlands correspondence for $\text{GL}_n(\mathbb{R})$ and $\text{GL}_{n'}(\mathbb{R})$ we have

$$\varrho({}^{\iota}\sigma_v) \ = \ \bigoplus_{i=1}^{n/2} I(\ell_i^\tau) \otimes |\ |^{-d},$$

$$\varrho({}^{\iota}\sigma_v'^{\vee}) \ = \ \begin{cases} \bigoplus_{j=1}^{(n'-1)/2} I(\ell_j'^\tau) \otimes |\ |^{d'} \oplus sgn^{\epsilon_v'} |\ |^{d'}, & \text{if } n' \text{ is odd}, \\ \bigoplus_{j=1}^{n'/2} I(\ell_j'^\tau) \otimes |\ |^{d'}, & \text{if } n' \text{ is even}. \end{cases} \tag{7.26}$$

(Here $\iota \circ \tau = \nu_v$.)

The following properties are readily checked:

(i) $I(k)$ is irreducible for $k \neq 0$, and $I(0) = 1\!\!1 \oplus sgn$;
(ii) $I(k) \simeq I(-k)$;
(iii) $I(k) \otimes sgn \simeq I(k)$; and
(iv) $I(k) \otimes I(k') = I(k+k') \oplus I(|k-k'|)$.

Under the assumption that σ_f and σ'_f have disjoint cuspidal parameters, if n' is odd then $\varrho({}^\iota\sigma_v) \otimes \varrho({}^\iota\sigma_v'^{\mathsf{v}})$ is given by

$$\bigoplus_{i=1}^{n/2} I(\ell_i^\tau) \otimes |\,|^{-a(\underline{\mu})} \bigoplus_{\substack{1\leq i\leq n/2 \\ 1\leq j\leq (n'-1)/2}} \left(I(\ell_i^\tau + \ell_j'^\tau) \otimes |\,|^{-a(\underline{\mu})} \oplus I(|\ell_i^\tau - \ell_j'^\tau|) \otimes |\,|^{-a(\underline{\mu})} \right), \tag{7.27}$$

(recall that $a(\underline{\mu}) = d - d'$), and if n' is even then

$$\varrho({}^\iota\sigma_v) \otimes \varrho({}^\iota\sigma_v'^{\mathsf{v}}) = \bigoplus_{\substack{1\leq i\leq n/2 \\ 1\leq j\leq n'/2}} \left(I(\ell_i^\tau + \ell_j'^\tau) \otimes |\,|^{-a(\underline{\mu})} \oplus I(|\ell_i^\tau - \ell_j'^\tau|) \otimes |\,|^{-a(\underline{\mu})} \right). \tag{7.28}$$

From the matching up of local L-factors on either side of the local Langlands correspondence (see Knapp [47, (3.6), (3.8)]), if n' is odd then from (7.27) we get

$$\begin{aligned} L_v(s, \iota, \sigma \times \sigma'^{\mathsf{v}}) &= \prod_{i=1}^{n/2} \Gamma_{\mathbb{C}}\left(s - a(\underline{\mu}) + \tfrac{\ell_i^\tau}{2}\right) \cdot \\ &\cdot \prod_{\substack{1\leq i\leq n/2 \\ 1\leq j\leq (n'-1)/2}} \Gamma_{\mathbb{C}}\left(s - a(\underline{\mu}) + \tfrac{\ell_i^\tau + \ell_j'^\tau}{2}\right) \Gamma_{\mathbb{C}}\left(s - a(\underline{\mu}) + \tfrac{|\ell_i^\tau - \ell_j'^\tau|}{2}\right), \end{aligned} \tag{7.29}$$

and if n' is even then from (7.28) we get

$$L_v(s, \iota, \sigma \times \sigma'^{\mathsf{v}}) = \prod_{\substack{1\leq i\leq n/2 \\ 1\leq j\leq n'/2}} \Gamma_{\mathbb{C}}\left(s - a(\underline{\mu}) + \tfrac{\ell_i^\tau + \ell_j'^\tau}{2}\right) \Gamma_{\mathbb{C}}\left(s - a(\underline{\mu}) + \tfrac{|\ell_i^\tau - \ell_j'^\tau|}{2}\right). \tag{7.30}$$

Proposition 7.7. *Suppose $\sigma_f \in \mathrm{Coh}_{!!}(G_n, \mu)$ and $\sigma'_f \in \mathrm{Coh}_{!!}(G_{n'}, \mu')$ have disjoint cuspidal parameters. (From Prop. 7.5, necessarily nn' is even.) If $N = n + n'$ then let $\epsilon_N \in \{0, 1\}$ be such that $N \equiv \epsilon_N \pmod 2$. Then the critical set* *for $L(s, \iota, \sigma_f \times \sigma_f'^{\mathsf{v}})$ is the finite set of half-integers:*

$$\left\{ m \in \tfrac{\epsilon_N}{2} + \mathbb{Z} \ : \ \frac{2 - \ell(\mu, \mu') + 2a(\underline{\mu})}{2} \ \le \ m \ \le \ \frac{\ell(\mu, \mu') + 2a(\underline{\mu})}{2} \right\}.$$

The critical set is centered around $\frac{1}{2} + a(\underline{\mu})$ and has cardinality equal to the cuspidal width $\ell(\mu, \mu')$.

Proof. The parity issue is dictated by Def. 7.3. We remind the reader that $\ell_i \equiv 2d + n - 1 \pmod 2$; hence $\ell(\mu, \mu') \pm 2a(\underline{\mu}) \equiv \epsilon_N \pmod 2$. Computing the critical set is an exercise involving (7.29) or (7.30). We will omit the details. □

The reader should note that the critical sets in Prop. 7.6 and Prop. 7.7 are compatible under the shift $(\mathbf{w} + \mathbf{w}' - 2a(\underline{\mu}))/2$ as in (7.23).

Corollary 7.8. *Suppose $\sigma_f \in \mathrm{Coh}_{!!}(G_n, \mu)$ and $\sigma_f' \in \mathrm{Coh}_{!!}(G_{n'}, \mu')$ have disjoint cuspidal parameters. (From Prop. 7.5, necessarily nn' is even.) Let $N = n + n'$. The points $-\frac{N}{2}$ and $1 - \frac{N}{2}$ are critical for $L(s, \iota, \sigma_f \times \sigma_f'^{\mathsf{v}})$ if and only if*

$$-\frac{N}{2} + 1 - \frac{\ell(\mu, \mu')}{2} \ \le \ a(\underline{\mu}) \ \le \ -\frac{N}{2} - 1 + \frac{\ell(\mu, \mu')}{2}.$$

Proof. Follows easily from the above proposition. □

Corollary 7.9. *The ratio of local L-values at any archimedean place $v \in \mathsf{S}_\infty$ is given by*

$$\frac{L_v\left(-\frac{N}{2}, \iota, \sigma \times \sigma'^{\mathsf{v}}\right)}{L_v\left(1 - \frac{N}{2}, \iota, \sigma \times \sigma'^{\mathsf{v}}\right)} \ = \ \frac{(2\boldsymbol{\pi})^{\frac{nn'}{2}}}{\prod_{i=1}^{n/2} \prod_{j=1}^{n'} \left(\frac{-N + 2(a(\underline{\mu})) + |\ell_i^\tau - \ell_j^\tau|}{2} \right)}.$$

In particular, the ratio of L-values at infinity is given by

$$\frac{L_\infty\left(-\frac{N}{2}, \iota, \sigma \times \sigma'^{\mathsf{v}}\right)}{L_\infty\left(1 - \frac{N}{2}, \iota, \sigma \times \sigma'^{\mathsf{v}}\right)} \ \approx_{\mathbb{Q}^\times} \ \boldsymbol{\pi}^{[F:\mathbb{Q}]nn'/2} \ = \ \boldsymbol{\pi}^{\dim(U_P)/2},$$

where $\approx_{\mathbb{Q}^\times}$ means up to a nonzero rational number.

Proof. Follows from (7.29) and (7.30), and the well-known relation $\Gamma(s+1) = s\Gamma(s)$. □

As a matter of notation, here and elsewhere, $\boldsymbol{\pi}$ will always denote the universal constant 3.14159..., and π will stand for a representation or a covering map, etc. This ratio of L-values will be relevant in the discussion around Thm. 7.25.

7.2.2.3 A digression for another look at the holomorphy of the Eisenstein series

Recall Thm. 6.4, which asserted that if we are on the right of the unitary axis then $\mathrm{Eis}_P(s, f)$ is holomorphic at $s = -N/2$. We will only be looking at the situation when the points $s = -N/2$ and $s = 1-N/2$ are critical for the Rankin–Selberg L-function attached to the pair $(\sigma, \sigma'^{\mathsf{v}})$. In this paragraph we provide another view on the holomorphy of the Eisenstein series if we asked for the criticality of these points. This is a purely analytic argument that uses some details from the Langlands–Shahidi machinery.

Proposition 7.10. *Notations are as in Sect. 5.3.7. Let $\sigma \in \mathrm{Coh}(G_n, \mu)$ (resp., $\sigma' \in \mathrm{Coh}(G_{n'}, \mu')$) be a cohomological cuspidal automorphic representation of $G_n(\mathbb{A})$ (resp., $G_{n'}(\mathbb{A})$). Suppose nn' is even. Assume that the points $-N/2$ and $1 - N/2$ are critical for the Rankin–Selberg L-function $L(s, \sigma \times \sigma'^{\mathsf{v}})$. Let $f = \otimes f_v \in I_P^G(s, \sigma \times \sigma')$ and $\tilde{f} = \otimes \tilde{f}_v \in I_Q^G(w, \sigma' \times \sigma)$. Then:*

1. *If $\frac{N}{2} + d - d' \leq 0$ then $\mathrm{Eis}_P(s, f)$ is holomorphic at $s = -N/2$,*
2. *If $\frac{N}{2} + d - d' \geq 0$ then $\mathrm{Eis}_Q(w, \tilde{f})$ is holomorphic at $w = N/2$.*

Proof of Prop. 7.10. The proof needs a few lemmas: 7.11, 7.12 and 7.13. We first state these lemmas and then prove them.

Lemma 7.11. *If $\mathrm{Eis}_P(s, f)$ has a pole at $s = -N/2$ then $L^{\mathsf{S}}(1-\frac{N}{2}, \sigma \times \sigma'^{\mathsf{v}}) = 0$, where $\mathsf{S} = \mathsf{S}_\infty \cup \mathsf{S}_f$ is a finite set of places that is the union of the set S_∞ of all the archimedean places and the set S_f of all the finite places where either σ or σ' is ramified.*

Lemma 7.12. *If $\mathrm{Eis}_Q(w, \tilde{f})$ has a pole at $w = N/2$ then $L^{\mathsf{S}}(1+\frac{N}{2}, \sigma' \times \sigma^{\mathsf{v}}) = 0$, where S is as in the previous lemma.*

Lemma 7.13. *Let v be a finite place of F, and let π and π' be irreducible admissible unitary generic representations of $\mathrm{GL}_n(F_v)$ and $\mathrm{GL}_{n'}(F_v)$, respectively. Suppose $L(s, \pi \times \pi')$ has a pole at $s = t_0 \in \mathbb{R}$, then $t_0 < 1$.*

Proofs of Lemmas 7.11 and 7.12

Suppose $f = \otimes_v f_v$ is a pure tensor in $I_P^G(s, \sigma \otimes \sigma')$, and for $v \notin \mathsf{S}$ suppose $f_v = f_v^0$ is the normalized spherical vector (normalized to take the value 1 on the identity), and similarly, $\tilde{f}_v^0$ is such a vector in the v-th component of $I_Q^G(-s, \sigma' \otimes \sigma)$, then from [65, Thm. 6.3.1] we have

$$\mathcal{F}^Q(\mathrm{Eis}_P(s, f)) = \frac{L^{\mathsf{S}}(s, \sigma \times \sigma'^{\mathsf{v}})}{L^{\mathsf{S}}(s+1, \sigma \times \sigma'^{\mathsf{v}})} \otimes_{v \notin \mathsf{S}} \tilde{f}_v^0 \otimes_{v \in \mathsf{S}} T_{\mathrm{st}}(s, \sigma_v \otimes \sigma'_v) f_v.$$

The constant term map $\mathcal{F}^Q$ is an integration over a compact space and will not affect the analytic discussion that follows. If $\mathrm{Eis}_P(s,f)$ has a pole at $s=-N/2$, then looking at the right-hand side, either

(1) $L^{\mathsf{S}}(s,\sigma\times\sigma'^{\vee})/L^{\mathsf{S}}(s+1,\sigma\times\sigma'^{\vee})$ has a pole at $-N/2$, or
(2) for some $v\in\mathsf{S}$, the local standard intertwining operator $T_{\mathrm{st}}(s,\sigma_v\otimes\sigma'_v)$ has a pole at $-N/2$.

In case (1), since the numerator $L^{\mathsf{S}}(s,\sigma\times\sigma'^{\vee})$ is entire, the only way the ratio can have a pole is if the denominator vanishes at that point, i.e., $L^{\mathsf{S}}(1-N/2,\sigma\times\sigma'^{\vee})=0$.

In case (2), if $T_{\mathrm{st}}(s,\sigma_v\otimes\sigma'_v)$ has a pole at $s=-N/2$ of order $k_v\geq 0$, and if $k=\sum_{v\in\mathsf{S}}k_v$, then we claim that the numerator $L^{\mathsf{S}}(s,\sigma\times\sigma'^{\vee})$ will have a zero of order at least k; i.e., all the poles of the local standard intertwining operators at bad places will cancel against zeros of the partial L-function appearing in the numerator; hence the only way $\mathrm{Eis}_P(s,f)$ can have a pole at $s=-N/2$ is when $L^{\mathsf{S}}(1-N/2,\sigma\times\sigma'^{\vee})=0$, which will prove the lemma.

For the proof of the claim made above: suppose $T_{\mathrm{st}}(s,\sigma_v\otimes\sigma'_v)$ has a pole at $s=-N/2$ of order $k_v\geq 0$. We appeal to Shahidi's results on local constants, which we briefly review here. Fix a nontrivial additive character ψ_v of the base field and let $\lambda_{\psi_v}(s,\sigma_v\times\sigma'_v)$ be the standard Whittaker functional (defined as a certain integral) on $I(s,\sigma_v\otimes\sigma'_v)$. We are using the fact here that cuspidal representations of GL_n are globally generic, and hence locally generic everywhere. We know that $\lambda_{\psi_v}(s,\sigma_v\times\sigma'_v)$ extends to an entire function that is nonzero for all v; see Shahidi [62, Prop. 3.1] for archimedean v and [63, Prop. 3.1] for finite v. By multiplicity one for Whittaker models there is a complex number $C_{\psi_v}(s,\sigma_v\times\sigma'_v)$, called a *local constant*, such that

$$C_{\psi_v}(s,\sigma_v\times\sigma'_v)\,(\lambda_{\psi_v}(-s,\sigma'_v\times\sigma_v)\circ T_{\mathrm{st}}(s,\sigma_v\otimes\sigma'_v)) \;=\; \lambda_{\psi_v}(s,\sigma_v\times\sigma'_v). \tag{7.31}$$

Furthermore, the local constant is related to local L- and ε-factors as

$$C_{\psi_v}(s,\sigma_v\times\sigma'_v) \;=\; \varepsilon(s,\psi_v,\sigma_v\times\sigma'^{\vee}_v)\,\frac{L(1-s,\sigma^{\vee}_v\times\sigma'_v)}{L(s,\sigma_v\times\sigma'^{\vee}_v)}. \tag{7.32}$$

See Shahidi [65, (5.1.4)]. Back to our proof of the above claim: suppose $T_{\mathrm{st}}(s,\sigma_v\otimes\sigma'_v)$ has a pole at $s=-N/2$ of order $k_v\geq 0$, since the right-hand side of (7.31) is holomorphic, the local constant $C_{\psi_v}(s,\sigma_v\times\sigma'_v)$ has a zero at $s=-N/2$ of order k_v. The epsilon factor is an exponential function (hence entire and nonvanishing), and local L-factors are nowhere vanishing; hence the denominator of the right-hand side of (7.32), i.e., $L(s,\sigma_v\times\sigma'^{\vee}_v)$ has a pole at $s=-N/2$ of order k_v. We conclude that $L_{\mathsf{S}}(s,\sigma\times\sigma'^{\vee})=\prod_{v\in\mathsf{S}}L(s,\sigma_v\times\sigma'^{\vee}_v)$ has a pole at $s=-N/2$ of order $k=\sum_{v\in\mathsf{S}}k_v$. But, $L(s,\sigma\times\sigma'^{\vee})=L_{\mathsf{S}}(s,\sigma\times\sigma'^{\vee})L^{\mathsf{S}}(s,\sigma\times\sigma'^{\vee})$ is entire (Thm. 7.1); hence $L^{\mathsf{S}}(s,\sigma\times\sigma'^{\vee})$ must have a zero at $s=-N/2$ of order at least k to cancel against the pole at $s=-N/2$ of order k of $L_{\mathsf{S}}(s,\sigma\times\sigma'^{\vee})$–as claimed above. Proof of Lem. 7.12 is identical to the above proof of Lem. 7.11.

Proof of Lemma 7.13

This is well-known and follows from the theory of zeta integrals. For example, in the case $n \neq n'$, see Prop. 6.2(i) and Prop. 6.3 in Cogdell [13]. More generally, it is embedded in the work of Jacquet, Piatetskii-Shapiro, and Shalika [40].

CONCLUSION OF PROOF OF PROP. 7.10

To prove (1), suppose $N/2 + d - d' \leq 0$, and if possible suppose also that $\mathrm{Eis}_P(s, f)$ has a pole at $s = -N/2$. From Lem. 7.11 we know that

$$L^S(1 - \tfrac{N}{2}, \sigma \times \sigma'^{\vee}) = L^S(1 - (\tfrac{N}{2} + d - d'), {}^\circ\sigma \times {}^\circ\sigma'^{\vee}) = 0,$$

where $\sigma = |\ |^{-d} \otimes {}^\circ\sigma$ with ${}^\circ\sigma$ being a unitary cuspidal representation; and similarly, $\sigma' = |\ |^{-d'} \otimes {}^\circ\sigma'$. Since $N/2 + d - d' \leq 0$, from Lem. 7.13, for any place v we know that $L(1 - (\frac{N}{2} + d - d'), {}^\circ\sigma_v \times {}^\circ\sigma_v'^{\vee})$ is finite. Hence, for the completed L-function we get $L(1 - (\frac{N}{2} + d - d'), {}^\circ\sigma \times {}^\circ\sigma'^{\vee}) = 0$; but this is not possible by (4) of Thm. 7.1 since $1 - (N/2 + d - d') \geq 1$; whence we conclude that $\mathrm{Eis}_P(s, f)$ is finite at $s = -N/2$. The proof of (2) is identical, and is left to the reader. □

This ends our digression, and we return to our discussion of the critical set for the Rankin–Selberg L-functions.

7.2.3 A combinatorial lemma

Assume that nn' is even, and suppose $\sigma_f \in \mathrm{Coh}_{!!}(G_n, \mu)$ and $\sigma'_f \in \mathrm{Coh}_{!!}(G_{n'}, \mu')$. Let $N = n + n'$. Recall (from Prop. 4.3 and Sect. 5.3.7) that the assumption on the existence of a balanced Kostant representative $w \in W^P$ such that $\lambda := w^{-1} \cdot (\mu + \mu')$ is dominant implies that various induced representations appear in boundary cohomology groups in prescribed degrees. The lemma below characterizes the existence of such a balanced Kostant representative as a purely combinatorial condition involving the cuspidal width $\ell(\mu, \mu')$ and $a(\underline{\mu})$. Amazingly, the same combinatorial condition says that the ratio of L-values

$$\frac{L\left(-\frac{N}{2}, \iota, \sigma_f \times \sigma_f'^{\vee}\right)}{L\left(1 - \frac{N}{2}, \iota, \sigma_f \times \sigma_f'^{\vee}\right)}$$

is in fact a ratio of *critical* L-values; this ratio will appear as a factor in $T_{\mathrm{Eis}}(\sigma, \sigma', \varepsilon')$ (see (6.16)), and we may ask whether a cohomological interpretation gives a rationality result for this term. We will elaborate further in Sect. 7.2.3.1.

Lemma 7.14. *Suppose $\sigma_f \in \mathrm{Coh}_{!!}(G_n, \mu)$ and $\sigma'_f \in \mathrm{Coh}_{!!}(G_{n'}, \mu')$ have disjoint cuspidal parameters. (From Prop. 7.5, necessarily nn' is even.) Let $\mu + \mu' \in X^*(T_N)$. Then the following are equivalent:*

1. *There is a balanced Kostant representative $w \in W^P$ such that $w^{-1} \cdot (\mu + \mu')$ is a dominant integral weight.*
2. $-\frac{N}{2} + 1 - \frac{\ell(\mu,\mu')}{2} \leq a(\underline{\mu}) \leq -\frac{N}{2} - 1 + \frac{\ell(\mu,\mu')}{2}$.
3. *The points $-\frac{N}{2}$ and $1 - \frac{N}{2}$ are critical for the automorphic Rankin–Selberg L-function $L(s, \iota, \sigma_f \times \sigma_f'^{\vee})$.*

Proof. The equivalence (1) $\iff$ (2) was conjectured in our announcement [33], and was later proved by Uwe Weselmann; his proof is given in Sect. 7.2.4. The equivalence (2) $\iff$ (3) is exactly the statement of Cor. 7.8. □

7.2.3.1 An arithmetic consequence of the combinatorial lemma

This monograph is about an application of Eisenstein cohomology to prove rationality results for ratios of successive L-values. The combinatorial lemma says that we can prove such rationality results for a ratio of L-values when these are critical values, and furthermore, the lemma also says that every ratio of critical values is captured by these techniques, and if either of the L-values is not critical then our methods do not apply. This may be justified as follows: we are able to prove a rationality result for the ratio

$$\frac{L^{\mathrm{aut}}\left(-\frac{N}{2}, \iota, \sigma_f \times \sigma_f'^{\vee}\right)}{L^{\mathrm{aut}}\left(1-\frac{N}{2}, \iota, \sigma_f \times \sigma_f'^{\vee}\right)} = \frac{L^{\mathrm{Coh}}\left(\iota, \sigma_f \times \sigma_f'^{\vee}, -\frac{N}{2} + \frac{(\mathbf{w}+\mathbf{w}')}{2} - a(\underline{\mu})\right)}{L^{\mathrm{Coh}}\left(\iota, \sigma_f \times \sigma_f'^{\vee}, 1-\frac{N}{2} + \frac{(\mathbf{w}+\mathbf{w}')}{2} - a(\underline{\mu})\right)}.$$

For brevity, let

$$\mathsf{m}_0 := -\frac{N}{2} + \frac{(\mathbf{w}+\mathbf{w}')}{2} - a(\underline{\mu}).$$

Now, given $\sigma_f \in \mathrm{Coh}_{!!}(G_n, \mu)$ and $\sigma'_f \in \mathrm{Coh}_{!!}(G_{n'}, \mu')$, let's take all possible Tate twists. Note that this means, we have fixed the semisimple parts $\mu^{(1)}$ and $\mu'^{(1)}$ and we are letting d and d', i.e., the abelian parts $\mu_{\mathrm{ab}} = d\boldsymbol{\delta}_n$ $\mu'_{\mathrm{ab}} = d'\boldsymbol{\delta}_{n'}$, to vary. The inequalities in (2) of Lem. 7.14 bounds $a(\underline{\mu})$ since the cuspidal width $\ell(\mu, \mu') = \ell(\mu^{(1)}, \mu'^{(1)})$ depends only on the semisimple parts of the weights. As $a(\underline{\mu})$ varies between its lower bound $-\frac{N}{2} + 1 - \frac{\ell(\mu,\mu')}{2}$ to its upper bound $-\frac{N}{2} - 1 + \frac{\ell(\mu,\mu')}{2}$, one may check using Prop. 7.6, that the ratio

$$\frac{L^{\mathrm{Coh}}\left(\iota, \sigma_f \times \sigma_f'^{\vee}, \mathsf{m}_0\right)}{L^{\mathrm{Coh}}\left(\iota, \sigma_f \times \sigma_f'^{\vee}, 1+\mathsf{m}_0\right)}$$

runs exactly over all possible successive critical values; no more and no less!

7.2.3.2 Comparison of notations with [33]

The combinatorial lemma, especially (1) $\iff$ (2), was conjectured by us [33, Conj. 5.1] in our announcement, and was proved by Uwe Weselmann; see Sect. 7.2.4 for his proof. In this short paragraph we just point to some notational differences: The quantity $p(\mu)$ of [33] is the largest critical point of the cohomological L-function which is the same as the motivic L-function for the effective motives. From Prop. 7.6 we can see that $p(\mu)$ is equal to our $(\ell(\mu,\mu')+\mathbf{w}+\mathbf{w}')/2$. It should now be clear that the above formulation of the combinatorial lemma is the same as in [33].

7.2.4 Proof of combinatorial lemma By Uwe Weselmann

We give a proof of the combinatorial Lem. 7.14. Let us suppose that the assumptions of Lem. 7.14 are satisfied. So n denotes an even integer, n' an arbitrary positive integer, and $N = n + n'$. For simplicity we will start to work inside the character group $X^*_{\mathbb{Q}}(T_0 \times_\tau E)$ for a fixed embedding $\tau : F \hookrightarrow E$. Since this can and will be identified with $\mathbb{Q}^n = X^*_{\mathbb{Q}}(T_0)$, we will drop the τ from all notations until the final proof. In $X^*_{\mathbb{Q}}(T_0)$ the half sum of the positive weights on GL_n will be denoted by $\boldsymbol{\rho}_n$, the determinant by $\boldsymbol{\delta}_n = \det_n$ and the i-th standard basis vector by e_i. Write

$$\mu + \boldsymbol{\rho}_n = (\bar{b}_1, \ldots, \bar{b}_n) \in \mathbb{Q}^n = X^*_{\mathbb{Q}}(T_0), \qquad \text{i.e., } \bar{b}_i \;=\; b_i + \frac{n+1}{2} - i$$

with the b_i from Sect. 2.2.4. Recall that the $\boldsymbol{\gamma}_i$, which are extensions of the fundamental weights, may be written in the form

$$\boldsymbol{\gamma}_i = \sum_{j=1}^{i} e_j - \frac{i}{n} \cdot \sum_{j=1}^{n} e_j \qquad \text{for } i = 1, \ldots, n-1.$$

Thus

$$\mu + \boldsymbol{\rho}_n = \sum_{i=1}^{n-1} a_i \boldsymbol{\gamma}_i + d \cdot \boldsymbol{\delta}_n = \sum_{j=1}^{n} \left(\sum_{i=j}^{n-1} a_i \right) e_j + \left(- \sum_{i=1}^{n-1} \frac{i}{n} a_i + d \right) \cdot \sum_{j=1}^{n} e_j.$$

If $\mu + \boldsymbol{\rho}_n$ is essentially self-dual, we have $a_{n-i} = a_i$ and thus

$$\begin{aligned}\sum_{i=1}^{n-1} \frac{i}{n} a_i &= \frac{1}{2} \cdot \left(\sum_{i=1}^{n-1} \frac{i}{n} a_i + \sum_{i=1}^{n-1} \frac{i}{n} a_{n-i} \right) \\ &= \frac{1}{2} \cdot \left(\sum_{i=1}^{n-1} \frac{i}{n} a_i + \sum_{i=1}^{n-1} \frac{n-i}{n} a_i \right) = \frac{1}{2} \cdot \left(\sum_{i=1}^{n-1} a_i \right).\end{aligned}$$

We deduce $\bar{b}_j = \beta_j + d$, where

$$\beta_j = \sum_{i=j}^{n-1} a_i - \frac{1}{2} \cdot \left(\sum_{i=1}^{n-1} a_i \right) = \frac{1}{2} \cdot \left(\sum_{i=j}^{n-1} a_i - \sum_{i=1}^{j-1} a_i \right).$$

Thus $\beta_j = \frac{l_j}{2}$ in the notations of (3.4). The weight being essentially self-dual now means $\beta_{n+1-j} = -\beta_j$. Observe $a_i = \beta_i - \beta_{i+1}$ and $\beta_1 > \beta_2 > \ldots > \beta_n$.

Similarly write $\mu' + \boldsymbol{\rho}_{n'} = (\bar{b}'_1, \ldots, \bar{b}'_{n'}) \in \mathbb{Q}^{n'}$ with $\bar{b}'_i = \beta'_i + d'$ and $\beta'_{n'+1-i} = -\beta'_i$.

The motivic weights are $\mathbf{w} = 2\beta_1$ and $\mathbf{w}' = 2\beta'_1$. The Hodge numbers of the effective motives are

$$\begin{aligned}&(\beta_1 + \beta_i, \beta_1 - \beta_i) \qquad i = 1, \ldots, n \qquad \text{for } \mathbf{M}_{\text{eff}} \\ &(\beta'_1 + \beta'_j, \beta'_1 - \beta'_j) \qquad j = 1, \ldots, n' \qquad \text{for } \mathbf{M}'_{\text{eff}}.\end{aligned}$$

Then $\mathbf{M}_{\text{eff}} \otimes \mathbf{M}'_{\text{eff}}$ has the Hodge numbers

$$(\beta_1 + \beta'_1 + \beta_i + \beta_j,\ \beta_1 + \beta'_1 - \beta_i - \beta_j) \quad \text{for } (i,j) \in S, \quad \text{where}$$

$$S = \{1, \ldots, n\} \times \{1, \ldots, n'\}.$$

In the following the characters of the split torus of GL_N will be written as pairs $\tilde{\lambda} = (\lambda, \lambda') \in \mathbb{Z}^n \times \mathbb{Z}^{n'}$; we especially write $\tilde{\mu} = (\mu, \mu')$. We introduce the half-integer

$$\begin{aligned}\tilde{p}(\tilde{\mu}) &:= \min \left\{ \beta_i + \beta'_j > 0 \mid (i,j) \in S \right\} \\ &= \min \left\{ \nu > 0 \mid \quad h^{\beta_1 + \beta'_1 + \nu, \beta_1 + \beta'_1 - \nu} \neq 0 \right\}.\end{aligned}$$

With the integer $p(\tilde{\mu}) = \min\{p > \frac{\mathbf{w}+\mathbf{w}'}{2} \mid h^{p, \mathbf{w}+\mathbf{w}'-p} \neq 0\}$ from [33, 3] we have

$$\tilde{p}(\tilde{\mu}) = p(\tilde{\mu}) - \frac{\mathbf{w} + \mathbf{w}'}{2} = p(\tilde{\mu}) - \beta_1 - \beta'_1.$$

Recall that it is an assumption of Lem. 7.14., that σ_f and σ'_f have disjoint cuspidal parameters. This is equivalent, by Prop. 7.5, to the fact that the middle Hodge number of the tensor motive is zero, which means $\beta_i + \beta'_j \neq 0$ for all $(i,j) \in S$. In view of the symmetries $\beta_{n+1-i} = -\beta_i$ and $\beta'_{n'+1-j} = -\beta'_j$ we thus get

$$\tilde{p}(\tilde{\mu}) = \min_{(i,j)\in S} |\beta_i - \beta'_j| = \min_{(i,j)\in S^+} |\beta_i - \beta'_j|, \tag{7.33}$$

where

$$S^+ = \left\{(i,j) \in S \mid 1 \leq i \leq \frac{n+1}{2},\ 1 \leq j \leq \frac{n'+1}{2}\right\}$$
$$= \left\{(i,j) \in S \mid \beta_i \geq 0, \beta'_j \geq 0\right\}.$$

We have

$$\tilde{\mu} + \boldsymbol{\rho}_N = (\mu, \mu') + \boldsymbol{\rho}_N = \left(\mu + \boldsymbol{\rho}_n + \frac{n'}{2} \cdot \boldsymbol{\delta}_n,\ \mu' + \boldsymbol{\rho}_{n'} - \frac{n}{2} \cdot \boldsymbol{\delta}_{n'}\right),$$

and if we write $\tilde{\mu} + \boldsymbol{\rho}_N = (\tilde{b}_1, \ldots, \tilde{b}_N)$, then we get

$$\tilde{b}_i = \bar{b}_i + \frac{n'}{2} = \beta_i + \tilde{d}, \quad \text{with } \tilde{d} = d + \frac{n'}{2}, \quad \text{for } 1 \leq i \leq n, \quad \text{and}$$

$$\tilde{b}_{j+n} = \bar{b}'_j - \frac{n}{2} = \beta'_j + \tilde{d}', \quad \text{with } \tilde{d}' = d' - \frac{n}{2}, \quad \text{for } 1 \leq j \leq n'.$$

Let

$$\tilde{a}(\tilde{\mu}) = a(\underline{\mu}) + \frac{n+n'}{2} = \tilde{d} - \tilde{d}'.$$

(Recall $a(\underline{\mu}) = d - d'$.)

Lemma 7.15. *If we have $|\tilde{a}(\tilde{\mu})| = |\beta_i - \beta'_j|$ for some $(i,j) \in S^+$, then there does not exist $w \in W^P$ such that $\tilde{\mu} + \boldsymbol{\rho}_N$ is of the form $w(\lambda + \boldsymbol{\rho}_N)$ with a dominant weight λ.*

Proof. If $|\tilde{a}(\tilde{\mu})| = |\beta_i - \beta'_j|$ then either $\tilde{d} - \tilde{d}' = \beta_i - \beta'_j$ which is equivalent to $\tilde{b}_{n+1-i} = \tilde{d} - \beta_i = \tilde{d}' - \beta'_j = \tilde{b}_{N+1-j}$ or we have $\tilde{d}' - \tilde{d} = \beta_i - \beta'_j$, which is equivalent to $\tilde{b}_i = \tilde{b}_{n+j}$. Thus at least two entries of $\tilde{\mu} + \boldsymbol{\rho}_N$ coincide. But if λ is dominant, then $\lambda + \boldsymbol{\rho}_N$ has strictly decreasing entries, and all entries of $w(\lambda + \boldsymbol{\rho}_N)$ are different for every $w \in W^P$. □

Lemma 7.16. *Assume $|\tilde{a}(\tilde{\mu})| \neq |\beta_i - \beta'_j|$ for all $(i,j) \in S^+$. Then there exists a unique $w \in W^P$, such that $\tilde{\mu} + \boldsymbol{\rho}_N = w(\lambda + \boldsymbol{\rho}_N)$ with a dominant weight $\lambda \in \mathbb{Z}^N$.*

(a) If $|\tilde{a}(\tilde{\mu})| < \tilde{p}(\tilde{\mu})$, *then* $l(w) = \frac{dim(U_P)}{2}$;
(b) If $\tilde{a}(\tilde{\mu}) > \tilde{p}(\tilde{\mu})$, *then* $l(w) < \frac{dim(U_P)}{2}$;

(c) If $\tilde{a}(\tilde{\mu}) < -\tilde{p}(\tilde{\mu})$, *then* $l(w) > \frac{dim(U_P)}{2}$.

Proof. The assumption implies, by the arguments of Lem. 7.15, that the entries of $\tilde{\mu} + \boldsymbol{\rho}_N$ are pairwise different. Therefore there exists a unique permutation $w \in W$ such that $w^{-1}(\tilde{\mu} + \boldsymbol{\rho}_N)$ is regular dominant (i.e., the entries form a strictly decreasing sequence). Since the entries differ by integers, we get an equation $\tilde{\mu} + \boldsymbol{\rho}_N = w(\lambda + \boldsymbol{\rho}_N)$ with a dominant λ. Since we already know $\tilde{b}_1 > \ldots > \tilde{b}_n$ and $\tilde{b}_{n+1} > \ldots > \tilde{b}_N$, we have $w \in W^P$.

Since the Weyl group W is the symmetric group $W = S_N$, the length can be computed by the formula

$$l(w) \quad = \quad \#\Big\{(i,j) \in \mathbb{N} \times \mathbb{N} \mid 1 \le i < j \le N, \quad \tilde{b}_i < \tilde{b}_j\Big\}.$$

But since the first n entries are already in a decreasing order as are the last n' entries, $l(w)$ is the cardinality of the set

$$S_< \;=\; \Big\{(i,j) \in S \mid \tilde{b}_i < \tilde{b}_{n+j}\Big\}, \quad \text{where } S \;=\; \{1,\ldots,n\} \times \{1,\ldots,n'\}.$$

Observe $\#(S) = n \cdot n' = dim(U_P)$. To calculate $l(w)$ we will determine the intersections $Q_{i,j} \cap S_<$, where

$$Q_{i,j} = \{(i,j),\ (n+1-i,j),\ (i,n'+1-j),\ (n+1-i,n'+1-j)\}$$

for $(i,j) \in S^+$. Observe that $S = \bigcup_{(i,j)\in S^+} Q_{i,j}$ is a disjoint partition in subsets of order 4 (for $j < \frac{n'+1}{2}$) respectively of order 2 for $j = \frac{n'+1}{2}$. Observe that $i \neq \frac{n+1}{2}$ for all i, since n is assumed to be even.

(a) If $|\tilde{a}(\tilde{\mu})| < \tilde{p}(\tilde{\mu})$ we have $|\tilde{d} - \tilde{d}'| < |\beta_i - \beta'_j|$ for all $(i,j) \in S^+$. In the case $\beta_i > \beta'_j \ge 0$ this means

$$\beta'_j - \beta_i < \tilde{d} - \tilde{d}' < \beta_i - \beta'_j, \quad \text{and consequently}$$

$$\tilde{d} - \beta_i = \tilde{b}_{n+1-i} \;<\; \tilde{b}_{n+n'+1-j} = \tilde{d}' - \beta'_j \;\le\; \tilde{b}_{n+j} = \tilde{d}' + \beta'_j \;<\; \tilde{b}_i = \tilde{d} + \beta_i.$$

Therefore $Q_{i,j} \cap S_< \;=\; \{(n+1-i,j),(n+1-i,n'+1-j)\}$. Thus $\#(Q_{i,j} \cap S_<) = \frac{1}{2}\cdot\#(Q_{i,j})$. (This means $2 = \frac{1}{2}\cdot 4$ in the case $j < n'+1-j$ and $1 = \frac{1}{2}\cdot 2$ in the case $\beta_j = 0 \Leftrightarrow j = n'+1-j$.) In the case $\beta'_j > \beta_i > 0$ we get similarly

$$\beta_i - \beta'_j < \tilde{d} - \tilde{d}' < \beta'_j - \beta_i \quad \text{and therefore} \quad \tilde{b}_{n+n'+1-j} < \tilde{b}_{n+1-i} < \tilde{b}_i < \tilde{b}_{n+j}.$$

Thus we have $Q_{i,j} \cap S_< \;=\; \{(i,j),(n+1-i,j)\}$ in this case, and again

$\#(Q_{i,j} \cap S_<) = \frac{1}{2} \cdot \#(Q_{i,j})$. Summing up over all indices we get

$$\begin{aligned} l(w) = \#(S_<) &= \sum_{(i,j)\in S^+} \#(Q_{i,j} \cap S_<) \\ &= \frac{1}{2} \sum_{(i,j)\in S^+} \#(Q_{i,j}) = \frac{1}{2}\#(S) = \frac{1}{2}\dim(U_P). \end{aligned}$$

(b) In this case we have $\tilde{d} - \tilde{d}' > |\beta_i - \beta'_j|$ for at least one pair $(i,j) \in S^+$. Consequently $\tilde{d} - \tilde{d}' > \beta_i - \beta'_j$ and $\tilde{d} - \tilde{d}' > \beta'_j - \beta_i$. This implies $\tilde{b}_i > \tilde{b}_{n+j} \geq \tilde{b}_{n+n'+1-j}$ and $\tilde{b}_{n+1-i} > \tilde{b}_{n+n'+1-j}$. Thus $Q_{i,j} \cap S_< \subset \{(n+1-i,j)\}$ in case $\beta'_j > 0$ and $Q_{i,j} \cap S_< = \emptyset$ in case $\beta'_j = 0$. In both cases we get

$$\#(Q_{i,j} \cap S_<) < \frac{1}{2} \cdot \#(Q_{i,j}).$$

For those pairs $(i,j) \in S^+$ for which we do not have $\tilde{d} - \tilde{d}' > |\beta_i - \beta'_j|$, we must have $\tilde{d} - \tilde{d}' = |\tilde{d} - \tilde{d}'| < |\beta_i - \beta'_j|$ and get $\#(Q_{i,j} \cap S_<) = \frac{1}{2} \cdot \#(Q_{i,j})$ by the calculations in (a). Summing up we conclude $l(w) = \#(S_<) < \frac{1}{2}\#(S) = \frac{1}{2}\dim(U_P)$.

(c) This case is dual to (b): If we have $\tilde{d} - \tilde{d}' < -|\beta_i - \beta'_j|$ for some pair $(i,j) \in S$, then we get $Q_{i,j} \cap S_< \supset \{((i,j), (n+1-i, n'+1-j), (n+1-i,j)\}$, and thus

$$\#(Q_{i,j} \cap S_<) > \frac{1}{2} \cdot \#(Q_{i,j}).$$

Since for all other pairs we have $\#(Q_{i,j} \cap S_<) \geq \frac{1}{2} \cdot \#(Q_{i,j})$ by the same reasoning respectively by (a), we conclude $l(w) = \#(S_<) > \frac{1}{2}\#(S) = \frac{1}{2}\dim(U_P)$. □

Proposition 7.17. *The existence of $w \in W^P$ of length $l(w) = \frac{dim(U_P)}{2}$, such that $\tilde{\mu} + \boldsymbol{\rho}_N = w(\lambda + \boldsymbol{\rho}_N)$ with a dominant weight $\lambda \in \mathbb{Z}^N$, is equivalent to*

$$-\tilde{p}(\tilde{\mu}) + 1 \leq \tilde{a}(\tilde{\mu}) \leq \tilde{p}(\tilde{\mu}) - 1.$$

Proof. This is an immediate consequence of the two lemmas, once we have proved that $|\tilde{a}(\tilde{\mu})| < \tilde{p}(\tilde{\mu})$ is equivalent to the inequality in the proposition. But this follows from the fact that the difference between the half integers $\tilde{a}(\tilde{\mu})$ and $\tilde{p}(\tilde{\mu})$ is an integer: Observe $2d = 2\bar{b}_1 - 2\beta_1 = 2b_1 + n - 1 - \mathbf{w} \equiv n - 1 - \mathbf{w} \bmod 2$, which implies $2\tilde{a}(\tilde{\mu}) = 2d - 2d' + N \equiv n - 1 - \mathbf{w} - (n' - 1 - \mathbf{w}') + n + n' \equiv \mathbf{w} + \mathbf{w}' \bmod 2$. Furthermore, $2\tilde{p}(\tilde{\mu}) = 2p(\tilde{\mu}) - (\mathbf{w} + \mathbf{w}') \equiv \mathbf{w} + \mathbf{w}' \bmod 2$, since $p(\tilde{\mu})$ is an integer. Consequently $2\tilde{a}(\tilde{\mu}) \equiv 2\tilde{p}(\tilde{\mu}) \bmod 2$. □

Corollary 7.18. *The conditions (1) and (2) of Lem. 7.14. are equivalent.*

Proof. One has to apply the proposition to all embeddings $\tau : F \to E$. Write $\tilde{\mu} = (\mu, \mu') = (\tilde{\mu}^\tau)_{\tau:F\to E} \in \prod_\tau X^*(T_0 \times_\tau E)$. Condition (1) means (see Def. 5.9) that for each embedding $\tau : F \to E$ there exists $w^\tau \in W^P$ of length $\dim(U_{P_0})/2$ such that $(w^\tau)^{-1}(\tilde{\mu}^\tau + \boldsymbol{\rho}_N) - \boldsymbol{\rho}_N$ is dominant.

After adding $\frac{N}{2}$ to the inequality (2) it reads: $-\frac{\ell(\mu,\mu')}{2} + 1 \leq \tilde{a}(\tilde{\mu}) \leq \frac{\ell(\mu,\mu')}{2} - 1.$ The relations $\beta_i^\tau = \frac{\ell_i^\tau}{2}$ imply, in view of (7.33) and Def. 7.4, that $\frac{\ell(\mu,\mu')}{2} = \min_{\tau:F\to E} \tilde{p}(\tilde{\mu}^\tau)$, so that condition (2) is equivalent to

$$-\tilde{p}(\tilde{\mu}^\tau) + 1 \;\leq\; \tilde{a}(\tilde{\mu}) \;\leq\; \tilde{p}(\tilde{\mu}^\tau) - 1 \qquad \text{for all } \tau : F \to E.$$

□

Corollary 7.19. *Conjecture 5.1. of [33] is true.*

Proof. This is clear from the proposition and the preceding corollary applied to $F = \mathbb{Q}$ which was the case in [33]. Observe that we have $\tilde{a}(\tilde{\mu}) = a(\tilde{\mu}) + \frac{N}{2}$, where $a(\tilde{\mu}) = -a(\underline{\mu})$ in the notations of loc. cit. □

Remark 7.20. It should be noted that the sets $\Phi = \{\beta_1, \ldots, \beta_n\}$ and $\Phi' = \{\beta_1', \ldots, \beta_n'\}$ are characteristic sets in the sense of [72]: The set Φ is either of type B_* (half integers) or of type D_* (integers). If $n = 2g$, then the weight satisfies $\mathbf{w} = 2\beta_1 \equiv a_g \bmod 2$. Thus for a_g even we have a characteristic set of type D_g, and for a_g odd the characteristic set is of type B_g.

If n' is odd, the set Φ' is of type C_* (i.e., $0 \in \Phi'$). If n' is even it may be of type B_* or of type D_*. (This subtle point where a parity dictates which classical group we need to consider also makes its appearance in [5].) If n' is odd and if the representations σ and σ' are self-equivalent, then the central Eisenstein class in Theorem 5.2 of [33] (i.e., where the usual Rankin–Selberg L-function is evaluated at $s = 0$ and at $s = 1$) is a twisted lift of an endoscopic L-packet on Sp_{N-1} by the results of [72].

7.3 THE MAIN RESULT ON SPECIAL VALUES OF L-FUNCTIONS

7.3.1 The main result on L-values

Given pure weights $\mu \in X_0^*(T_n)$ and $\mu' \in X_0^*(T_{n'})$, and for $\sigma_f \in \mathrm{Coh}_{!!}(G_n, \mu)$ and $\sigma_f' \in \mathrm{Coh}_{!!}(G_{n'}, \mu')$, we put $\underline{\sigma}_f = \sigma_f \times \sigma_f'$. Let $E/\mathbb{Q}$ be a finite Galois extension over which $\underline{\sigma}_f$ is defined. Assume that nn' is even. In this situation we define an array of periods $\Omega({}^\iota\underline{\sigma})$, using the definition of the relative periods in Sect. 5.2.3. If both n and n' are even, then we put

$$\Omega({}^\iota\underline{\sigma}) \ := \ 1.$$

(This is independent of the choice of the signatures ε and ε' which can appear with ${}^\iota\sigma_f$ and ${}^\iota\sigma'_f$, respectively.) If one of the two numbers is odd, say n' is odd and n is even, then we put

$$\Omega({}^\iota\underline{\sigma}_f) \ := \ \Omega^{\varepsilon'}({}^\iota\sigma_f).$$

In other words, since n is even, we have two periods for σ_f, but the signature character ε' for σ'_f tells us which of the two periods gives us $\Omega({}^\iota\underline{\sigma}_f)$.

Theorem 7.21. *Let $\mu \in X_0^*(T_n)$ (resp., $\mu' \in X_0^*(T_{n'})$) be a pure weight. Suppose $\sigma_f \in \mathrm{Coh}_{!!}(G_n, \mu)$ and $\sigma'_f \in \mathrm{Coh}_{!!}(G_{n'}, \mu')$ have disjoint cuspidal parameters; in particular nn' is even. Let $\mathsf{m}_0 = -\frac{N}{2} + \frac{(\mathbf{w}+\mathbf{w}')}{2} - a(\underline{\mu})$, and assume m_0 and $1+\mathsf{m}_0$ are critical. Let the rest of the notations be as in Sect. 5.3.7.*

1. *Suppose for some $\iota : E \to \mathbb{C}$, we have $L^{\mathrm{Coh}}(\iota, \sigma \times \sigma'^{\mathsf{v}}, 1+\mathsf{m}_0) = 0$. Then,*

$$1 + \mathsf{m}_0 = (1 + \mathbf{w} + \mathbf{w}')/2$$

   is the center of symmetry for the cohomological L-function; furthermore, for every $\iota : E \to \mathbb{C}$, we have $L^{\mathrm{Coh}}(\iota, \sigma \times \sigma'^{\mathsf{v}}, 1+\mathsf{m}_0) = 0$.

2. *Given $\iota : E \to \mathbb{C}$, suppose $L^{\mathrm{Coh}}(\iota, \sigma \times \sigma'^{\mathsf{v}}, 1+\mathsf{m}_0) \neq 0$; then we have*

$$\frac{1}{\Omega({}^\iota\underline{\sigma}_f)} \frac{L^{\mathrm{Coh}}(\iota, \sigma \times \sigma'^{\mathsf{v}}, \mathsf{m}_0)}{L^{\mathrm{Coh}}(\iota, \sigma \times \sigma'^{\mathsf{v}}, 1+\mathsf{m}_0)} \ \in \ \iota(E).$$

   Moreover, for any $\tau \in Gal(\bar{\mathbb{Q}}/\mathbb{Q})$, we have

$$\tau\left(\frac{1}{\Omega({}^\iota\underline{\sigma}_f)} \frac{L^{\mathrm{Coh}}(\iota, \sigma \times \sigma'^{\mathsf{v}}, \mathsf{m}_0)}{L^{\mathrm{Coh}}(\iota, \sigma \times \sigma'^{\mathsf{v}}, 1+\mathsf{m}_0)}\right) = \frac{1}{\Omega({}^{\tau\circ\iota}\underline{\sigma}_f)} \frac{L^{\mathrm{Coh}}(\tau\circ\iota, \sigma \times \sigma'^{\mathsf{v}}, \mathsf{m}_0)}{L^{\mathrm{Coh}}(\tau\circ\iota, \sigma \times \sigma'^{\mathsf{v}}, 1+\mathsf{m}_0)}.$$

7.3.2 Proof of Thm. 7.21.

The notations are as in Sect. 5.3.7. If we are on the right of the unitary axis, then we go from P to Q as suggested by Prop. 7.10, and we will present the details of the proof that follows, and we will discuss the situation of being on the left of the unitary axis in Sect. 7.3.2.5. If we are on the right, then from

Rem. 7.2 it follows that $L^{\mathrm{Coh}}(\iota, \sigma \times \sigma'^{\vee}, 1+\mathsf{m}_0) \neq 0$ for any ι. Since m_0 and $1+\mathsf{m}_0$ are critical we know (Combinatorial Lemma) that there is a balanced Kostant representative $w \in W^P$ such that $\lambda = w^{-1} \cdot (\mu + \mu')$ is dominant. By Thm. 6.2, the subspace $\mathfrak{I}^b(\sigma_f, \sigma'_f, \varepsilon')$, which is the image of global cohomology in the $2\mathsf{k}$-dimensional E-vector space $I_b^{\mathsf{S}}(\sigma_f, \sigma'_f, \varepsilon')_{P,w} \oplus I_b^{\mathsf{S}}(\sigma'_f(n), \sigma_f(-n'), \varepsilon')_{Q,w'}$, is a k-dimensional E-subspace; and furthermore, from the proof of that theorem, it is clear that we get an intertwining operator $T_{\mathrm{Eis}}(\sigma, \sigma', \varepsilon')$ defined over E (see (6.16)) such that

$$\mathfrak{I}^b(\sigma_f, \sigma'_f, \varepsilon') = \left\{(\xi, T_{\mathrm{Eis}}(\sigma, \sigma', \varepsilon')(\xi)) \mid \xi \in I_b^{\mathsf{S}}(\sigma_f, \sigma'_f, \varepsilon')_{P,w}\right\}.$$

The idea of the proof is to take $T_{\mathrm{Eis}}(\sigma, \sigma', \varepsilon')$ to a transcendental level, use the constant term theorem which gives L-values, and after introducing the relative periods, to descend to an arithmetic level, giving a rationality result for the said L-values divided by the relative periods.

7.3.2.1 *The arithmetic of local intertwining operators*

We return to the modules described in Sect. 5.3.7, especially as we are working over the field E and work with algebraic induction. As before, we identify M_P and M_Q via the conjugation by w_P. This conjugation also provides a bijection between the isomorphism types of representations $\underline{\sigma}_f = (\sigma_f \times \sigma'_f) \mapsto \underline{\sigma}'_f = \sigma'_f \times \sigma_f$. More precisely, we get an identification $H_P^b(\sigma_f \times \sigma'_f) = H_Q^b(\sigma'_f \times \sigma_f)$; notation as in (5.22). Our next goal is to construct, using the integrals in Sect. 6.3.3, an intertwining operator between the Hecke modules

$$T_{\mathrm{st}}^{\mathrm{ar}}(\sigma_f \times \sigma'_f)(1) : I_b^{\mathsf{S}}(\sigma_f \times \sigma'_f) \to I_b^{\mathsf{S}}((\sigma'_f \times \sigma_f) \otimes |\boldsymbol{\delta}_Q|).$$

Here we have to be alert, because the integrals are infinite sums.

Recall from Sect. 5.3.6 the isomorphism $T_{\mathrm{arith}}(\underline{\sigma}_f, \pm) : H_P(\underline{\sigma}_f) \xrightarrow{\sim} H_Q(\underline{\sigma}'_f)$. As in (2.16), write $H_P(\underline{\sigma}_f)$ as a tensor product of local Hecke modules

$$H_P(\underline{\sigma}_f) = \bigotimes_v{}' H_P(\underline{\sigma}_v),$$

where for $v \notin \mathsf{S}$, the module $H(\underline{\sigma}_v) = E$. Then our intertwining operator will be constructed from a tensor product of local normalized intertwining operators:

$$\bigotimes_{v \in \mathsf{S}} T_{\mathrm{norm}}^{\mathrm{ar}}(\underline{\sigma}_v)(1) \otimes \bigotimes_{v \notin \mathsf{S}}{}' T_{\mathrm{loc}}^{\mathrm{ar}}(\underline{\sigma}_v)(1);$$

hence we have to construct the local operators $T_{\mathrm{loc}}^{\mathrm{ar}}(\underline{\sigma}_v)(1)$ and $T_{\mathrm{norm}}^{\mathrm{ar}}(\underline{\sigma}_v)(1)$.

To achieve this we pick a finite place v and consider the ring of Laurent polynomials $E[q_v^{-z}, q_v^z]$; i.e., we consider q_v^{-z} as a formal variable. Let $\{q_v^{-z}\}_{P_0}$ be the free $E[q_v^{-z}, q_v^z]$ module of rank one on which $M_{P_0}(F_v)$ acts by the character

$m \mapsto (q_v^{-z})^{ord_v(\boldsymbol{\gamma}_{P_0}(m))}$. Similarly, define $\{q_v^{-z}\}_{Q_0}$. Consider the module $H(\underline{\sigma}_v)\otimes E[q_v^{-z}, q_v^z]$ and define the twisted induced module

$$\begin{aligned}&I_b^{\mathsf{S}}(\underline{\sigma}_v)\otimes\{q_v^{-z}\}_P\\ &= \{f : G_0(F_v)/K_v \to H(\underline{\sigma}_v)\otimes E[q_v^{-z}, q_v^z] \mid f(pg) = \underline{\sigma}_v(\bar{p})f(g)(q_v^{-z})^{-ord_v(\boldsymbol{\delta}_{P_0}(p))}\},\end{aligned}$$

where $g \in G_0(F_v), p \in P_0(F_v)$, and $\bar{p} = p \mod U_0(F_v) \in M_0(F_v)$. Now we have the following proposition.

Proposition 7.22. *The integral in (6.12)*

$$f(g) \mapsto T_{\mathrm{arith}}(\underline{\sigma}_f)(f)(g) = \Big(\int_{U_0(F_v)} f(w_{P_0}ug)du\Big)$$

defines an intertwining operator

$$T_{\mathrm{st}}^{\mathrm{ar}}(\underline{\sigma}_v)(q_v^{-z}) : \; I_b^{\mathsf{S}}(\underline{\sigma}_v)\otimes\{q_v^{-z}\}_P \longrightarrow (I_b^{\mathsf{S}}(\underline{\sigma}'_v)\otimes\{q_v^{z-N}\}_Q)\otimes E[q_v^z][[q_v^{-z}]].$$

Proof. This is a standard argument. Write $w_{P_0}u = p(u)k(u)$; then the integrand becomes

$$f(p(u)k(u)g) = \underline{\sigma}_v(p(u))f(k(u)g)(q_v^{-z})^{ord_v(\boldsymbol{\delta}_{P_0}(\overline{p(u)}))}.$$

Now the sets

$$\{u \in U_0(F_v) \mid w_{P_0}u = p(u)k(u);\ \mathrm{ord}_v(u) \le T\}$$

are compact. Since our functions are locally constant it follows that

$$T_{\mathrm{st}}^{\mathrm{ar}}(\underline{\sigma}_v)(q_v^{-z}) = \sum_{i=0}^{\infty} A_i(\underline{\sigma}_v)(q_v^{-z})^i,$$

where $A_i(\underline{\sigma}_v) \in \mathrm{Hom}_{G_0(\mathcal{O}_v)}(I_b^{\mathsf{S}}(\underline{\sigma}_v), I_b^{\mathsf{S}}(\underline{\sigma}'_v)|\boldsymbol{\delta}_Q|_v)$. □

A rationality result of Waldspurger [70, Thm. IV.1.1] asserts that this power series is actually a rational function; i.e., there exists a polynomial $P(q_v^{-z}) = 1 + a_1(\underline{\sigma}_v)q_v^{-z} + \cdots + a_m(\underline{\sigma}_v)q_v^{-mz} \in E[q_v^{-z}]$, (where m depends on $\underline{\sigma}_v$) such that

$$T_{\mathrm{st}}^{\mathrm{ar}}(\underline{\sigma}_v)(q_v^{-z}) = \frac{\sum_{i=0}^M B_i(\underline{\sigma}_v)(q_v^{-z})^i}{P(q_v^{-z})} = \frac{\sum_{i=0}^M B_i(\underline{\sigma}_v)(q_v^{-z})^i}{\prod(1-\omega_\nu(\underline{\sigma}_v)q_v^{-z})}, \tag{7.34}$$

with $B_i(\underline{\sigma}_v) \in \mathrm{Hom}_{G_0(\mathcal{O}_v)}(I_b^{\mathsf{S}}(\underline{\sigma}_v, I_b^{\mathsf{S}}(\underline{\sigma}'_v)|\boldsymbol{\delta}_Q|_v)$. It is clear that we may assume that there is no cancellation in the above expression; i.e., we can assume

$$\sum_{i=0}^M B_i(\underline{\sigma}_v)\omega_\nu(\underline{\sigma}_v)^i \neq 0 \text{ for all } \nu;$$

if this assumption is satisfied then we can evaluate the numerator and we know for sufficiently small level K_v the numerator-interwining operators evaluated at any $u_0 = q_v^{-s_0}$ are nonzero, i.e., $\sum_{i=0}^{M} B_i(\underline{\sigma}_v)u_0^i \neq 0$.

Consider the local operator in (6.12) at an unramified place $\underline{\sigma}_v$, then $H(\underline{\sigma}_v) = H(\underline{\sigma}'_v) = E$ and we have the *normalized spherical* vectors $f_v^0 \in I_b^{\mathsf{S}}(\underline{\sigma}_v)$ and $f_v^{\prime,0} \in I_b^{\mathsf{S}}(\underline{\sigma}'_v)$ which are normalized to take the value 1 at the identity. Define the local arithmetic intertwining operator by

$$T_{\mathrm{loc}}^{\mathrm{ar}}(\underline{\sigma}_v)(q_v^{-z})(f_v^0) \;=\; f_v^{\prime,0}. \tag{7.35}$$

In this case we have the fundamental formula of Langlands for the intertwining operator. For any $\iota : E \to \mathbb{C}$, let $\mathrm{diag}(\vartheta_{1,v}, \dots, \vartheta_{n,v})$ (resp., $\mathrm{diag}(\vartheta'_{1,v}, \dots, \vartheta'_{n',v})$) be the Satake parameters of ${}^{\iota}\sigma_v$ (resp., ${}^{\iota}\sigma'_v$). Recall from (7.21), the local *automorphic* Rankin–Selberg L-function:

$$L(s, {}^{\iota}\sigma_v \times {}^{\iota}\sigma_v^{\prime\mathsf{v}}) \;=\; \prod_{\substack{1\le i\le n \\ 1\le j\le n'}} (1 - \vartheta_{i,v}\vartheta_{j,v}^{\prime -1} q_v^{-s})^{-1},$$

and Langlands' formula [49] yields

$$T_{\mathrm{st}}^{\mathrm{ar}}(\underline{\sigma}_v)(q_v^{-z}) \otimes_{E,\iota} \mathbb{C} = \frac{L(z - N/2, {}^{\iota}\sigma_v \times {}^{\iota}\sigma_v^{\prime\mathsf{v}})}{L(z + 1 - N/2, {}^{\iota}\sigma_v \times {}^{\iota}\sigma_v^{\prime\mathsf{v}})} T_{\mathrm{loc}}^{\mathrm{ar}}(\underline{\sigma}_v)(q_v^{-z}) \otimes_{E,\iota} \mathbb{C}. \tag{7.36}$$

The elementary symmetric functions in n (resp., n') variables evaluated at

$$\{q_v^{(n-1)/2}\vartheta_{1,v}, \dots, q_v^{(n-1)/2}\vartheta_{n,v}\}, \;\; (\text{resp., } \{q_v^{(n'-1)/2}\vartheta_{1,v}^{\prime -1} \dots, q_v^{(n'-1)/2}\vartheta_{n',v}^{\prime -1}\})$$

are eigenvalues of Hecke operators on $H(\underline{\sigma}_v)$ and hence are in $\iota(E)$. This implies

$$L(z - N/2, {}^{\iota}\sigma_v \times {}^{\iota}\sigma_v^{\prime\mathsf{v}})^{-1} \;\in\; \iota(E)[q_v^{-z}].$$

Moreover, there exists $L(z - N/2, \sigma_v \times \sigma_v^{\prime\mathsf{v}})^{-1} \in E[q_v^{-z}]$ such that after applying ι to the coefficients to the polynomial over E, we get the local Rankin-Selberg L-factor, i.e.,

$$\iota(L(z - N/2, \sigma_v \times \sigma_v^{\prime\mathsf{v}})) = L(z - N/2, {}^{\iota}\sigma_v \times {}^{\iota}\sigma_v^{\prime\mathsf{v}}).$$

This is true at all finite places; i.e., for any place $v \notin \mathsf{S}_\infty$, there exists a local factor defined at an arithmetic level: $L(z - N/2, \sigma_v \times \sigma_v^{\prime\mathsf{v}})^{-1} \in E[q_v^{-z}]$ such that after applying ι to the coefficients in E of this polynomial we get the above equality of local Rankin–Selberg L-factor; this may be seen as in the proof of [53, Prop. 3.17], together with a Galois descent argument. Now we formally define a normalized arithmetic intertwining operator:

$$T^{\mathrm{ar}}_{\mathrm{norm}}(\underline{\sigma}_v)(q_v^{-z}) = \left(\frac{L(z-N/2,\sigma_v\times\sigma_v'^{\vee})}{L(1+z-N/2,\sigma_v\times\sigma_v'^{\vee})}\right)^{-1} T^{\mathrm{ar}}_{\mathrm{st}}(\underline{\sigma}_v)(q_v^{-z}). \tag{7.37}$$

At the point of evaluation $z=0$, we have the following proposition.

Proposition 7.23. *If (μ,μ') is on the right of the unitary axis, i.e., $-a(\underline{\mu}) \geq N/2$, then for all $v \in \mathsf{S}_f$, $T^{\mathrm{ar}}_{\mathrm{norm}}(\underline{\sigma}_v)(1)$ is a nonzero intertwining operator over E.*

Proof. We apply an ι, and then see using the fact ([52, Prop. I.10]) that normalized intertwining operator is holomorphic and nonzero at the point of evaluation; hence at an arithmetic level, a fortiori, we have $T^{\mathrm{ar}}_{\mathrm{norm}}(\underline{\sigma}_v)(1)$ is nonzero. □

Given $\underline{\sigma}_f = \sigma_f \times \sigma'_f$ and $\iota : E \to \mathbb{C}$, let ${}^\iota\sigma = {}^\iota\sigma_\infty \times {}^\iota\sigma_f$ and ${}^\iota\sigma' = {}^\iota\sigma'_\infty \times {}^\iota\sigma'_f$, then recall from (6.11) the global intertwining operator $T^{\mathrm{ar}}_{\mathrm{st}}(z, {}^\iota\sigma \times {}^\iota\sigma')$ which we write as

$$T^{\mathrm{ar}}_{\mathrm{st}}(z, {}^\iota\sigma \times {}^\iota\sigma') : {}^a\mathrm{Ind}_{P(\mathbb{A})}^{G(\mathbb{A})}\left(({}^\iota\sigma \times {}^\iota\sigma') \otimes |\gamma_P|^z\right) \\ \to {}^a\mathrm{Ind}_{P(\mathbb{A})}^{G(\mathbb{A})}\left(({}^\iota\sigma' \times {}^\iota\sigma) \otimes |\gamma_P|^{N-z}\right).$$

The global intertwining operator is a tensor product of local intertwining operators; the set of all places are grouped together as the set S_∞ of all archimedean places, a finite set S_f of finite places containing all the ramified places, and all the remaining ($v \notin \mathsf{S} = \mathsf{S}_\infty \cup \mathsf{S}_f$) necessarily consisting of finite and unramified places. At $z=0$ we see that $T^{\mathrm{ar}}_{\mathrm{st}}(0, {}^\iota\sigma \otimes {}^\iota\sigma')$ may be written as

$$\frac{L(-\frac{N}{2}, {}^\iota\sigma_f \times {}^\iota\sigma_f'^{\vee})}{L(1-\frac{N}{2}, {}^\iota\sigma_f \times {}^\iota\sigma_f'^{\vee})} \cdot \bigotimes_{v\in\mathsf{S}_\infty} T^{\mathrm{ar}}_{\mathrm{st}}(0, {}^\iota\underline{\sigma}_v) \\ \otimes \left(\bigotimes_{v\in\mathsf{S}_f} T^{\mathrm{ar}}_{\mathrm{norm}}(\underline{\sigma}_v)(1) \otimes \bigotimes_{v\notin\mathsf{S}} T^{\mathrm{ar}}_{\mathrm{loc}}(\underline{\sigma}_v)(1)\right) \otimes_{E,\iota} \mathbb{C}.$$

Let's translate this into normalized induction. Put $z = s + N/2$, then

$$I_P^G(s, {}^\iota\sigma \otimes {}^\iota\sigma') = {}^a\mathrm{Ind}_{P(\mathbb{A})}^{G(\mathbb{A})}({}^\iota\sigma \times {}^\iota\sigma')|\gamma_P|^z$$

and the intertwining operator $T_{\mathrm{st}}(s, {}^\iota\sigma \otimes {}^\iota\sigma') : I_P^G(s, {}^\iota\sigma \otimes {}^\iota\sigma') \to I_P^G(-s, {}^\iota\sigma' \otimes {}^\iota\sigma)$

at the point of evaluation, which is now at $s = -N/2$, is given by

$$\frac{L(-\frac{N}{2}, {}^{\iota}\sigma_f \times {}^{\iota}\sigma_f'^{\vee})}{L(1-\frac{N}{2}, {}^{\iota}\sigma_f \times {}^{\iota}\sigma_f'^{\vee})} \cdot \bigotimes_{v \in \mathsf{S}_\infty} T_{\mathrm{st}}(-\tfrac{N}{2}, {}^{\iota}\underline{\sigma}_v)$$
$$\otimes \Big(\big(\bigotimes_{v \in \mathsf{S}_f} T_{\mathrm{norm}}^{\mathrm{ar}}(\underline{\sigma}_v)(1) \otimes \bigotimes_{v \notin \mathsf{S}} T_{\mathrm{loc}}^{\mathrm{ar}}(\underline{\sigma}_v)(1) \Big) \otimes_{E,\iota} \mathbb{C}. \quad (7.38)$$

The expression inside $(\cdots)$ is a nonzero operator over E. It remains to understand the standard intertwining operator at the archimedean places.

7.3.2.2 The local intertwining operator at the infinite places

We begin by showing that $T_{\mathrm{st}}(-\frac{N}{2}, {}^{\iota}\sigma_\infty \times {}^{\iota}\sigma'_\infty)^{\bullet}(\varepsilon')$ is a nonzero isomorphism between one-dimensional vector spaces.

Proposition 7.24. *The induced representations*

$${}^{\mathrm{a}}\mathrm{Ind}_{P(\mathbb{R})}^{G(\mathbb{R})}({}^{\iota}\sigma_\infty \otimes {}^{\iota}\sigma'_\infty) \quad \text{and} \quad {}^{\mathrm{a}}\mathrm{Ind}_{Q(\mathbb{R})}^{G(\mathbb{R})}({}^{\iota}\sigma'_\infty(n) \otimes {}^{\iota}\sigma_\infty(-n'))$$

are irreducible representations of $G(\mathbb{R})$. *The operator* $T_{\mathrm{st}}(-\frac{N}{2}, {}^{\iota}\sigma \times {}^{\iota}\sigma')_\infty$ *is an isomorphism between these two induced representations, and hence induces an isomorphism* $T_{\mathrm{st}}(-\frac{N}{2}, {}^{\iota}\sigma \times {}^{\iota}\sigma')^{\bullet}_\infty(\varepsilon')$ *between the one-dimensional* $\mathbb{C}$*-vector spaces:*

$$H^{b_N^F}(\mathfrak{g}, K_\infty^0, {}^{\mathrm{a}}\mathrm{Ind}_{P(\mathbb{R})}^{G(\mathbb{R})}({}^{\iota}\sigma_\infty \otimes {}^{\iota}\sigma'_\infty) \otimes \mathcal{M}_{{}^{\iota}\lambda,\mathbb{C}})(\varepsilon')$$
$$\longrightarrow H^{b_N^F}(\mathfrak{g}, K_\infty^0, {}^{\mathrm{a}}\mathrm{Ind}_{Q(\mathbb{R})}^{G(\mathbb{R})}({}^{\iota}\sigma'_\infty(n) \otimes {}^{\iota}\sigma_\infty(-n')) \otimes \mathcal{M}_{{}^{\iota}\lambda,\mathbb{C}})(\varepsilon').$$

Proof. Irreducibility of the induced representations follows by applying results of Birgit Speh; see [51, Thm. 10b]. We briefly sketch the details in one case. (Let's introduce some convenient and well-known notation. For any local field $\mathbb{F}$, suppose π_1 and π_2 are representations of $\mathrm{GL}_{n_1}(\mathbb{F})$ and $\mathrm{GL}_{n_2}(\mathbb{F})$, then by $\pi_1 \times \pi_2$ we denote the representation of $\mathrm{GL}_{n_1+n_2}$ obtained by normalized parabolical induction from the representation $\pi_1 \otimes \pi_2$ of the Levi subgroup $\mathrm{GL}_{n_1}(\mathbb{F}) \times \mathrm{GL}_{n_2}(\mathbb{F})$.) Translating to normalized induction, and using transitivity of normalized parabolic induction, we may write ${}^{\mathrm{a}}\mathrm{Ind}_{P(\mathbb{R})}^{G(\mathbb{R})}({}^{\iota}\sigma_\infty \otimes {}^{\iota}\sigma'_\infty)$ as

$$\mathbb{D}_{{}^{\iota}\ell_1}(-d-\tfrac{n'}{2}) \times \cdots \times \mathbb{D}_{{}^{\iota}\ell_{n/2}}(-d-\tfrac{n'}{2}) \times \mathbb{D}_{{}^{\iota}\ell'_1}(-d'-\tfrac{n}{2}) \times \cdots$$
$$\cdots \times \mathbb{D}_{{}^{\iota}\ell_{(n'-1)/2}}(-d'-\tfrac{n}{2}) \times {}^{\iota}\varepsilon'(-d'-\tfrac{n}{2}).$$

To check irreducibility, by (ii) and (iii) of [51, Thm. 10b], we need to check, for

$1 \le i \le n/2$ and $1 \le j \le (n'+1)/2$, that

$$-\frac{|{}^\iota\ell_i - {}^\iota\ell'_j|}{2} + |d - d' + \tfrac{N}{2}| \notin \{1, 2, 3, \dots\}.$$

We leave it to the reader to see that this follows from the inequalities in (ii) of Lem.7.14. It is also easy to see along the same lines that the irreducibility holds if both n and n' are even.

Now, we have the standard intertwining operator $T_{\rm st}(s, {}^\iota\sigma \times {}^\iota\sigma')_\infty$, at $s = -N/2$ between two irreducible representations; but we need to show that this operator is finite and nonzero at $s = -N/2$. For this we appeal to Shahidi's results on local constants that was briefly reviewed in Sect. 7.2.2.3. Recall (7.32):

$$C_{\psi_v}(s, \sigma_v \times \sigma'_v) \;=\; \varepsilon(s, \psi_v, \sigma_v \times \sigma'^{\vee}_v)\,\frac{L(1-s, \sigma^{\vee}_v \times \sigma'_v)}{L(s, \sigma_v \times \sigma'^{\vee}_v)}.$$

The ε-factor is a constant and is irrelevant to this discussion. Now, if $T(-N/2) := T_{\rm st}(s, \sigma_v \times \sigma'_v)|_{s=-N/2}$ is 0 then from (7.31) we see that $L(1+N/2, \sigma^{\vee}_v \times \sigma'_v)$ has to be a pole to guarantee nonvanishing of $\lambda_{\psi_v}(-N/2, \sigma_v \times \sigma'_v)$, and similarly, if $T(-N/2)$ is a pole then the L-value in the denominator $L(-N/2, \sigma_v \times \sigma'^{\vee}_v)$ has to be a pole to guarantee finiteness of $\lambda_{\psi_v}(-N/2, \sigma_v \times \sigma'_v)$; but neither case is possible since $-N/2$ is a critical point. Whence, $T(-N/2)$ is finite and nonzero, and indeed is an isomorphism between the two irreducible modules. □

Let's note en passant that from a very general result contained in Casselman–Shahidi [12] we get irreducibility of the induced representation as a consequence of criticality of the L-values at hand.

7.3.2.3 Choice of bases at infinity

Fix a place $v \in \mathsf{S}_\infty$, and let's suppress the ι from the notation as it is fixed for this discussion. Then $\underline{\sigma}_v = \mathbb{D}_{\underline{\mu}_v}(\varepsilon') = \mathbb{D}_{\mu_v} \otimes (\mathbb{D}_{\mu'_v} \times \varepsilon')$ if n even and n' is odd, and is $\mathbb{D}_{\mu_v} \otimes \mathbb{D}_{\mu'_v}$ when both n and n' are even. The details that follow are for n even and n' odd, and the reader can make the necessary changes when both are even. Delorme's Lemma [9, Thm. III.3.3] describes the relative Lie algebra cohomology of a parabolically induced representation in terms of the cohomology of the inducing data. This allows us to make the following identifications:

$$\begin{aligned}
&H^{b^F_N}(\mathfrak{g}, K^0_v, {}^{\rm a}{\rm Ind}^{G(\mathbb{R})}_{P(\mathbb{R})}\mathbb{D}_{\underline{\mu}_v}(\varepsilon') \otimes \mathcal{M}_{\lambda_v,\mathbb{C}})(\varepsilon') \\
&\xrightarrow{\sim}\; H^{b^F_n}(\mathfrak{g}_n, K^0_{n,v}, \mathbb{D}_{\mu_v} \otimes \mathcal{M}_{\mu_v,\mathbb{C}})(\varepsilon') \otimes H^{b^F_{n'}}(\mathfrak{g}_{n'}, K^0_{n',v}, \mathbb{D}_{\mu'_v} \times \varepsilon' \otimes \mathcal{M}_{\mu'_v,\mathbb{C}}),
\end{aligned} \tag{7.39}$$

and

$$H^{b_N^F}(\mathfrak{g}, K_v^0, {}^{\mathrm{a}}\mathrm{Ind}_{Q(\mathbb{R})}^{G(\mathbb{R})}\mathbb{D}_{\underline{\mu}_v'-N\gamma_Q}(\varepsilon')\otimes\mathcal{M}_{\lambda,\mathbb{C}})(\varepsilon') \xrightarrow{\sim}$$
$$H^{b_{n'}^F}(\mathfrak{g}_{n'}, K_{n',v}^0, \mathbb{D}_{\mu_v'-n\delta_{n'}}\times\varepsilon'\otimes\mathcal{M}_{\mu_v',\mathbb{C}})\otimes H^{b_n^F}(\mathfrak{g}_n, K_{n,v}^0, \mathbb{D}_{\mu_v-n'\delta_n}\otimes\mathcal{M}_{\mu_v,\mathbb{C}})(\varepsilon'). \tag{7.40}$$

We fixed basis elements in Sect. 5.2.2:

$$w^{\varepsilon'}(\mu_v)\in \quad H^{b_n^F}(\mathfrak{g}_n, K_{n,v}^0, \mathbb{D}_{\mu_v}\otimes\mathcal{M}_{\mu_v,\mathbb{C}})(\varepsilon'),$$
$$w^{\varepsilon'}(\mu_v')\in \quad H^{b_{n'}^F}(\mathfrak{g}_{n'}, K_{n',v}^0, \mathbb{D}_{\mu_v'}\times\varepsilon'\otimes\mathcal{M}_{\mu_v',\mathbb{C}}),$$

and hence we have basis elements

$$w^{\varepsilon'}(\underline{\mu}_v) = w^{\varepsilon'}(\mu_v)\otimes w^{\varepsilon'}(\mu_v') \in H^{b_N^F}(\mathfrak{g}, K_v^0, {}^{\mathrm{a}}\mathrm{Ind}_{P(\mathbb{R})}^{G(\mathbb{R})}\mathbb{D}_{\underline{\mu}_v}(\varepsilon')\otimes\mathcal{M}_{\lambda,\mathbb{C}})(\varepsilon')$$

and

$$w^{\varepsilon'}(\underline{\mu}_v+N\gamma_Q) = w^{\varepsilon'}(\mu_v'-n\delta_{n'})\otimes w^{\varepsilon'}(\mu_v+n'\delta_n)$$
$$\in H^{b_N^F}(\mathfrak{g}, K_v^0, {}^{\mathrm{a}}\mathrm{Ind}_{Q(\mathbb{R})}^{G(\mathbb{R})}\mathbb{D}_{\underline{\mu_v}'-N\gamma_Q}(\varepsilon')\otimes\mathcal{M}_{\lambda,\mathbb{C}})(\varepsilon').$$

We define the local intertwining operator

$$T_{\mathrm{loc}}(\underline{\sigma}_v) : {}^{a}\mathrm{Ind}_{P(\mathbb{R})}^{G(\mathbb{R})}\mathbb{D}_{\underline{\mu}_v}(\varepsilon') \to {}^{a}\mathrm{Ind}_{Q(\mathbb{R})}^{G(\mathbb{R})}\mathbb{D}_{\underline{\mu}_v'}(\varepsilon')\otimes|\gamma_Q|^{-N}$$

by the requirement that

$$T_{\mathrm{loc}}(\underline{\sigma}_v)^{\bullet}(w^{\varepsilon'}(\underline{\mu}_v)) = i^{n/2}w^{\varepsilon'}(\underline{\mu}_v+N\gamma_Q). \tag{7.41}$$

The reader should parse $T_{\mathrm{loc}}(\underline{\sigma}_v)^{\bullet}$ as: take Tate twists

$$w^{\varepsilon'}(\mu_v) \mapsto w^{-\varepsilon'}(\mu_v+n'\delta_n), \quad \text{and} \quad w^{\varepsilon'}(\mu_v') \mapsto w^{\varepsilon'}(\mu_v'-n\delta_{n'}),$$

(we have used the parities on n and n' here), and further compose the former Tate twist by $T^{-\varepsilon'}(\mu_v+n'\delta_n)$; see (5.9).

The following result is proved in Thm. 8.7 (see also Sect. 9.6.12), and in combination with Cor. 7.9 it yields the following theorem.

Theorem 7.25. *There is a nonzero rational number $c(\underline{\sigma}_v) = c(\sigma_v\times\sigma_v')$ such that*

$$T_{\mathrm{st}}(\sigma_v\otimes\sigma_v')^{\bullet}(\varepsilon') = c(\underline{\sigma}_v)i^{nn'/2}\cdot\frac{L_v^{\mathrm{Coh}}(\sigma_v\times\sigma_v'^{\vee}, \mathsf{m}_0)}{L_v^{\mathrm{Coh}}(\sigma_v\times\sigma_v'^{\vee}, 1+\mathsf{m}_0)}T_{\mathrm{loc}}(\underline{\sigma}_v). \tag{7.42}$$

In Chap. 8 we raise the question whether $c(\underline{\sigma}_v)$ could be a very *simple* number, for instance could it be simply ± 1? In the special case $N = 3$ this has been proved in [29] with the help of Zagier [74], and in the general case this is proven by Weselmann in Chap. 9; see especially Sect. 9.6.12.

Now consider all places in S_∞, and we reintroduce the ι to get basis elements

$$\omega_\infty^{\varepsilon'}({}^\iota\underline{\mu}) = \bigotimes_{v\in\mathsf{S}_\infty} w^{\varepsilon'}({}^\iota\underline{\mu}_v) \in H^{b_N^F}(\mathfrak{g}, K_\infty^0, {}^{\mathrm{a}}\mathrm{Ind}_{P(\mathbb{R})}^{G(\mathbb{R})}\mathbb{D}_{{}^\iota\underline{\mu}}(\varepsilon') \otimes \mathcal{M}_{{}^\iota\lambda,\mathbb{C}})(\varepsilon'),$$

$$\omega_\infty^{\varepsilon'}(\underline{\mu}' - N\gamma_Q) = \bigotimes_{v\in\mathsf{S}_\infty} w^{\varepsilon'}(\underline{\mu}'_v - N\gamma_Q)$$
$$\in H^{b_N^F}(\mathfrak{g}, K_\infty^0, {}^{\mathrm{a}}\mathrm{Ind}_{Q(\mathbb{R})}^{G(\mathbb{R})}\mathbb{D}_{\underline{\mu}'-N\gamma_Q}(\varepsilon') \otimes \mathcal{M}_{{}^\iota\lambda,\mathbb{C}})(\varepsilon'),$$

and we now have (in self-explanatory notation) that

$$T_{\mathrm{st}}(-\tfrac{N}{2}, {}^\iota\underline{\sigma}_\infty)^\bullet(\omega_\infty^{\varepsilon'}({}^\iota\underline{\mu}))$$
$$= c({}^\iota\underline{\sigma}_\infty)\cdot i^{rnn'/2}\cdot \frac{L_\infty^{\mathrm{Coh}}({}^\iota\sigma_\infty \times {}^\iota\sigma_\infty'^{\mathsf{v}}, \mathsf{m}_0)}{L_\infty^{\mathrm{Coh}}({}^\iota\sigma_\infty \times {}^\iota\sigma_\infty'^{\mathsf{v}}, 1+\mathsf{m}_0)}\, w^{\varepsilon'}(\underline{\mu}'_\infty - N\gamma_Q),$$

for a nonzero rational number $c({}^\iota\underline{\sigma}_\infty)$.

7.3.2.4 Conclusion of the proof of Thm. 7.21 when on the right of the unitary axis

We now introduce the periods into the play. The module

$$\mathbb{D}_{{}^\iota\underline{\mu}}(\varepsilon') \otimes H_P({}^\iota\underline{\sigma}_f) \otimes_{E,\iota} \mathbb{C}$$

being a cuspidal automorphic representation, we get a commutative diagram (see Sect. 5.2.2):

$$\begin{array}{ccc}
\mathbb{D}_{{}^\iota\underline{\mu}}(\varepsilon') \otimes H_P({}^\iota\underline{\sigma}_f) \otimes_{E,\iota} \mathbb{C} & \xrightarrow{\quad\Phi\quad} & \mathcal{A}_{\mathrm{cusp}}(M_P(\mathbb{Q})\backslash M_P(\mathbb{A})) \\
\Big\downarrow {\scriptstyle |\gamma_Q|_\infty^N \otimes T_{\mathrm{arith}}} & & \Big\downarrow {\scriptstyle w_P \otimes |\gamma_Q|_f^N} \\
\mathbb{D}_{{}^\iota\underline{\mu}+N\gamma_Q}(-\varepsilon') \otimes H_P({}^\iota\underline{\sigma}'_f \otimes |\gamma_{\mathbb{Q}}|^N) \otimes_{E,\iota} \mathbb{C} & \xrightarrow{\quad\Phi_N\quad} & \mathcal{A}_{\mathrm{cusp}}(M_Q(\mathbb{Q})\backslash M_Q(\mathbb{A}))
\end{array}$$

where the second vertical arrow is defined by

$$w_P \otimes |\gamma_Q|_f^N(h)(\underline{m}) = h(w_P\underline{m}w_P^{-1})|\gamma_Q|_f^N(\underline{m}).$$

The choice $\omega_\infty^{\varepsilon'}({}^\iota\underline{\mu})$ (resp., $\omega_\infty^{\varepsilon'}({}^\iota\underline{\mu}' - N\gamma_Q)$) provides an endomorphism, defined on a transcendental level, of $H_P(\underline{\sigma}_f) \otimes_{E,\iota} \mathbb{C}$ (resp., $H_P(\underline{\sigma}'_f \otimes |\gamma_{\mathbb{Q}}|^N) \otimes_{E,\iota} \mathbb{C}$,) defined by

$$\Phi^\bullet(\omega_\infty^{\varepsilon'}(\underline{\mu})) : h \;\mapsto\; \Phi^\bullet(\omega_\infty^{\varepsilon'}(\underline{\mu}) \otimes h),$$

and respectively by

$$\Phi_N^\bullet(\omega_\infty^{\varepsilon'}(\underline{\mu}' + N\gamma_Q)) : h' \;\mapsto\; \Phi_N^\bullet(\omega_\infty^{\varepsilon'}(\underline{\mu}' + N\gamma_Q) \otimes h').$$

Since these are endomorphisms of absolutely irreducible Hecke modules, these endomorphisms are given by numbers

$$\Omega({}^\iota\underline{\sigma}_f, \omega_\infty^{\varepsilon'}({}^\iota\underline{\mu}), \Phi), \quad \text{resp.,} \quad \Omega({}^\iota\underline{\sigma}'_f, \omega_\infty^{\varepsilon'}({}^\iota\underline{\mu}' + N\gamma_Q), \Phi_N).$$

If we take the ratio then the choice of Φ cancels out and by definition

$$\Omega({}^\iota\underline{\sigma}_f) = \frac{\Omega({}^\iota\underline{\sigma}_f), \omega_\infty^{\varepsilon'}({}^\iota\underline{\mu}), \Phi)}{\Omega({}^\iota\underline{\sigma}'_f, \omega_\infty^{\varepsilon'}({}^\iota\underline{\mu}' + N\gamma_Q), \Phi_N)}; \tag{7.43}$$

see the beginning of Sect. 7.3.1. Of course this number depends on the choice of $T_{\text{arith}}(\underline{\sigma}_f)$ and it is unique only up to an element in $E^\times$.

We return to $T_{\text{Eis}}(\sigma, \sigma', \varepsilon')$ defined over E (see (6.16)), and taking the base extension by an $\iota : E \hookrightarrow \mathbb{C}$, we get $T_{\text{Eis}}(\sigma, \sigma', \varepsilon') \otimes_{E,\iota} \mathbb{C} = T_{\text{st}}(-\frac{N}{2}, {}^\iota\sigma \otimes {}^\iota\sigma')^\bullet$ and that

$$\mathfrak{I}^b({}^\iota\sigma_f, {}^\iota\sigma'_f, \varepsilon') \;=\; \left\{\left(\xi,\, T_{\text{st}}(-\tfrac{N}{2}, {}^\iota\sigma \otimes {}^\iota\sigma')^\bullet\xi\right) \;:\; \xi \in I_b^{\mathsf{S}}({}^\iota\sigma_f, {}^\iota\sigma'_f, \varepsilon')_{P,{}^\iota w}\right\}.$$

Using (7.38) we get that $T_{\text{st}}(-\frac{N}{2}, {}^\iota\sigma \otimes {}^\iota\sigma')^\bullet$ is given by

$$\frac{1}{\Omega({}^\iota\underline{\sigma}_f)} \frac{L^{\text{Coh}}(\iota, \sigma \times \sigma'^{\mathsf{v}}, \mathsf{m}_0)}{L^{\text{Coh}}(\iota, \sigma \times \sigma'^{\mathsf{v}}, 1 + \mathsf{m}_0)} \cdot \Big(c_\infty({}^\iota\underline{\sigma}_\infty) \bigotimes_{v \in \mathsf{S}_f} T^{\text{ar}}_{\text{norm}}(\underline{\sigma}_v)(1) \otimes \bigotimes_{v \notin \mathsf{S}} T^{\text{ar}}_{\text{loc}}(\underline{\sigma}_v)\Big) \otimes_{E,\iota} \mathbb{C}.$$

The proof of Theorem (7.21) follows if (μ, μ') is on the right of the unitary axis; we have used the fact that the expression in $\Big(\cdots\Big)$ is a nonzero linear operator over E (see Sect. 7.3.2.1).

7.3.2.5 *Conclusion of the proof of Thm. 7.21*

Now suppose that the data (μ, μ') is on the left of the unitary axis; i.e., that $-a(\underline{\mu}) < N/2$. Suppose that for a given $\iota : E \to \mathbb{C}$, $L^{\text{Coh}}(\iota, \sigma \times \sigma'^{\mathsf{v}}, 1 + \mathsf{m}_0) \neq 0$. Then the Eisenstein series $\text{Eis}_P(s, f)$ (see Thm. 7.10) is holomorphic at $s = -N/2$ and the above proof goes through verbatim. The Eisenstein series picks up a pole only in an exceptional case that is excluded by the condition that σ and σ' have disjoint cuspidal parameters.

It follows from the preceding paragraph that if $L^{\mathrm{Coh}}(\iota, \sigma \times \sigma'^{\mathsf{v}}, 1 + \mathsf{m}_0)$ is nonvanishing for one ι then it is nonvanishing for every ι; i.e., we have:

$$\text{the condition } L^{\mathrm{Coh}}(\iota, \sigma \times \sigma'^{\mathsf{v}}, 1 + \mathsf{m}_0) = 0 \text{ is independent of } \iota.$$

Indeed, if for an ι, we have $L^{\mathrm{Coh}}(\iota, \sigma \times \sigma'^{\mathsf{v}}, 1 + \mathsf{m}_0) = 0$, then it means that the image $\mathfrak{I}^b({}^\iota\sigma_f, {}^\iota\sigma'_f, \varepsilon')$ is the set of all classes of the form $\{(0, \tilde{\xi}) : \tilde{\xi} \in I_b^{\mathsf{S}}({}^\iota\sigma_f, {}^\iota\sigma'_f, \varepsilon')_{Q, {}^\iota w'}\}$ and it is clear that the structure of this image is independent of ι.

The reader should view the foregoing property of (non)vanishing of an L-value being independent of ι as an verifying an instance of Gross's conjecture stated in Deligne [15, Conj. 2.7(ii)].

7.3.2.6 Remarks on Thm. 7.21 concerning the parities of n and n'

It might help the reader to know the subtle difference of the two cases (i) n is even and n' is odd, and (ii) both n and n' are even. In case (i), the representation σ appears with a Tate twist of n' that is odd (as a constituent of the inducing data from the associate parabolic subgroup Q) and this causes the appearance of the relative period $\Omega^{\varepsilon'}(\sigma_f)$ in the theorem. However, in case (ii), both the constituents σ and σ' appear with even Tate twists, which are rationally defined, whence, the successive critical L-values differ only by an element of $\iota(E)$.

Our argument depends on the fact that we find balanced elements $w \in W^P$ and having a very precise understanding of certain parts of the Eisenstein cohomology related to these elements. We are working with the strongly inner cohomology of maximal parabolics and if n, n' are odd there are no balanced elements. On the other hand, Prop. 7.26 says that the critical points have to satisfy an additional parity condition (that may be construed as coming from the parity by which complex conjugation acts on the middle Hodge type of the tensor product motive). We want to raise the question whether an understanding of other parts of the Eisenstein cohomology could give results on ratios of L-values where the arguments differ by 2. This strategy has actually been used in [25], where the Eisenstein cohomology coming from the Borel subgroup is investigated.

7.3.2.7 Remarks on Thm. 7.21 and twisting by characters

Given σ_f and σ'_f as in the theorem, and given a finite-order Dirichlet character χ of F with coefficients in E, observe that for any $\iota : E \to \mathbb{C}$, we have

$$L(s, ({}^\iota\sigma \otimes {}^\iota\chi) \times {}^\iota\sigma') = L(s, {}^\iota\sigma \times ({}^\iota\sigma' \otimes {}^\iota\chi));$$

i.e., in a Rankin–Selberg L-function the twisiting character could be absorbed into either of the two constituents σ or σ'. Compatibility of Thm. 7.21 with respect to twisting by χ follows from Prop. 5.5 and the observation that ε' changes to $\varepsilon'\varepsilon_\chi$ upon twisting σ' by χ.

7.4 SOME REMARKS

7.4.1 The case when nn' is odd
By Chandrasheel Bhagwat and A. Raghuram

After the announcement [33], the authors of this subsection proved certain period relations in [3] for the periods of a tensor product $\mathbf{M} \otimes \mathbf{M}'$ of two pure motives of ranks n and n'. This was generalized in Bhagwat [2], and subsequently put in to a very general framework in Deligne–Raghuram [16]. If the parities of n and n' are different then these period relations together with Deligne's conjecture [15] exactly predict the main result on L-values Thm. 7.21. It was already observed by Yoshida [73] that if n and n' are both even, then the two periods $c^{\pm}$ of Deligne for the tensor product motive are equal. In the case when both n and n' are odd, in [3], an interesting period relation for the ratio c^+/c^- for $\mathbf{M} \otimes \mathbf{M}'$ is proved; this suggests that there should be a result analogous to Thm. 7.21; however, it turns out that the methods of the paper break down in this situation. The aim of this subsection is to discuss the combinatorial lemma (Lem. 7.14) in the situation when the Levi subgroup is $\mathrm{GL}_n \times \mathrm{GL}_{n'}$ with both n and n' being odd positive integers. We will see that certain simple combinatorial conditions on the highest weights μ and μ' assure us of the existence of two successive critical points; however, the same conditions will give that there is no element of the Weyl group that makes the character $\mu + \mu'$ dominant. This is already seen when the base field $F = \mathbb{Q}$, and so we will work over $\mathbb{Q}$ for notational simplicity. We also take $E = \mathbb{C}$ and suppress it from the notations.

Proposition 7.26. *Let $\sigma \in \mathrm{Coh}_{!!}(\mathrm{GL}_n, \mu)$ and $\sigma' \in \mathrm{Coh}_{!!}(\mathrm{GL}_{n'}, \mu')$ for pure weights μ and μ' respectively. Assume that n and n' are odd positive integers. Let $\epsilon, \epsilon' \in \{0,1\}$ be such that $\sigma_\infty = \mathbb{D}_\mu \otimes \mathrm{sgn}^\epsilon$ and $\sigma'_\infty = \mathbb{D}_{\mu'} \otimes \mathrm{sgn}^{\epsilon'}$. Put $\epsilon_0 = \epsilon + \epsilon'$ (mod 2). Let ℓ_i and ℓ'_j denote the cuspidal parameters as in (3.4); recall that all the ℓ_i and ℓ'_j are even, and $\ell_{(n+1)/2} = \ell'_{(n'+1)/2} = 0$; and because of the last property, we modify the definition the cuspidal width as*

$$\ell^+(\mu, \mu') \ := \ \min\{|\ell_i - \ell'_j| \ : \ 1 \leq i \leq (n-1)/2, \ 1 \leq j \leq (n'-1)/2\}.$$

Then we have:

1. *If $\ell^+(\mu, \mu') = 0$ then $L(s, \sigma \times \sigma'^{\mathsf{v}})$ has no critical points.*
2. *If $\ell := \ell^+(\mu, \mu') > 0$ and $\epsilon = \epsilon'$, i.e., $\epsilon_0 = 0$, then the critical set of integers*

for $L(s, \sigma \times \sigma^{\vee})$ is given by

$$\left\{-a(\underline{\mu}) - [\tfrac{\ell}{2}]_e - 1, \ \ldots, \ -a(\underline{\mu}) - 3, \ -a(\underline{\mu}) - 1; \right.$$
$$\left. -a(\underline{\mu}) + 2, \ -a(\underline{\mu}) + 4, \ \ldots, \ -a(\underline{\mu}) + [\tfrac{\ell}{2}]_e \right\},$$

where $[\ell/2]_e$ is the largest even integer less than or equal to $\ell/2$.

3. *If $\ell = \ell^+(\mu, \mu') > 0$ and $\epsilon \neq \epsilon'$, i.e., $\epsilon_0 = 1$, then the critical set of integers for $L(s, \sigma \times \sigma^{\vee})$ is given by*

$$\left\{-a(\underline{\mu}) - [\tfrac{\ell}{2}]_o - 1, \ \ldots \ , -a(\underline{\mu}) - 2, \ -a(\underline{\mu}); \right.$$
$$\left. -a(\underline{\mu}) + 1, \ -a(\underline{\mu}) + 3, \ \ldots, \ -a(\underline{\mu}) + [\tfrac{\ell}{2}]_o \right\},$$

where $[\ell/2]_o$ is the largest odd integer less than or equal to $\ell/2$.

4. *The points $s = -\frac{N}{2}$ and $s = 1 - \frac{N}{2}$ are critical for $L(s, \ \sigma \times \sigma'^{\vee})$ if and only if $\ell^+(\mu, \mu') > 0$, $\epsilon \neq \epsilon'$ and $d - d' = -\frac{N}{2}$.*

5. *Suppose $d - d' = -N/2$ then there does not exist an element $w \in W$ such that $w^{-1} \cdot (\mu + \mu')$ is dominant.*

Proof. For the proofs of (1), (2), and (3), recall that the critical set for the Rankin–Selberg L-function $L(s, \sigma \times \sigma'^{\vee})$ is the set of all integers m such that both $L_\infty(s, \sigma \times \sigma'^{\vee})$ and $L_\infty(1-s, \sigma^{\vee} \times \sigma')$ are regular at $s = m$. The Langlands parameters for σ_∞ and σ'_∞ are given by

$$\rho(\sigma) = \bigoplus_{i=1}^{(n-1)/2} I(\ell_i) \otimes |\ |^{-d} \ \oplus \ \mathrm{sgn}^{\epsilon} \otimes |\ |^{-d},$$

and

$$\rho(\sigma') = \bigoplus_{j=1}^{(n'-1)/2} I(\ell'_j) \otimes |\ |^{-d'} \ \oplus \ \mathrm{sgn}^{\epsilon'} \otimes |\ |^{-d'}.$$

If $\ell^+(\mu, \mu') > 0$ then the local L-factor $L_\infty(s, \sigma \times \sigma'^{\vee})$ at infinity for $\sigma \times \sigma'^{\vee}$ is given by

$$\Gamma_{\mathbb{R}}\left(s + \epsilon_0 + a(\underline{\mu})\right) \cdot \prod_{i=1}^{(n-1)/2} \Gamma_{\mathbb{C}}\left(s + d' - d + \frac{\ell_i}{2}\right) \prod_{j=1}^{(n'-1)/2} \Gamma_{\mathbb{C}}\left(s + d' - d + \frac{\ell'_j}{2}\right)$$
$$\cdot \prod_{\substack{1 \leq i \leq (n-1)/2 \\ 1 \leq j \leq (n'-1)/2}} \Gamma_{\mathbb{C}}\left(s + d' - d + \frac{\ell_i + \ell'_j}{2}\right) \Gamma_{\mathbb{C}}\left(s + d' - d + \frac{|\ell_i - \ell'_j|}{2}\right).$$

Now, $L_\infty(s, \sigma \times \sigma'^{\vee})$ is regular at $s = m$ if and only if for all indices $1 \leq i \leq (n+1)/2$ and $1 \leq j \leq (n'+1)/2$ (except when both $i = (n+1)/2$, $j = (n'+1)/2$), we have

$$\begin{aligned} m - (-a(\underline{\mu})) + |\ell_i - \ell'_j|/2 &\geq 1, \\ m - (-a(\underline{\mu})) + \epsilon_0 &\notin \{0, -2, -4, -6, \dots\}. \end{aligned} \tag{7.44}$$

Similarly, the factor $L_\infty(1-s,\ \sigma^\vee \times \sigma')$ is regular at $s = m$ if and only if for all indices $1 \leq i \leq (n+1)/2,\ \ 1 \leq j \leq (n'+1)/2$ (except when $i = (n+1)/2,\ j = (n'+1)/2$), we have

$$\begin{aligned} -m - a(\underline{\mu}) + |\ell_i - \ell'_j|/2 &\geq 0, \\ 1 - m - a(\underline{\mu}) + \epsilon_0 &\notin \{0, -2, -4, -6, \dots\}. \end{aligned} \tag{7.45}$$

We leave it to the reader to check that (2) and (3) follow from (7.44) and (7.45). For (1), suppose $\ell^+(\mu, \mu') = 0$ then $\ell_i = \ell'_j$ for some i, j; the local factor at infinity for $\sigma \times \sigma'^\vee$ will have factors $\Gamma((s - a(\underline{\mu}))/2)\Gamma((s - a(\underline{\mu}) + 1)/2)$ and it is easy to see then that there are no critical points. Also, (4) easily follows from (1), (2) and (3).

We now prove (5). Suppose $\mu = (\mu_1, \mu_2, \dots, \mu_n)$ and $\mu' = (\mu'_1, \mu'_2, \dots, \mu'_{n'})$ with nonincreasing sequences μ_i and μ'_j. By definition, we have

$$\mu + \mu' = (\mu_1,\ \mu_2,\ \dots,\ \mu_n,\ \mu'_1,\ \mu'_2,\ \dots,\ \mu'_{n'}).$$

Let $\boldsymbol{\rho}_N = \left(\frac{N-1}{2},\ \frac{N-3}{2},\ \dots,\ \frac{1-N}{2}\right)$ be the half sum of positive roots. It is clear that

$$(\mu + \mu' + \boldsymbol{\rho}_N)_j = \begin{cases} \mu_j + \frac{N+1}{2} - j & \text{if } \ 1 \leq j \leq n, \\ \\ \mu'_{j-n} + \frac{N+1}{2} - j & \text{if } \ 1 + n \leq j \leq N. \end{cases}$$

By definition of the twisted action of the Weyl group we have $w \cdot (\mu + \mu') := w(\mu + \mu' + \boldsymbol{\rho}_N) - \boldsymbol{\rho}_N$. The weight $w^{-1} \cdot (\mu + \mu')$ is dominant if and only if

$$(w^{-1} \cdot (\mu + \mu'))_j \ \geq\ (w^{-1} \cdot (\mu + \mu'))_{j+1} \quad \forall\, j \geq 1.$$

Equivalently,

$$(w^{-1}(\mu + \mu' + \boldsymbol{\rho}_N))_j \ \geq\ (w^{-1}(\mu + \mu' + \boldsymbol{\rho}_N))_{j+1} + 1 \quad \forall\, j \geq 1.$$

In particular, for $w^{-1} \cdot (\mu + \mu')$ to be dominant, it is necessary that $(\mu + \mu' + \boldsymbol{\rho}_N)_j$ are all distinct as $1 \leq j \leq N$. However, observe that

$$(\mu + \mu' + \boldsymbol{\rho}_N)_{\frac{n+1}{2}} \ =\ d + n'/2 \ =\ d' - n/2 \ =\ (\mu + \mu' + \boldsymbol{\rho}_N)_{n + \frac{n'+1}{2}},$$

since $\mu_{(n+1)/1} = d$, $\mu'_{(n'+1)/1} = d'$, and by hypothesis $d - d' = -N/2$. In other words, there is no element w of the Weyl group W so that the weight $w^{-1} \cdot (\mu + \mu')$ is dominant. □

To give a feel for the critical sets described in (2) and (3) of the proposition, let's consider a very classical example: Suppose χ is a Dirichlet character, i.e., we have a homomorphism $\chi : (\mathbb{Z}/N\mathbb{Z})^\times \to \mathbb{C}^\times$, and we consider the Dirichlet L-function $L(s, \chi)$. Then one knows:

(e) If χ is even, i.e., $\chi(-1) = 1$, then the critical set for $L(s, \chi)$ is

$$\{\dots, -5, -3, -1, 2, 4, 6, \dots\}.$$

(o) If χ is odd, i.e., $\chi(-1) = -1$, then the critical set for $L(s, \chi)$ is

$$\{\dots, -4, -2, 0, 1, 3, 5, \dots\}.$$

If $n = n' = 1$, and $\mu = \mu' = 0$, then σ and σ' are Dirichlet characters; let's put $\chi = \sigma \otimes \sigma'^{-1}$. The cuspidal parameters is an empty set and so $\ell(\mu, \mu')$ is to interpreted as infinity–the minimum of an empty set. If $\epsilon = \epsilon'$ then χ is even, and Prop. 7.26, (2), is exactly (e). Similarly, if $\epsilon \neq \epsilon'$ then χ is odd, and Prop. 7.26, (3) is exactly (o) above.

Back to the general situation. If we hope to use the theory of Eisenstein cohomology to prove a result on successive L-values, it is clear from the main body of this monograph that we need the points $-N/2$ and $1-N/2$ to be critical. However, in the situation when n and n' are odd, the conditions (given in (4) of the above proposition) which ensure criticality of these two points also say (in (5)) that the relevant induced representation does not occur in the boundary cohomology of GL_N for any coefficient system λ, because, if there were such a dominant integral weight λ, then this would mean $\lambda = w^{-1} \cdot (\mu + \mu')$ is dominant, contradicting (5) above.

7.4.2 Example: Hilbert modular forms

Highest weight. Let $\mathbf{k} = (k_\tau)_{\tau : F \to E}$ be a r_F-tuple of integers. Let $k_0 = \max\{k_\tau\}$. Let $m \in \mathbb{Z}$. Consider the weight $\lambda \in X^*(T_2 \times E)$ given by

$$\lambda \;=\; (\lambda^\tau), \quad \lambda^\tau \;=\; (k_\tau - 2)\boldsymbol{\rho}_2 + (-m - \tfrac{k_0}{2})\boldsymbol{\delta}_2.$$

In the standard basis we have

$$\lambda^\tau \;=\; \left(\frac{k_\tau - k_0}{2} - 1 - m, \, \frac{-k_\tau - k_0}{2} + 1 - m \right).$$

Then λ is dominant if and only if $k_\tau \geq 2$ for all τ, and λ is integral if and only if $k_\tau \equiv k_0 \pmod 2$. Furthermore, by definition, it satisfies the algebraicity condition and, clearly, λ is also pure. The motivic weight of λ is $\mathbf{w} = \mathbf{w}_\lambda = k_0 - 1$. The abelian part of λ is $d \cdot \boldsymbol{\delta}_2$ where $d = d_\lambda = -m - k_0/2$.

Strongly inner Hecke module. Let $\sigma_f \in \mathrm{Coh}_{!!}(G_2, \lambda)$. Take $\iota : E \to \mathbb{C}$.

There is a holomorphic Hilbert modular cusp form ${}^{\iota}\mathbf{f}$ of weight $\mathbf{k}$, such that if $\pi({}^{\iota}\mathbf{f})$ is the associated cuspidal automorphic representation of $\mathrm{GL}_2(\mathbb{A}_F)$ then

$$ {}^{\iota}\sigma = \pi({}^{\iota}\mathbf{f}) \otimes |\ |^{m+\frac{k_0}{2}}. $$

(This can be seen from Raghuram–Tanabe [57, Thm. 1.4].) The cuspidal parameters of σ_f are $\ell = (\ell^{\tau})$ with $\ell^{\tau} = (\ell_1^{\tau}, \ell_2^{\tau})$ with $\ell_1^{\tau} = k_{\tau} - 1$ and $\ell_2^{\tau} = -\ell_1^{\tau} = 1 - k_{\tau}$.

Motives and their Hodge types. Let $\mathbf{M}(\sigma_f)$ be the motive attached to σ_f. Its Hodge types are:

$$ \mathcal{H}od(\nu, {}^{\iota}\mathbf{M}(\sigma_f)) = \left\{ \left(\frac{k_{\tau} - k_0 - 2m}{2}, \frac{2 - k_{\tau} - k_0 - 2m}{2} \right), \left(\frac{2 - k_{\tau} - k_0 - 2m}{2}, \frac{k_{\tau} - k_0 - 2m}{2} \right) \right\}, $$

where $\tau = \iota^{-1} \circ \nu$. The weight of $\mathbf{M}(\sigma_f)$ is $\tilde{\mathbf{w}} = 2d + n - 1 = -2m - k_0 + 1$. The effective motive $\mathbf{M}_{\text{eff}}(\sigma_f)$ attached to σ_f has Hodge types:

$$ \mathcal{H}od(\nu, {}^{\iota}\mathbf{M}_{\text{eff}}(\sigma_f)) = \left\{ \left(\frac{k_{\tau} + k_0 - 2}{2}, \frac{-k_{\tau} + k_0}{2} \right), \left(\frac{-k_{\tau} + k_0}{2}, \frac{k_{\tau} + k_0 - 2}{2} \right) \right\}, $$

which are independent of the abelian part of λ. The purity weight of the effective motive $\mathbf{M}_{\text{eff}}(\sigma_f)$ is the motivic weight $\mathbf{w} = k_0 - 1$ of λ.

Relations between various L-functions. We leave it to the reader to find his or her own comfort zone with the following three objects: a holomorphic cuspidal Hilbert modular form, an effective regular rank-2 simple motive, and a cuspidal automorphic representation depicted in the following triangle:

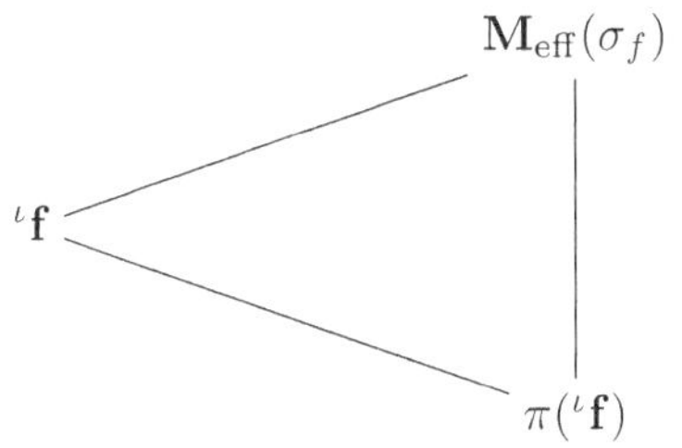

Each object lives naturally in its own habitat and within that framework there is a well-working anaytic theory of L-functions:

1. Hecke–Shimura theory for Hilbert modular form, with the Mellin transform of ${}^{\iota}\mathbf{f}$ giving the Hecke L-function $L^{\text{Hecke}}(s, {}^{\iota}\mathbf{f})$.
2. Jacquet–Langlands theory for automorphic forms on $\mathrm{GL}(2)$ giving the automorphic L-function: $L_f^{\text{aut}}(s, \pi({}^{\iota}\mathbf{f}))$.

3. The Artin L-function attached to the Galois representation on the ℓ-adic realization of the effective motive giving us $L(\iota, \mathbf{M}_{\text{eff}}(\sigma_f), s)$.

These L-functions are related to each other as follows:

$$\begin{aligned} L(\iota, \mathbf{M}_{\text{eff}}(\sigma_f), s) &= L^{\text{aut}}\left(s + \frac{2d - \mathbf{w}}{2}, {}^{\iota}\sigma_f\right) \\ &= L_f^{\text{aut}}\left(s + \frac{-k_0 - 2m - (k_0 - 1)}{2}, \pi({}^{\iota}\mathbf{f}) \otimes |\ |^{m + \frac{k_0}{2}}\right) \\ &= L_f^{\text{aut}}\left(s - \frac{k_0 - 1}{2}, \pi({}^{\iota}\mathbf{f})\right) \\ &= L^{\text{Hecke}}(s, {}^{\iota}\mathbf{f}), \end{aligned}$$

where the last equality follows from [57, Thm. 1.4, (1)]. Note that irrespective of the integer m, we always end up with same L-function for the effective motive, which is why the abelian part of the highest weight is not so interesting. Furthermore, in the triangle above, we remind the reader that the relations between the duals of the various objects (in their respective habitats) is reflected in the functional equation for the various L-functions, and these look like:

1. $L^{\text{Hecke}}(s, {}^{\iota}\mathbf{f}) \sim L^{\text{Hecke}}(k_0 - s, {}^{\iota}\mathbf{f}^{\vee})$,
   where ${}^{\iota}\mathbf{f}^{\vee}$ is obtained by applying the relevant Atkin-Lefner involution to ${}^{\iota}\mathbf{f}$.
2. $L^{\text{aut}}(s, \pi({}^{\iota}\mathbf{f})) \sim L^{\text{aut}}(1 - s, \pi({}^{\iota}\mathbf{f})^{\vee})$,
   where $\pi({}^{\iota}\mathbf{f})^{\vee}$ is the contragredient representation of $\pi({}^{\iota}\mathbf{f})$.
3. $L(\iota, \mathbf{M}_{\text{eff}}(\sigma_f), s) \sim L(\iota, \mathbf{M}_{\text{eff}}(\sigma_f)^{\vee}, 1 - s)$,
   where $\mathbf{M}_{\text{eff}}(\sigma_f)^{\vee}$ is the motive dual to $\mathbf{M}_{\text{eff}}(\sigma_f)$ defined by taking duals at the level of Betti, de Rham, and ℓ-adic realizations.

In the above functional equations $\sim$ means equality up to a suitable epsilon factor.

Critical points of Hecke L-function. To compute the critical set, let's denote $k^0 = \min(k_\tau)$ and observe that

$$p_{\max} = \frac{k_0 - k^0}{2}, \quad q_{\min} = \frac{k_0 + k^0}{2} - 1.$$

The critical set for $L(\iota, \mathbf{M}_{\text{eff}}(\sigma_f), s)$ is

$$\left\{m \in \mathbb{Z} \ : \ \frac{k_0 - k^0}{2} < m < \frac{k_0 + k^0}{2}\right\}.$$

Examples of special values of L-functions. The reader should work out the case of $\mathrm{GL}_2 \times \mathrm{GL}_1$ (resp., $\mathrm{GL}_2 \times \mathrm{GL}_2$) explicating our Thm. 7.21 using the well-known results of Shimura on the special values of the standard L-function attached to a Hilbert modular form and a Hecke character of F [66, Thm. 4.3] (resp., the special values of the Rankin–Selberg L-function attached to a pair of Hilbert modular forms [66, Thm. 4.2]). Some more explicit examples may be produced using symmetric power transfers of Hilbert modular forms and special value results in Raghuram [53], [55], and the various other examples described in Grobner–Raghuram [23, Sect. 8].

Chapter Eight

Harish-Chandra Modules over $\mathbb{Z}$

By Günter Harder

This chapter was written during the preparation of this book, when we wanted to provide a proof of the rationality of the intertwining operator and we also wanted to formulate a conjecture concerning the arithmetic properties of this intertwining operator. (See Quest. 8.82.) In the meanwhile, Uwe Weselmann gave a positive answer to this question by explicit computation of the effect of this intertwining operator on some very special vectors (see Chap. 9), making this current chapter a bit redundant. However, we believe that there are some interesting rationality and integrality aspects that are discussed in this chapter that can prove to be useful for later applications and possible generalizations to other reductive groups.

8.1 INTRODUCTION

Harish-Chandra modules play an important role in the theory of representations of semisimple Lie groups over $\mathbb{R}$; in a certain sense they are the algebraic skeleton of a certain class of representations of semisimple real Lie groups.

In this chapter we show that certain classes of Harish-Chandra modules have in a natural way a structure over $\mathbb{Z}$. The Lie group is replaced by a split reductive group scheme $G/\mathbb{Z}$, its Lie algebra is denoted by $\mathfrak{g}_{\mathbb{Z}}$. On the group scheme $G/\mathbb{Z}$ we have a Cartan involution Θ that acts by $t \mapsto t^{-1}$ on the split maximal torus. The fixed points of $G/\mathbb{Z}$ under Θ is a flat group scheme $\mathcal{K}/\mathbb{Z}$. A Harish-Chandra module over $\mathbb{Z}$ is a $\mathbb{Z}$-module $\mathcal{V}$ that comes with an action of the Lie algebra $\mathfrak{g}_{\mathbb{Z}}$, an action of the group scheme $\mathcal{K}$, and we require some compatibility conditions between these two actions. Finally, we require $\mathcal{K}$-finiteness, which is that $\mathcal{V}$ is a union of finitely generated $\mathbb{Z}$ modules $\mathcal{V}_I$ that are $\mathcal{K}$-invariant. The definitions imitate the definition of a Harish-Chandra modules over $\mathbb{R}$ or over $\mathbb{C}$. (See, for instance, [9, 0.2.5], where these modules are called $(\mathfrak{g}, K)$ modules.) We will set up the basics of these integral Harish-Chandra modules in Sect. 8.2.

For these $(\mathfrak{g}_{\mathbb{Z}}, \mathcal{K})$ modules $\mathcal{V}$ we define cohomology modules $H^\bullet(\mathfrak{g}_{\mathbb{Z}}, \mathcal{K}, \mathcal{V})$

and these will be finitely generated $\mathbb{Z}$ modules provided the module $\mathcal{V}$ satisfies suitable finiteness conditions. We construct some simple examples in Sect. 8.3. In particular, we construct the $\mathbb{Z}$-version of the discrete series representations of $\mathrm{GL}_2(\mathbb{R})$ and compute their cohomology and the effect of a certain standard intertwining operator on these cohomology groups.

In Sect. 8.4 we discuss the process of induction: For a parabolic subgroup $P/\mathbb{Z}$ and a $(\mathfrak{m}, \mathcal{K}^M)$-module $\mathcal{V}$ for its reductive quotient $M/\mathbb{Z}$ we define the induced module $\mathfrak{Ind}_P^G\mathcal{V}$. Then we study intertwining operators between some specific induced Harish-Chandra modules $\mathfrak{Ind}_P^G\mathbb{D}_\mu, \mathfrak{Ind}_Q^G\mathbb{D}_{\mu'}$ where P and Q are maximal parabolic subgroups of $\mathrm{GL}_N/\mathbb{Z}$. Here we have to introduce some twisting which is achieved by extending the scalars from $\mathbb{Z}$ to the function field $\mathbb{Q}(s)$ and define $\mathfrak{Ind}_P^G\mathbb{D}_\mu \otimes s$ over $\mathbb{Q}(s)$. Then our intertwining operators are defined as integrals. We cannot expect that they are defined over $\mathbb{Q}(s)$. But it turns out (and this is certainly not surprising) that they can be written down in terms of the form $\Gamma(s)R(s)$ with $R(s) \in \mathbb{Q}(s)$ and $\Gamma(s)$ is of course the Γ-function. If the intertwining operator is holomorphic at $s = 0$ we can evaluate at $s = 0$ and it turns out that our intertwining operator, which is defined by the transcendental process of integration, is essentially a power of π times a nonzero rational number. See Thm. 8.7, which is the main theorem of this chapter.

Such a rationality result was stated as Thm. 7.25 and used in the proof of Thm. 7.21. It can be formulated without reference to rational integral structures on Harish-Chandra modules; we just have to choose the *right* basis in certain one-dimensional vector spaces.

The main reason why we develop these concepts is an intriguing question concerning the cohomology of these modules and its behavior under the intertwining operators. It turns out that the cohomology in certain situations is a free module of rank one over a small ring R (like $\mathbb{Z}$ or $\mathbb{Z}[i, \frac{1}{2}]$, etc.). Then the intertwining operator divided by the appropriate power of π induces an isomorphism between these cohomology modules after we tensor them by the quotient field of R. This isomorphism depends on some data, for instance, some highest weights. Our question is whether this isomorphism is already an isomorphism over the basic ring R independently of the data. This question has been investigated in [29] in a special case and reduced to an combinatorial identity, which then was proved by D. Zagier (see [74]) in an appendix to [29]. This gives a positive answer to the preceding question in this special case, suggesting an inherent naturality to the question.

Let's end the introduction with a speculative note. In this chapter we work with certain specific choices of Cartan involutions. Such a choice provides the so-called maximal definite group schemes $\mathcal{K}/\mathbb{Z}$; these group schemes are flat over $\mathrm{Spec}(\mathbb{Z})$ and they are even reductive if we invert the prime 2. But we can also choose other maximal definite group schemes $\mathcal{K}'$ that are reductive at the prime 2 and perhaps nonreductive at some other places. This suggests that we should consider sheaves of Harish-Chandra modules over $\mathrm{Spec}(\mathbb{Z})$.

The reader is also referred to Harris [34] and Januszewski [41] for similar considerations about rational structures on Harish-Chandra modules and with similar motivations of controlling the archimedean contribution to a global rationality result of special values of L-functions.

8.2 HARISH-CHANDRA MODULES OVER $\mathbb{Z}$

8.2.1 The general setup

For any affine group scheme $H/\mathrm{Spec}(\mathbb{Z})$ we denote by $A(H)$ its algebra of regular functions. The affine algebra of the multiplicative group scheme $\mathbb{G}_m$ is $A(\mathbb{G}_m) = \mathbb{Z}[x, x^{-1}]$, i.e., we choose the generator γ_1 of the character module $X^*(\mathbb{G}_m)$ to be given by the identity map $\mathbb{G}_m \to \mathbb{G}_m$, and γ_1 as an element of $A(\mathbb{G}_m)$ corresponds to x in $\mathbb{Z}[x, x^{-1}]$. Let $\mathbb{G}_a$ be the one-dimensional additive group scheme for which we have $A(\mathbb{G}_a) = \mathbb{Z}[x]$.

Let $G/\mathrm{Spec}(\mathbb{Z})$ be a reductive connected group scheme. We assume that the derived group $G^{(1)}/\mathrm{Spec}(\mathbb{Z})$ is a simply connected Chevalley scheme, and that the maximal central torus $C/\mathrm{Spec}(\mathbb{Z})$ is split. Let $\mathfrak{g}_{\mathbb{Z}}$, $\mathfrak{g}_{\mathbb{Z}}^{(1)}$ and $\mathfrak{c}_{\mathbb{Z}}$ be the Lie algebras of $G/\mathrm{Spec}(\mathbb{Z})$, $G^{(1)}/\mathrm{Spec}(\mathbb{Z})$ and C, respectively. We have the maximal split torus $T/\mathrm{Spec}(\mathbb{Z})$, and let $T^{(1)}/\mathrm{Spec}(\mathbb{Z}) = T \cap G^{(1)}$. We choose a Borel subgroup $B/\mathrm{Spec}(\mathbb{Z}) \supset T/\mathrm{Spec}(\mathbb{Z})$. As usual, we denote the character module $\mathrm{Hom}(T, \mathbb{G}_m)$ by $X^*(T)$, and we have the direct sum decomposition

$$X^*_{\mathbb{Q}}(T) = X^*(T) \otimes \mathbb{Q} = X^*_{\mathbb{Q}}(T^{(1)}) \oplus X^*_{\mathbb{Q}}(C), \qquad (8.1)$$

under which we will always write $\gamma = \gamma^{(1)} + \delta$ for the decomposition of a character $\gamma \in X^*_{\mathbb{Q}}(T)$ into its semisimple and its abelian part.

Let $\Delta \subset X^*(T)$ (resp., $\Delta^+ \subset X^*(T)$) be the set of roots (resp., positive roots), and let $\Pi = \{\alpha_1, \alpha_2, \ldots, \alpha_r\} \subset \Delta$ be the set of simple positive roots. Let $\gamma_1, \gamma_2, \ldots, \gamma_r \in X^*(T^{(1)})$ be the dominant fundamental weights; we extend them to elements in $X^*_{\mathbb{Q}}(T)$ by putting the abelian part equal to zero. The element $\rho \in X_{\mathbb{Q}}(T)$ is the half sum of positive roots. For any root α we have the root subgroup scheme $U_\alpha/\mathrm{Spec}(\mathbb{Z})$, and assume that for all simple roots we have fixed an isomorphism

$$\tau_\alpha : \mathbb{G}_a/\mathrm{Spec}(\mathbb{Z}) \xrightarrow{\sim} U_\alpha/\mathrm{Spec}(\mathbb{Z});$$

i.e., we have selected a generator e_α of the abelian group $U_\alpha(\mathbb{Z}) \xrightarrow{\sim} \mathbb{Z}$.

For a simple root α, we also get a subgroup scheme $H_\alpha \subset G^{(1)}/\mathrm{Spec}(\mathbb{Z})$ which is "generated" by U_α, $U_{-\alpha}$ and which is isomorphic to $\mathrm{SL}_2/\mathrm{Spec}(\mathbb{Z})$. It has a maximal torus $T_\alpha/\mathrm{Spec}(\mathbb{Z}) \subset T^{(1)}/\mathrm{Spec}(\mathbb{Z})$ which is the intersection of the kernels of the fundamental weights γ_β where $\beta \neq \alpha$. The choice of τ_α is the same as the choice of an isomorphism

$$\tilde{\tau}_\alpha : \mathrm{SL}_2/\mathrm{Spec}(\mathbb{Z}) \to H_\alpha$$

which sends the diagonal torus to T_α and on the $\mathbb{Z}$-valued points it sends

$$\begin{pmatrix} 1 & 1 \\ 0 & 1 \end{pmatrix} \to e_\alpha.$$

The derivative of τ_α defines a generator $E_\alpha \in \mathrm{Lie}(U_\alpha)$. Finally, we define the coroot $\alpha^\vee : \mathbb{G}_m \xrightarrow{\sim} T_\alpha$ by the rule $\langle \alpha^\vee, \alpha \rangle = 2$.

Let Θ be the unique automorphism of $G^{(1)}/\mathrm{Spec}(\mathbb{Z})$ that induces $t \mapsto t^{-1}$ on $T^{(1)}$ and restricted to H_α and composed with $\tilde{\tau}_\alpha^{-1}$ is the inner automorphism of $\mathrm{SL}_2/\mathrm{Spec}(\mathbb{Z})$ given by the element

$$s_\alpha = \begin{pmatrix} 0 & 1 \\ -1 & 0 \end{pmatrix}.$$

We call the pair $(G^{(1)}, \Theta)$ an Arakelow Chevalley scheme. The automorphism restricted to $G^{(1)}(\mathbb{R})$ is a Cartan involution and the fixed point set $G^{(1)}(\mathbb{R})^\Theta = K_\infty^{(1)}$ is a maximal compact subgroup of $G^{(1)}(\mathbb{R})$. Here we use this automorphism to give the structure of a group scheme over $\mathrm{Spec}(\mathbb{Z})$ to $K_\infty^{(1)}$. To be more precise: The group scheme of fixed points $\mathcal{K}^{(1)}/\mathrm{Spec}(\mathbb{Z}) = (G^{(1)})^\Theta/\mathrm{Spec}(\mathbb{Z})$ is a flat group scheme over $\mathrm{Spec}(\mathbb{Z})$, it is smooth and connected over $\mathrm{Spec}(\mathbb{Z}[\frac{1}{2}])$. We call $\mathcal{K}^{(1)}$ a *maximal definite connected* subgroup scheme of $G^{(1)}/\mathbb{Z}$. We denote by $\mathfrak{k}_{\mathbb{Z}[\frac{1}{2}]}$ its Lie algebra over $\mathbb{Z}[\frac{1}{2}]$. We put $\mathfrak{k}_\mathbb{Z} = \mathfrak{g}_\mathbb{Z} \cap \mathfrak{k}_{\mathbb{Z}[\frac{1}{2}]}$. Then $\mathfrak{k}_\mathbb{Z}$ is a maximal subalgebra for which the restriction of the Killing form is negative definite, justifying the terminology.

If we have an extension of the Cartan involution to $G/\mathbb{Z}$ then we can also look at the fixed point scheme $G^\Theta/\mathbb{Z}$ and define $\mathcal{K}/\mathbb{Z} = G^\Theta$. We are mostly interested in cases where this extension induces $t \mapsto t^{-1}$ on $C/\mathbb{Z}$, then $\mathcal{K}^{(1)}$ is the connected component of the identity of $\mathcal{K}/\mathbb{Z}$. In general we denote by $\mathcal{K}$ a group scheme lying between $\mathcal{K}^{(1)}$ and G^Θ. Then $\mathcal{K}/\mathcal{K}^{(1)}$ is a finite constant group scheme which is isomorphic to $(\mathbb{Z}/2\mathbb{Z})^s$. We also consider larger subschemes of the form $\tilde{\mathcal{K}} = \mathcal{K}^{(1)} \cdot C'$, where C' is any subtorus of C. We call them *essentially maximal definite* subgroup schemes. They are also smooth over $\mathbb{Z}[\frac{1}{2}]$ and the Lie algebra is denoted by $\tilde{\mathfrak{k}}_{\mathbb{Z}[\frac{1}{2}]}$. Again we define $\tilde{\mathfrak{k}}_\mathbb{Z} = \tilde{\mathfrak{k}}_{\mathbb{Z}[\frac{1}{2}]} \cap \mathfrak{g}_\mathbb{Z}$.

8.2.2 Definition of integral Harish-Chandra modules

For any commutative ring $R \supset \mathbb{Z}$ we define the notion of a Harish-Chandra module over R, or equivalently a $(\mathfrak{g}_\mathbb{Z}, \mathcal{K})$- module over R as follows:

1. This will be a projective R-module $\mathcal{V}$ that is the union of finitely generated projective submodules $\mathcal{V}_I, I \in \mathcal{I}$ such that $\mathcal{V}/\mathcal{V}_I$ is torsion free. We have an action of $\mathcal{K}$ on $\mathcal{V}$ that respects the $\mathcal{V}_I$.

2. If L is the quotient field of R then every irreducible finite-dimensional representation ϑ of $\mathcal{K} \times L$ occurs with finite multiplicity in this module and we have the isotypical decomposition

$$\mathcal{V} \otimes L = \bigoplus \mathcal{V}(\vartheta)$$

where $\mathcal{V}(\vartheta)$ is the ϑ isotypical component.

3. We have a Lie algebra action of $\mathfrak{g}_{\mathbb{Z}} \otimes R$ on $\mathcal{V}$.

4. The group scheme $\mathcal{K}$ acts by the adjoint action on $\mathfrak{g}_{\mathbb{Z}}$ and the R-module homomorphism

$$(\mathfrak{g}_{\mathbb{Z}} \otimes R) \otimes \mathcal{V} \to \mathcal{V}$$

given by (3) is $\mathcal{K}$ invariant.

5. The restriction of the Lie algebra action of $\mathfrak{g}_{\mathbb{Z}}$ to the Lie algebra $\mathfrak{k}_{\mathbb{Z}} = \mathrm{Lie}(\mathcal{K})$ is the differential of the action of $\mathcal{K}$.

6. For any $I \in \mathcal{I}$ we find an $I_1 \in \mathcal{I}$ such that $\mathcal{V}_I \subset \mathcal{V}_{I_1}$, and such that the Lie algebra action of $\mathfrak{g}_{\mathbb{Z}}$ on $\mathcal{V}$ induces an R-bilinear map:

$$\mathfrak{g}_{\mathbb{Z}} \times \mathcal{V}_I \to \mathcal{V}_{I_1}. \tag{8.2}$$

We say that the $(\mathfrak{g}_{\mathbb{Z}}, \mathcal{K})$-module has a central character if the Lie algebra $\mathfrak{c}_{\mathbb{Z}} = \mathrm{Lie}(C)$ of the central torus C acts by a linear map $z_V : \mathfrak{c}_{\mathbb{Z}} \to R$.

8.2.3 Some comments

This is almost the same as the usual definition of a Harish-Chandra module except that the field of scalars $\mathbb{C}$ has been replaced by the ring R and the action of the maximal compact group K_∞ is replaced by the action of the group scheme $\mathcal{K}$.

We want to remind the reader what it means for the group scheme $\mathcal{K}/\mathrm{Spec}(\mathbb{Z})$ to act upon $\mathcal{V}$ and $\mathcal{V}_I$. We recall that by definition $\mathcal{K}/\mathrm{Spec}(\mathbb{Z})$ is a functor from the category of affine schemes $Y \to \mathrm{Spec}(R)$ to the category of groups. This means that for any commutative ring R_1 containing R we get an abstract group of R_1-valued points $G(R_1)$ which depends functorially on R_1. Then the action of $\mathcal{K}/\mathrm{Spec}(\mathbb{Z})$ on the R module $\mathcal{V}$ provides for any R_1 an action of $\mathcal{K}(R_1)$ on the R_1 module $\mathcal{V} \otimes_R R_1$. We require that for all our finitely generated submodules the module $\mathcal{V}_I \otimes R_1$ is invariant under $\mathcal{K}(R_1)$.

In all examples that will be discussed in the text that follows we take for $\mathcal{I}$ the set of finite sets of isomorphism classes of irreducible representations of the group scheme $\mathcal{K}$. If $I = \{\vartheta_1, \dots, \vartheta_r\}$ then

$$\mathcal{V}_I = \mathcal{V} \cap \oplus_{\nu=1}^{r} \mathcal{V}(\vartheta_\nu). \tag{8.3}$$

In this case the requirement (6) is superfluous.

We call $\mathcal{V}$ irreducible if $\mathcal{V} \otimes L$ does not contain a proper nontrivial $(\mathfrak{g}_\mathbb{Z}, \mathcal{K})$ submodule, and we call it absolutely irreducible if $\mathcal{V} \otimes L_1$ stays irreducible for any finite extension L_1/L.

We saw already that we have some flexibility in the choice of $\mathcal{K}$. If we replace $\mathcal{K}$ by the connected component of the identity $\mathcal{K}^{(1)}$ then we can restrict the $(\mathfrak{g}_\mathbb{Z}, \mathcal{K})$ module to $(\mathfrak{g}_\mathbb{Z}, \mathcal{K}^{(1)})$. It may happen that the restriction of an irreducible module is not irreducible anymore.

8.2.4 Motivation for this concept

Take a dominant weight $\lambda \in X^*(T)$ and construct a highest weight module $\mathcal{M}_{\lambda,\mathbb{Z}}$. This highest weight module has a central character $\zeta_\lambda \in X^*(C)$. We are looking for absolutely irreducible Harish-Chandra modules $\mathcal{V}$ (over $\mathbb{Z}$ or a slightly larger ring) having the central character $z_V = -d\zeta_\lambda$, and that have nontrivial $(\mathfrak{g}_\mathbb{Z}, \mathcal{K})$-cohomology with coefficients in $\mathcal{M}_{\lambda,\mathbb{Z}}$. The cohomology is defined as the cohomology of the complex

$$\mathrm{Hom}_\mathcal{K}(\Lambda^\bullet(\mathfrak{g}_\mathbb{Z}/\mathfrak{k}_\mathbb{Z}), \mathcal{V} \otimes \mathcal{M}_{\lambda,\mathbb{Z}}),$$

where the definition of the complex is exactly the same as in the traditional situation. Hence we define

$$H^\bullet(\mathfrak{g}_\mathbb{Z}, \mathcal{K}, \mathcal{V} \otimes \mathcal{M}_{\lambda,\mathbb{Z}}) = H^\bullet(\mathrm{Hom}_\mathcal{K}(\Lambda^\bullet(\mathfrak{g}_\mathbb{Z}/\mathfrak{k}_\mathbb{Z}), \mathcal{V} \otimes \mathcal{M}_{\lambda,\mathbb{Z}})). \tag{8.4}$$

It easy to see that only the semisimple component is relevant for the computation of the cohomology, and we have

$$H^\bullet(\mathfrak{g}_\mathbb{Z}, \mathcal{K}, \mathcal{V} \otimes \mathcal{M}_{\lambda,\mathbb{Z}}) = H^\bullet(\mathfrak{g}_\mathbb{Z}^{(1)}, \mathcal{K}^{(1)}, \mathcal{V} \otimes \mathcal{M}_{\lambda,\mathbb{Z}})^{\mathcal{K}/\mathcal{K}^{(1)}} \otimes \Lambda^\bullet(\mathfrak{c}_\mathbb{Z}) \tag{8.5}$$

We will see that that factor $\Lambda^\bullet(\mathfrak{c}_\mathbb{Z})$ is rather uninteresting. If we replace $\mathcal{K}$ by a larger group $\tilde{\mathcal{K}} = \mathcal{K}^{(1)} \cdot C'$ then we define more generally

$$H^\bullet(\mathfrak{g}_\mathbb{Z}, \tilde{\mathcal{K}}, \mathcal{V} \otimes \mathcal{M}_{\lambda,\mathbb{Z}}) = H^\bullet(\mathrm{Hom}_{\mathcal{K}^{(1)}}(\Lambda^\bullet(\mathfrak{g}_\mathbb{Z}/\tilde{\mathfrak{k}}_\mathbb{Z}), \mathcal{V} \otimes \mathcal{M}_{\lambda,\mathbb{Z}})). \tag{8.6}$$

(Observe the subscript at the Hom is $\mathcal{K}^{(1)}$ and we need not write it as $\tilde{\mathcal{K}}$ since the adjoint action of C' is trivial on both the arguments of Hom.) If we choose $C' = C$ and replace $\mathcal{K}$ in (8.5) by $\tilde{\mathcal{K}}$ then the factor $\Lambda^\bullet(\mathfrak{c}_\mathbb{Z})$ is replaced by $\Lambda^0(\mathfrak{c}_\mathbb{Z}) = \mathbb{Z}$.

We will be mainly concerned with the group scheme $G = \mathrm{GL}_n/\mathrm{Spec}(\mathbb{Z})$. The involution Θ will be the usual involution $g \mapsto {}^t g^{-1}$. Our first aim will be to construct for a given highest weight module $\mathcal{M}_{\lambda,\mathbb{Z}}$ a very specific absolutely irreducible $(\mathfrak{g}_\mathbb{Z}, \mathcal{K})$ module $\mathbb{D}_\lambda$ that has nontrivial cohomology. More precisely: The lowest degree where we find nontrivial cohomology is $b_n = [\frac{n^2}{4}]$ and

$$H^{b_n}(\mathfrak{g}_\mathbb{Z}, \mathcal{K}^{(1)}, \mathbb{D}_\lambda \otimes \mathcal{M}_{\lambda,\mathbb{Z}}) \xrightarrow{\sim} \begin{cases} \mathbb{Z}[\frac{1}{2}]\omega_\lambda^+ \oplus \mathbb{Z}[\frac{1}{2}]\omega_\lambda^- & n \text{ even} \\ \mathbb{Z}[\frac{1}{2}]\omega_\lambda & n \text{ odd} \end{cases} \otimes \Lambda^\bullet(\mathfrak{c}_\mathbb{Z}). \quad (8.7)$$

There is an action of $\mathcal{K}/\mathcal{K}^{(1)} = \mathbb{Z}/2\mathbb{Z}$ $(= \pi_0(\mathrm{GL}_n(\mathbb{R}))$ on the cohomology. This action is nontrivial if n is even and the cohomology decomposes in a $+$ and a $-$ eigenspace (see Sect. 8.4.4) that is relevant for the definition of the relative periods as in Sect.5.2.

If we take the tensor product $\mathbb{D}_\lambda \otimes \mathbb{C}$ then we get the usual Harish-Chandra modules over $\mathbb{C}$, which was previously denoted by $\mathbb{D}_\lambda$ as in Sect. 3.1. In this chapter we will call these modules over $\mathbb{C}$ the transcendental Harish-Chandra modules. These special transcendental modules will be the only essentially tempered modules that have cohomology, and they contribute to the cuspidal cohomology.

8.3 FIRST EXAMPLES

8.3.1 The case of the torus $\mathbb{G}_m$

For the multiplicative group scheme $\mathbb{G}_m/\mathbb{Z}$ we have $\mathrm{Lie}(\mathbb{G}_m)_\mathbb{Z} = \mathbb{Z} \cdot H$, for a generator H. We may choose for the group scheme $\mathcal{K}$ simply the subscheme $\mathcal{K} = \mu_2$ of second roots of unity. Then we can construct a $(\mathbb{Z}H, \mathcal{K})$-module $\mathbb{Z}[\gamma \otimes m]$ for any pair (γ, m) where $\gamma \in X^*(\mathbb{G}_m)$ and where m is an integer modulo two. If $\gamma = x^n$ then the generator H of $\mathrm{Lie}(\mathbb{G}_m)$ acts by multiplication by n and the action of $\mathcal{K}(\mathbb{Z})$ is given by the sign character $-1 \mapsto (-1)^m$. Therefore it is clear that these modules $\mathbb{Z}[\gamma \otimes m]$ are the absolutely irreducible $(\mathrm{Lie}(\mathbb{G}_m)_\mathbb{Z}, \mathcal{K})$ modules. The pairs (γ, m) are called the characters of Hecke type $-\gamma$; if $m = 0$ then these are the rational characters. We can do essentially the same for any split torus C; for any pair $\gamma \in X^*(C)$ and any $\epsilon : \mathcal{K} = C(\mathbb{Z}) \to \{\pm\}$ we can construct the $(\mathrm{Lie}(C)_\mathbb{Z}, \mathcal{K})$ module $\mathbb{Z}[\gamma \otimes \epsilon]$.

8.3.2 The case of $\mathrm{GL}_2/\mathbb{Z}$

We consider the special case $G = \mathrm{GL}_2/\mathrm{Spec}(\mathbb{Z})$ with $\tilde{\tau}_\alpha = Id$. The group $\mathrm{GL}_2(\mathbb{R})$ has its discrete series representations and the resulting $(\mathfrak{g}_\mathbb{R}, K_\infty)$-modules. We want to show that these discrete series representations are base extensions of Harish-Chandra modules over $\mathbb{Z}$.

Inside G we have the subgroup scheme

$$\tilde{\mathcal{K}} = \{\begin{pmatrix} a & b \\ -b & a \end{pmatrix}\} \subset \mathrm{GL}_2.$$

The affine algebra of $\tilde{\mathcal{K}}$ is $A(\tilde{\mathcal{K}}) = \mathbb{Z}[a, b, 1/(a^2+b^2)]$. Let $\mathcal{O} = \mathbb{Z}[i]$, where $i^2 = -1$. We define the flat group scheme $R_{\mathcal{O}/\mathbb{Z}}(\mathbb{G}_m)$, its R valued points are $R_{\mathcal{O}/\mathbb{Z}}(\mathbb{G}_m)(R) = (\mathcal{O} \otimes_{\mathbb{Z}} R)^\times$. Choose an isomorphism

$$j : \tilde{\mathcal{K}} \xrightarrow{\sim} R_{\mathcal{O}/\mathbb{Z}}(\mathbb{G}_m)$$

defined by the rule

$$j : \begin{pmatrix} 0 & 1 \\ -1 & 0 \end{pmatrix} \mapsto i.$$

The group scheme $\tilde{\mathcal{K}}/\mathbb{Z}$ is not smooth, but the embedding $\mathbb{Z}[i] \otimes \mathbb{Z}[i] \to \mathbb{Z}[i] \oplus \mathbb{Z}[i]$ induces an embedding

$$\tilde{\mathcal{K}} \times \mathrm{Spec}(\mathbb{Z}[i]) \hookrightarrow \mathbb{G}_m \times \mathbb{G}_m. \tag{8.8}$$

This embedding yields an inclusion of affine algebras

$$\mathbb{Z}[i][x, x^{-1}] \otimes \mathbb{Z}[i][y, y^{-1}] \ \hookrightarrow \ A(\tilde{\mathcal{K}}) \otimes \mathbb{Z}[i].$$

Here $y = \bar{x}$ is the complex conjugate of x. Then we get

$$a = \frac{1}{2}(x+y), \ b = \frac{1}{2i}(x-y). \tag{8.9}$$

This inclusion becomes an isomorphism if we invert 2. We observe that we have the obvious inclusion $i : \mathbb{G}_m \hookrightarrow \tilde{\mathcal{K}}$ and we have the restriction of the determinant $\det : \tilde{\mathcal{K}} \to \mathbb{G}_m$. The kernel of det is the group scheme

$$\mathcal{K}^{(1)} = \{\begin{pmatrix} a & b \\ -b & a \end{pmatrix} \subset \mathrm{GL}_2 | \ a^2 + b^2 = 1\}. \tag{8.10}$$

The character module $X^*(\mathcal{K}^{(1)} \times \mathbb{Z}[i]) = \mathbb{Z}e$, where

$$e : \{\begin{pmatrix} a & b \\ -b & a \end{pmatrix}\} \mapsto (a + bi). \tag{8.11}$$

The matrix

$$c_2 = \begin{pmatrix} 1 & 1 \\ -i & i \end{pmatrix} \in \mathrm{GL}_2(\mathbb{Z}[i, \frac{1}{2}]) \tag{8.12}$$

conjugates the standard diagonal torus $T \times \mathbb{Z}[i, \frac{1}{2}]$ into $\mathcal{K}^{(1)} \cdot \mathbb{G}_m \times \mathbb{Z}[i, \frac{1}{2}]$.

Take a weight $\lambda = l\gamma_1 + d\delta$, where $l \equiv 2d \pmod 2$. Consider the space of regular functions on G that satisfy

$$A_\lambda(B\backslash G) = \{f \in A(G) \mid f(bg) = \lambda(b)f(g)\}, \tag{8.13}$$

where B is the Borel subgroup scheme of G of all upper-triangular matrices. On this space of sections we have the action of G by right translations. The following is rather obvious and well known.

Proposition 8.1. *The module $A_\lambda(B\backslash G)$ of regular functions is trivial if $l > 0$. If $l \leq 0$ it realizes the module $\mathcal{M}_{\lambda^-,\mathbb{Z}}$ of highest weight $\lambda^- = -l\gamma_1 + d\delta$.*

We can also say that λ defines a line bundle $\mathcal{L}_\lambda$ on $B\backslash G$ and that

$$A_\lambda(B\backslash G) = H^0(B\backslash G, \mathcal{L}_\lambda). \tag{8.14}$$

(In greater generality this has already been discussed in Sect. 2.4 on integral cohomology.) The algebra of regular functions is embedded into the larger algebra $A(G)_e$, consisting of all functions on G which are regular at the identity element, it is the localization of $A(G)$ at e. Again we define

$$A_\lambda(B\backslash G)_e = \{f \in A(G)_e \mid f(bg) = \lambda(b)f(g)\}. \tag{8.15}$$

On this module we do not have an action of G, but it is clear that we still have an action of $\mathfrak{g}_\mathbb{Z}$.

Consider the morphism of schemes $m : B \times \tilde{\mathcal{K}} \to G$ given by multiplication. The intersection $B \cap \tilde{\mathcal{K}} = C = \mathbb{G}_m$ is embedded into the product $t \mapsto (t, t^{-1})$. The fiber of the morphism m are torsors under the action of C. Then m induces a homomorphism of affine algebras

$$A(G) \hookrightarrow (A(B) \otimes A(\tilde{\mathcal{K}}))^C \hookrightarrow A(G)_e. \tag{8.16}$$

The character λ defines the rank one module $\mathbb{Z}[\lambda]$ of $A(B)$, where we let the unipotent radical of B act trivially. Let λ_C be the restriction of λ to the center C. Then the above embedding defines an inclusion

$$\mathbb{Z}[\lambda] \otimes A_{\lambda_C}(\tilde{\mathcal{K}}) \hookrightarrow A(G)_e[\lambda], \tag{8.17}$$

where $A_{\lambda_C}(\tilde{\mathcal{K}})$ consists of those elements of $A(\tilde{\mathcal{K}})$ on which C acts via λ_C. The left-hand side is a $\tilde{\mathcal{K}}$-module, which is invariant under the action of the Lie algebra $\mathfrak{g}_\mathbb{Z}$. This allows us to define the induced module

$$\mathfrak{Ind}_B^G \mathbb{Z}[\lambda] = \mathbb{Z}[\lambda] \otimes A_{\lambda_C}(\tilde{\mathcal{K}}). \tag{8.18}$$

For $\nu \equiv l \pmod 2$, let's define the elements

$$\Phi_{d,\nu} := \frac{(a+bi)^\nu}{(a^2+b^2)^{\frac{\nu-2d}{2}}} \in A_{\lambda_C}(\tilde{\mathcal{K}}) \otimes \mathbb{Z}[i]. \tag{8.19}$$

We get an inclusion

$$A_{\lambda_C}(\tilde{\mathcal{K}}) \otimes \mathbb{Z}[i] \supset \bigoplus \mathbb{Z}[i]\Phi_{d,\nu}, \tag{8.20}$$

which becomes an isomorphism if we invert 2. Then we get a decomposition into eigenspaces under the action of $\mathcal{K}$. The $\Phi_{d,\nu}$ are characters. The complex conjugation $\mathbf{c}$ (the nontrivial element in $\mathrm{Gal}(\mathbb{Q}(i)/\mathbb{Q})$) acts on the modules above and let's note:

$$\mathbf{c}(\Phi_{d,\nu}) = \Phi_{d,-\nu}.$$

Define a submodule

$$A'_{\lambda_C}(\tilde{\mathcal{K}}) = \Big(\bigoplus_{\nu\equiv l \pmod 2} \mathbb{Z}[i]\Phi_{d,\nu}\Big) \cap A_{\lambda_C}(\tilde{\mathcal{K}}),$$

and again, if we invert 2 it becomes isomorphic to $A_{\lambda_C}(\tilde{\mathcal{K}})$.

The Lie algebra $\mathfrak{g}_{\mathbb{Z}}^{(1)}$ is a direct sum $\mathfrak{g}_{\mathbb{Z}}^{(1)} = \mathbb{Z}H \oplus \mathbb{Z}E_+ \oplus \mathbb{Z}E_-$, where

$$H = \begin{pmatrix} 1 & 0 \\ 0 & -1 \end{pmatrix}, \quad E_+ = \begin{pmatrix} 0 & 1 \\ 0 & 0 \end{pmatrix}, \quad E_- = \begin{pmatrix} 0 & 0 \\ 1 & 0 \end{pmatrix}. \tag{8.21}$$

We introduce some more notation

$$\begin{aligned} V &:= E_+ + E_-, \quad & Y &:= E_+ - E_-, \\ P_+ &:= H + i \otimes V, \quad & P_- &:= H - i \otimes V, \end{aligned} \tag{8.22}$$

where the elements in the first row are in $\mathfrak{g}_{\mathbb{Z}}^{(1)}$ and the elements in the second row are in $\mathfrak{g}_{\mathbb{Z}[i]}^{(1)}$. Under the adjoint action of $\tilde{\mathcal{K}}$ the elements P_+ and P_- are eigenvectors. We have

$$\mathrm{Ad}(k)P_+ = \Phi_{0,2}(k)P_+, \quad \mathrm{Ad}(k)P_- = \Phi_{0,-2}(k)P_- \tag{8.23}$$

Some elementary computations yield

$$Y\Phi_{d,\nu} = i\nu\Phi_{d,\nu}, \quad P_+\Phi_{d,\nu} = (l+\nu)\Phi_{d,\nu+2}, \quad P_-\Phi_{d,\nu} = (l-\nu)\Phi_{d,\nu-2}. \tag{8.24}$$

It's clear that $\mathfrak{g}_{\mathbb{Z}}/\mathfrak{k}_{\mathbb{Z}} \otimes \mathbb{Z}[\frac{1}{2}]$ is generated by

$$H = \frac{1}{2}(P_+ + P_-), \quad V = \frac{1}{2i}(P_+ - P_-),$$

and because of the parity conditions it is clear $A'_{\lambda_C}(\tilde{\mathcal{K}})$ is a $(\mathfrak{g}_{\mathbb{Z}}, \tilde{\mathcal{K}})$-module and hence we get that

$$\mathfrak{Ind}_B^G \mathbb{Z}[\lambda] \otimes \mathbb{Z}[\frac{1}{2}] \text{ is a } (\mathfrak{g}_{\mathbb{Z}}, \tilde{\mathcal{K}}) \text{ module} \tag{8.25}$$

Now it becomes clear that $\mathfrak{Ind}_B^G \mathbb{Z}[\lambda] \otimes \mathbb{Z}[i, \frac{1}{2}]$ is never irreducible. We have two cases:

Let us first assume $l \leq 0$, then we get from our formulas (8.24) that

$$P_+ \Phi_{d,-l} = 0, \quad P_- \Phi_{d,l} = 0$$

and we find a nontrivial invariant submodule:

$$\bigoplus_{\substack{l \leq \nu \leq -l \\ \nu \equiv l \pmod 2}} \mathbb{Z}[i, \frac{1}{2}] \Phi_{d,\nu},$$

and if we look a little bit more closely then we see that this is the module $\mathcal{M}_{\lambda^-, \mathbb{Z}[i,\frac{1}{2}]}$. The quotient by this submodule decomposes into a direct sum; i.e., we get an exact sequence

$$0 \to \mathcal{M}_{\lambda^-, \mathbb{Z}[i,\frac{1}{2}]} \to \mathfrak{Ind}_B^G \mathbb{Z}[\lambda] \otimes \mathbb{Z}[i, \frac{1}{2}] \to \mathcal{D}_\lambda^+ \oplus \mathcal{D}_\lambda^- \to 0, \tag{8.26}$$

where

$$\mathcal{D}_\lambda^+ = \bigoplus_{\substack{\nu \geq -l+2 \\ \nu \equiv l \pmod 2}} \mathbb{Z}[i, \frac{1}{2}] \, \Phi_{d,\nu} \; ; \quad \mathcal{D}_\lambda^- = \bigoplus_{\substack{\nu \leq l-2, \\ \nu \equiv l \pmod 2}} \mathbb{Z}[i, \frac{1}{2}] \Phi_{d,\nu} \tag{8.27}$$

is a decomposition into two invariant submodules.

Consider now the second case where $l \geq 0$. In this case let's look at the induced module

$$\mathfrak{Ind}_B^G \mathbb{Z}[\lambda + 2\rho].$$

Here 2ρ is the sum of the positive roots; of course $2\rho = \alpha$. In (8.24) replace l by $l + 2$. We have

$$P_- \Phi_{d,l+2} = 0, \quad P_+ \Phi_{d,-l-2} = 0$$

and hence we see that the two modules in (8.27) are invariant submodules and we get an exact sequence

$$0 \to (\mathcal{D}_\lambda^+ \oplus \mathcal{D}_\lambda^-) \otimes \mathbb{Z}[i, \frac{1}{2}] \to \mathfrak{Ind}_B^G \mathbb{Z}[\lambda + 2\rho] \otimes \mathbb{Z}[i, \frac{1}{2}] \to \mathcal{M}_\lambda \otimes \mathbb{Z}[i, \frac{1}{2}] \to 0. \tag{8.28}$$

For any λ the modules $\mathcal{D}_\lambda^\pm \otimes \mathbb{C}$ are the familiar discrete series modules for $\mathfrak{gl}_2(\mathbb{R})$. If we consider the two weights $\lambda = l\gamma_1 + d\delta$ and $\lambda^- = -l\gamma_1 + d\delta$, then the two discrete series $\mathcal{D}_\lambda^\pm$ and $\mathcal{D}_{\lambda^-}^\pm$ are not isomorphic but if we take the tensor product with the rationals then we find isomorphisms:

$$\Psi_{d,l}^\pm : \mathcal{D}_\lambda^\pm \otimes \mathbb{Q} \;\to\; \mathcal{D}_{\lambda^-}^\pm \otimes \mathbb{Q} \tag{8.29}$$

which are uniquely defined by the condition $\Psi_{d,l}^\pm(\Phi_{d,\pm(l+2)}) = \Phi_{d,\pm(l+2)}$.

In our notation the discrete series Harish-Chandra modules for GL_2 are parametrized by a pair (λ, sign) where $\lambda = l\gamma_1 + d\delta$, $l \geq 0$ is a highest weight. The exact sequences above (which are in fact not split exact sequences) tell us that

$$\mathrm{Ext}^1(\mathcal{M}_{\lambda,\mathbb{Z}[i]}, \mathcal{D}_\lambda^\pm) \neq 0. \tag{8.30}$$

If we restrict the action of $\mathcal{K}$ on $\mathcal{D}_\lambda^+$ to $\mathcal{K}^{(1)}$ then we get a decomposition into $\mathcal{K}^{(1)}$–types

$$\mathcal{D}_\lambda^+ = \bigoplus_{\substack{\nu \geq l+2 \\ \nu \equiv l \pmod 2}} \mathbb{Z}[i][\nu],$$

where $\mathcal{K}^{(1)}$ acts by νe on $\mathbb{Z}[i][\nu]$. The character $(l+2)e$ is called the *minimal* $\mathcal{K}^{(1)}$ *type* in $\mathcal{D}_\lambda^+$. The character $-(l+2)e$ is also called the minimal $\mathcal{K}^{(1)}$ type in $\mathcal{D}_\lambda^-$. Consider now the module $\mathcal{M}_{\lambda,\mathbb{Z}[i]}$. Let us assume that we realized $\mathcal{M}_{\lambda,\mathbb{Z}}$ as the module of homogeneous polynomials of degree l in two variables U, V. (This is actually the module $H^0(B\backslash G, \mathcal{L}_{\lambda^-})$.) We consider the action of $\mathcal{K}^{(1)} \times \mathbb{Z}[i]$ on it and we have the decomposition into eigenspaces

$$\mathbb{Z}[i, \frac{1}{2}](U - iV)^l \oplus \;\dots\; \oplus \mathbb{Z}[i, \frac{1}{2}](U + iV)^l. \tag{8.31}$$

We are interested only in the highest and lowest weight vectors. We abbreviate

$$(U - iV)^l \;=:\; e_{-l}, \quad (U + iV)^l \;=:\; e_l.$$

We also put

$$\mathcal{D}_\lambda = \mathcal{D}_\lambda^+ \oplus \mathcal{D}_\lambda^-.$$

The relative Lie algebra cohomology of $\mathcal{D}_\lambda \otimes \mathcal{M}_{\lambda,\mathbb{Z}[i,\frac{1}{2}]}$ is the cohomology of the complex

$$\mathrm{Hom}_{\mathcal{K}^{(1)}}(\Lambda^\bullet(\mathfrak{g}_\mathbb{Z}/\tilde{\mathfrak{k}}_\mathbb{Z}) \otimes \mathbb{Z}[i, \frac{1}{2}], \mathcal{D}_\lambda \otimes \mathcal{M}_{\lambda,\mathbb{Z}[i,\frac{1}{2}]}). \tag{8.32}$$

Here we observe that $\Lambda^0((\mathfrak{g}_\mathbb{Z}/\tilde{\mathfrak{k}}_\mathbb{Z}) \otimes \mathbb{Z}[i, \frac{1}{2}]) = \mathbb{Z}[i, \frac{1}{2}]$ and also $\Lambda^2((\mathfrak{g}_\mathbb{Z}/\tilde{\mathfrak{k}}_\mathbb{Z}) \otimes \mathbb{Z}[i, \frac{1}{2}]) = \mathbb{Z}[i, \frac{1}{2}]$, where we choose $P_+ \wedge P_-$ as the generator. Since $\mathcal{D}_\lambda \otimes \mathcal{M}_{\lambda,\mathbb{Z}[i,\frac{1}{2}]}$ does not contain the trivial $\mathcal{K}^{(1)}$ module this complex is zero in degrees 0 and 2. In degree 1 we have

$$\Lambda^1(\mathfrak{g}_\mathbb{Z}/\tilde{\mathfrak{k}}_\mathbb{Z}) \otimes \mathbb{Z}[i, \frac{1}{2}]) \;=\; \mathfrak{g}_\mathbb{Z}/\tilde{\mathfrak{k}}_\mathbb{Z} \otimes \mathbb{Z}[i, \frac{1}{2}] \;=\; \mathbb{Z}[i, \frac{1}{2}]P_+ \oplus \mathbb{Z}[i, \frac{1}{2}]P_-. \tag{8.33}$$

We denote by $P_+^\vee, P_-^\vee \in \mathrm{Hom}((\mathfrak{g}_\mathbb{Z}/\tilde{\mathfrak{k}}_\mathbb{Z}) \otimes \mathbb{Z}[i,\frac{1}{2}], \mathbb{Z}[i,\frac{1}{2}])$ the dual basis.

Proposition 8.2.

$$H^1(\mathfrak{g}_\mathbb{Z}, \tilde{\mathcal{K}}, \mathcal{D}_\lambda \otimes \mathcal{M}_{\lambda,\mathbb{Z}[i,\frac{1}{2}]}) = \mathbb{Z}[i,\frac{1}{2}]P_+^\vee \otimes \Phi_{d,l+2} \otimes e_{-l} \bigoplus \mathbb{Z}[i,\frac{1}{2}]P_-^\vee \otimes \Phi_{d,-l-2} \otimes e_l.$$

Proof. Obvious! □

If we replace $\tilde{\mathcal{K}}$ by $\mathcal{K} = \mathcal{K}^{(1)}$, then we get the same, but we have to multiply the right-hand side by $\Lambda^\bullet \mathfrak{c}_\mathbb{Z}$.

It is clear from the construction that the element $\mathbf{c}$ in the Galois group acts on $\mathcal{D}_\lambda$ and more precisely we have $\mathbf{c}(P_+) = P_-$, $\mathbf{c}(\Phi_{d,\nu}) = \Phi_{d,-\nu}$, $\mathbf{c}(e_l) = e_{-l}$. Then we put

$$\Omega_{d,l} := P_+^\vee \otimes \Phi_{d,l+2} \otimes e_{-l}, \quad \bar{\Omega}_{d,l} := P_-^\vee \otimes \Phi_{d,-l-2} \otimes e_l.$$

We may think of these elements as arithmetic versions of holomorphic and antiholomorphic 1-forms. Now we define

$$\mathcal{D}_{\lambda,\mathbb{Z}} := \text{the } (\mathfrak{g}_\mathbb{Z}, \mathcal{K}) \text{ module of elements in } \mathcal{D}_\lambda \text{ fixed by } \mathbf{c}.$$

Then

$$\mathrm{Hom}_\mathcal{K}(\Lambda^1(\mathfrak{g}_\mathbb{Z}/\tilde{\mathfrak{k}}_\mathbb{Z}), \mathcal{D}_{\lambda,\mathbb{Z}} \otimes \mathcal{M}_{\lambda,\mathbb{Z}}) = \mathbb{Z}(\Omega_{d,l} + \bar{\Omega}_{d,l}) \oplus \mathbb{Z}(i\Omega_{d,l} - i\bar{\Omega}_{d,l}). \quad (8.34)$$

We introduce some abbreviations

$$\omega_{d,l}^{(1)} := \Omega_{d,l} + \bar{\Omega}_{d,l}, \quad \omega_{d,l}^{(2)} := i\Omega_{d,l} - i\bar{\Omega}_{d,l}, \quad \eta = \begin{pmatrix} -1 & 0 \\ 0 & 1 \end{pmatrix}.$$

We still have the action of $\mathrm{O}(2)/\mathrm{SO}(2) = \mathbb{Z}/2\mathbb{Z} = \pi_0(G(\mathbb{R}))$, where the nontrivial element is represented by η. Under this action, $\mathrm{Hom}_{\mathcal{K}^{(1)}}(\Lambda^1(\mathfrak{g}_\mathbb{Z}/\tilde{\mathfrak{k}}_\mathbb{Z}), \mathcal{D}_{\lambda,\mathbb{Z}} \otimes \mathcal{M}_{\lambda,\mathbb{Z}})$ decomposes into a + and a − eigenspace. A straightforward computation shows that

$$\eta(P_\pm^\vee) = P_\mp^\vee, \quad \eta(\Phi_{d,\nu}) = \Phi_{d,-\nu}, \quad \eta(e_{\pm l}) = (-1)^{\frac{2d-l}{2}} e_{\mp l}.$$

Proposition 8.3. *The elements* $\omega_{d,l}^{(1)}, \omega_{d,l}^{(2)} \in \mathrm{Hom}_\mathcal{K}(\Lambda^1(\mathfrak{g}_\mathbb{Z}/\tilde{\mathfrak{k}}_\mathbb{Z}), \mathcal{D}_{\lambda,\mathbb{Z}} \otimes \mathcal{M}_{\lambda,\mathbb{Z}})$ *are generators of the* ± *eigenspaces (possibly up to a power of* 2*). We have*

$$\eta(\omega_{d,l}^{(1)}) = (-1)^{\frac{2d-l}{2}} \omega_{d,l}^{(1)}, \qquad \eta(\omega_{d,l}^{(2)}) = -(-1)^{\frac{2d-l}{2}} \omega_{d,l}^{(2)}.$$

Proof. Obvious! □

Recall that $d \in \frac{1}{2}\mathbb{Z}$ and satisfies $2d \equiv l \mod 2$, hence it is well defined modulo $\mathbb{Z}$. Which of the two generators $\omega_{d,l}^{(1)}, \omega_{d,l}^{(2)}$ is the generator of the +

eigenspace depends on d and they change role if we replace d by $d+1$. This flip plays a decisive role in the definition of the periods in Chap. 5.

The module $\mathcal{D}_{\lambda,\mathbb{Z}}$ is irreducible, but its base extension $\mathcal{D}_{\lambda,\mathbb{Z}}\otimes\mathbb{Z}[i]$ is reducible, since it decomposes into $\mathcal{D}^+_\lambda \oplus \mathcal{D}^-_\lambda$. If we enlarge $\mathcal{K}^{(1)}$ to $\mathcal{K} = \mathcal{K}^{(1)} \rtimes <\eta>$ then $\mathcal{D}_{\lambda,\mathbb{Z}} \otimes \mathbb{Z}[i]$ becomes an absolutely irreducible $(\mathfrak{g}_\mathbb{Z}, \mathcal{K})$-module.

Then it is easy to see that in the case λ regular (i.e., $l \neq 0$) $\mathcal{D}_{\lambda,\mathbb{Z}}$ is the only irreducible $(\mathfrak{g}_\mathbb{Z}, \tilde{\mathcal{K}})$–module that has nontrivial cohomology with coefficients in $\mathcal{M}_{\lambda,\mathbb{Q}}$. If $l = 0$ then the trivial one-dimensional $(\mathfrak{g}_\mathbb{Z}, \mathcal{K}^{(1)})$–module $\mathbb{Z}$ has nontrivial cohomology in degree 0 and 2 and this module completes the list of modules that have nontrivial cohomology with coefficients in some $\mathcal{M}_{\lambda,\mathbb{Z}}$.

8.3.3 The intertwining operator

We come back to our highest weight $\lambda = l\gamma_1 + d\delta$, with $l \geq 0$. In (8.29) we wrote down an intertwining isomorphism $\Psi_{d,l}$ between the two discrete series representation. If we look at the inverse of this operator and observe that $\mathcal{D}^+_{\lambda^-} \oplus \mathcal{D}^-_{\lambda^-}$ is a quotient of $\mathfrak{Ind}_B^G\mathbb{Z}[\lambda^-]$ and $\mathcal{D}^+_\lambda \oplus \mathcal{D}^-_\lambda$ is a submodule of $\mathfrak{Ind}_B^G\mathbb{Z}[\lambda+2\rho]$ then our isomorphism provides an intertwining operator

$$T^{alg}_{\lambda^-} : \mathfrak{Ind}_B^G\mathbb{Z}[\lambda^-] \otimes \mathbb{Q} \;\to\; \mathfrak{Ind}_B^G\mathbb{Z}[\lambda+2\rho] \otimes \mathbb{Q}, \tag{8.35}$$

which is unique up to a scalar and normalized by fixing its value on a lowest $\mathcal{K}$–type.

By the same token we get an operator in the opposite direction:

$$T^{alg}_{\lambda+2\rho} : \mathfrak{Ind}_B^G\mathbb{Z}[\lambda+2\rho] \otimes \mathbb{Q} \;\to\; \mathfrak{Ind}_B^G\mathbb{Z}[\lambda^-] \otimes \mathbb{Q}. \tag{8.36}$$

In this direction the space of homomorphisms is of rank one. The homomorphisms factor over a finite-dimensional quotient.

In our situation here the maximal torus $T = \mathbb{G}_m \times \mathbb{G}_m$ and so far we discussed only the modules that are induced from rational characters. In this case we also may induce characters $\lambda \otimes \epsilon$, where $\epsilon : \tilde{\mathcal{K}}^T = \mu_2 \times \mu_2 \to \mu_2$ is a sign character, it is the form $(\pm 1, \pm 1) \mapsto (\pm 1)^{m_1}(\pm 1)^{m_2}$. Then the induced module $\mathfrak{Ind}_B^G\mathbb{Z}[\lambda \otimes \epsilon]$ is still reducible if the sign character $\epsilon = \underline{m}$ factors over the determinant; i.e., we have $m_1 = m_2$. But if the sign character does not factor over the determinant then the induced module $\mathfrak{Ind}_B^G\mathbb{Z}[\lambda \otimes \epsilon]$ is in fact irreducible.

8.3.4 Transcendental Harish-Chandra modules

We return briefly to the transcendental theory of Harish-Chandra modules. If we tensor by $\mathbb{C}$ then our modules become Harish-Chandra modules in the traditional sense. The group scheme $\mathcal{K}^{(1)}$ is replaced by the group $\mathrm{SO}(2) = K_\infty = \mathcal{K}^{(1)}(\mathbb{R})$. The following is of course well known.

The evaluation of the highest weight λ on $T(\mathbb{R})$ provides an (algebraic) character $\lambda_{\mathbb{R}} : T(\mathbb{R}) \to \mathbb{R}^{\times}$. We define a larger class of (analytic) characters $\chi : T(\mathbb{R}) \to \mathbb{C}^{\times}$ that are of the form

$$\chi = (z, d, \underline{m}) : \begin{pmatrix} t_1 & 0 \\ 0 & t_2 \end{pmatrix} \mapsto (|\frac{t_1}{t_2}|)^{z/2} \, |t_1 t_2|^d \, (\frac{t_1}{|t_1|})^{m_1} (\frac{t_2}{|t_2|})^{m_2}, \tag{8.37}$$

where z is a complex variable and $\underline{m} = (m_1, m_2)$ is a pair of integers modulo 2. The central contribution given by the half integer d should be fixed. For us it is adequate to distinguish between the character $\lambda \in X^*(T)$ and its evaluation $\lambda_{\mathbb{R}}$. For $\lambda = l\gamma_1 + d\delta$ we have

$$\lambda_{\mathbb{R}} = \chi = (z, d, \underline{m}) \iff z = l \text{ and } m_1 \equiv \frac{l}{2} + d, \; m_2 \equiv -\frac{l}{2} + d \pmod 2.$$

We call such a χ algebraic, and we will say that χ is cohomological if $l \neq 0$. We say that χ is of algebraic type if z is an integer but the parity conditions may fail.

Define the induced representation

$$I_B^{G,\infty}\chi = \{f \in \mathcal{C}^{\infty}(\mathrm{GL}_2(\mathbb{R}), \mathbb{C}) \,|\, f(\begin{pmatrix} t_1 & u \\ 0 & t_2 \end{pmatrix} g) = \chi(t) f(g)\},$$

which is a $\mathrm{GL}_2(\mathbb{R})$-module with the group action given by right translations. (Note that this is algebraic or unnormalized parabolic induction.) The submodule of K_∞ finite functions in $I_B^{G,\infty}\chi$ is the induced Harish-Chandra module $I_B^G\chi$. For $\chi = \lambda_{\mathbb{R}}$ we have

$$I_B^G \lambda_{\mathbb{R}} = \mathfrak{Ind}_B^G \lambda \otimes \mathbb{C}.$$

Let $m = m_1 + m_2$ mod 2. Then the module is a direct sum

$$I_B^G\chi = \bigoplus_{\nu \equiv m \bmod 2} \Phi_\nu^\chi,$$

where

$$\Phi_\nu^\chi\left(\begin{pmatrix} t_1 & u \\ 0 & t_2 \end{pmatrix} \cdot \begin{pmatrix} \cos(\phi) & \sin(\phi) \\ -\sin(\phi) & \cos(\phi) \end{pmatrix}\right) = (|\frac{t_1}{t_2}|)^{z/2} \, |t_1 t_2|^d \, (\frac{t_1}{|t_1|})^{m_1} (\frac{t_2}{|t_2|})^{m_2} \, e^{2\pi i \nu \phi}.$$

We have essentially the same formulae for the action of the Lie algebra:

$$Y\Phi_\nu^\chi = i\nu\Phi_\nu^\chi, \quad P_+\Phi_\nu^\chi = (z + \nu)\Phi_{\nu+2}^\chi, \quad P_-\Phi_\nu^\chi = (z - \nu)\Phi_{\nu-2}^\chi. \tag{8.38}$$

Note that the parity of ν is equal to the parity of $m_1 + m_2$.

For $\chi = (z, d, \underline{m})$ we put $\chi' = (-z, d, \underline{m}')$, where $\underline{m}' = (m_2, m_1)$. Then we can write down the classical (standard) intertwining operator

$$T^{st}_\chi : I_B^G \chi \to I_B^G(\chi' \otimes \rho^2), \tag{8.39}$$

which is defined by

$$T^{st}_\chi(f)(g) = \int_{-\infty}^{\infty} f(\begin{pmatrix} 0 & 1 \\ -1 & 0 \end{pmatrix} \cdot \begin{pmatrix} 1 & u \\ 0 & 1 \end{pmatrix} g)du, \tag{8.40}$$

where du is the Lebesgue measure on $\mathbb{R}$. This integral converges for $\Re(z) \gg 0$ and has an meromorphic continuation into the entire z-plane. We need to locate the poles and we want to show that this operator is never identically zero.

We introduce the notation $\chi^\dagger = \chi' \otimes \rho^2$. We evaluate it at the smallest K_∞ type, which is Φ_0^χ if $m_1 + m_2$ is even, and Φ_1^χ if $m_1 + m_2$ is odd. Let us put $\epsilon(\underline{m}) = 0$ if $m_1 + m_2$ is even and $\epsilon(\underline{m}) = 1$ else. Then an easy computation (see, for example, Sect. 9.2.3) shows

$$T^{st}_\chi(\Phi^\chi_{\epsilon(\underline{m})}) = \frac{\Gamma(\frac{z+\epsilon(\underline{m})-1}{2})\Gamma(\frac{1}{2})}{\Gamma(\frac{z+\epsilon(\underline{m})}{2})} \Phi^{\chi^\dagger}_{\epsilon(\underline{m})}. \tag{8.41}$$

Then we can evaluate T^{st}_χ on any element Φ^χ_ν, using the recursion provided by the formulae in (8.38). We get for $n \geq 1$

$$P^n_+(\Phi^\chi_{\epsilon(\underline{m})}) = (z + \epsilon(\underline{m})) \dots (z + \epsilon(\underline{m}) + 2(n-1))\Phi^\chi_{\epsilon(\underline{m})+2n}, \tag{8.42}$$

and on the other side (we have to replace z by $2 - z$)

$$P^n_+(\Phi^{\chi^\dagger}_{\epsilon(\underline{m})}) = (2 - z + \epsilon(\underline{m})) \dots (2 - z + \epsilon(\underline{m}) + 2(n-1))\Phi^{\chi^\dagger}_{\epsilon(\underline{m})+2n}. \tag{8.43}$$

Therefore, if $\nu = \epsilon(\underline{m}) + 2n$

$$T^{st}_\chi(\Phi^\chi_\nu) = \frac{(2 - z + \epsilon(\underline{m})) \dots (2 - z + \epsilon(\underline{m}) + \nu - 2))}{(z + \epsilon(\underline{m})) \dots (z + \nu - 2)} \frac{\Gamma(\frac{z+\epsilon(\underline{m})-1}{2})\Gamma(\frac{1}{2})}{\Gamma(\frac{z+\epsilon(\underline{m})}{2})} \Phi^{\chi^\dagger}_\nu. \tag{8.44}$$

Note that here $\nu > 1$. The product is empty if $\nu = 0, 1$ and hence it has value 1 if this is the case. Of course we get a corresponding formula for $\nu \leq 0$. We say that the intertwining operator T^{st}_χ is holomorphic at $\chi = (z_0, d, \underline{m})$ if for all $\nu \equiv \epsilon(m) \pmod 2$, the function $T^{st}_\chi(\Phi^\chi_{d,\nu})/\Phi^{\chi^\dagger}_{d,\nu}$ is holomorphic at z_0. Otherwise we say that $T^{st}_\chi(\Phi_{d,\nu})$ has a pole at z_0.

Proposition 8.4. *The intertwining operator T^{st}_χ has its poles at the arguments $z_0 = 1 - \epsilon(\underline{m}), -1 - \epsilon(\underline{m}), \dots$ and these are first-order poles. At these arguments $T^{st}_\chi(\Phi^\chi_\nu)$ has a pole for all values ν.*

Proof. This is an easy exercise in using the properties of the Γ-function. Look at the denominator of the expression in (8.44). We have

$$(z+\epsilon(\underline{m}))\dots(z+\nu-2)\Gamma(\frac{z+\epsilon(\underline{m})}{2}) = 2^n\Gamma(\frac{z+\epsilon(\underline{m})+2}{2}+n-1).$$

The Γ-function has no zeros; hence the denominator does not contribute to poles. The Γ-factor in the numerator has its poles exactly at the above list; these are first-order poles and they do not cancel against the product of linear factors in front of the Γ-factor. □

We can form the composite $T^{st}_{\chi^\dagger}\circ T^{st}_\chi$ and this is an endomorphism of $I_B^G\chi$. Since for a general value of z the module is irreducible the operator must be a scalar $\Lambda(\chi)$ and it is not too difficult to write down this scalar. Define $a(\underline{m})=+1$ if $\epsilon(\underline{m})=0$ otherwise $a(\underline{m})=-1$. The scalar is given by

$$\Lambda(\chi) = \frac{\Gamma(\frac{z-1+\epsilon(\underline{m})}{2})\Gamma(\frac{1-z+\epsilon(\underline{m})}{2})}{\Gamma(\frac{z+\epsilon(\underline{m})}{2})\Gamma(\frac{2-z+\epsilon(\underline{m})}{2})} = \frac{2}{z-1}(\frac{\sin(\frac{\pi}{2}z)}{\cos(\frac{\pi}{2}z)})^{a(\underline{m})}.$$

For us the important arguments for χ are the values $\chi=(l+2,m)$ and $\chi=(-l,m)$ where $l\geq 0$ is an integer and $l\equiv m \mod 2$, and for such values we had called χ to be cohomological. Our proposition tells us that

T^{st}_χ is holomorphic at cohomological arguments.

But we also see that $\Lambda(\chi)$ vanishes at these arguments, i.e., $T^{st}_{\chi'}\circ T^{st}_\chi=0$. Since it is clear that the linear map T^{st}_χ is never zero it follows that T^{st}_χ maps $I_B^G\chi$ to the kernel of $T^{st}_{\chi^\dagger}$.

This is of course consistent with the results in Sect. 8.3.2; if we tensor the two exact sequences (8.26) and (8.28) by $\mathbb{C}$, and apply our intertwining operator to the terms in the middle of the exact sequences then for $\chi=\lambda_{\mathbb{R}}, \lambda=l\gamma_1+d\delta, l\geq 0$ we get

$$T^{st}_{\chi\otimes\rho^2_{\mathbb{R}}}: I_B^G\chi\otimes\rho^2_{\mathbb{R}} = \mathfrak{Ind}_B^G\mathbb{Z}[\lambda+2\rho]\otimes\mathbb{C} \longrightarrow \mathcal{M}_{\lambda,\mathbb{C}}\subset\mathfrak{Ind}_B^G\mathbb{Z}[\lambda^-]\otimes\mathbb{C} \quad (8.45)$$

and

$$T^{st}_{\chi^\dagger}: I_B^G\chi^\dagger = \mathfrak{Ind}_B^G\mathbb{Z}[\lambda^-]\otimes\mathbb{C} \longrightarrow (\mathcal{D}^+_\lambda\oplus\mathcal{D}^-_\lambda)\otimes\mathbb{C}\subset\mathfrak{Ind}_B^G\mathbb{Z}[\lambda+2\rho]\otimes\mathbb{C}. \quad (8.46)$$

These two intertwining operators are of course multiples of our earlier operators $T^{alg}_{\lambda+2\rho}\otimes\mathbb{C}$ and $T^{alg}_{\lambda^-}\otimes\mathbb{C}$, which had been normalized such that they gave the "identity" on certain K_∞ types. For the operator $T^{alg}_{\lambda+2\rho}\otimes\mathbb{C}$ the K_∞-type Φ_l, and for $T^{alg}_{\lambda^-} \otimes \mathbb{C}$ it is $\Phi_{d,l+2}$. For $\chi = \lambda_{\mathbb{R}}$, a straightforward computation yields

$$\begin{aligned} T^{st}_{\chi \otimes \rho^2} &= \left(2^{\frac{3l-\epsilon(m)}{2}} (-1)^{\frac{l-\epsilon(m)}{2}} \pi\right) T^{alg}_{\lambda+2\rho} \\ T^{st}_{\chi^\dagger} &= \left(2^{\frac{\epsilon(m)-l-2}{2}} (-1)^{\frac{l-\epsilon(m)}{2}} \pi\right) T^{alg}_{\lambda^-}. \end{aligned} \tag{8.47}$$

This tells us that the operators

$$\frac{1}{\pi} T^{st}_{\chi^\dagger} \quad \text{and} \quad \frac{1}{\pi} T^{st}_{\chi \otimes \rho^2}$$

evaluated at cohomological arguments are defined over $\mathbb{Q}(i)$. They even induce isomorphisms between the $\mathbb{Z}[i, \frac{1}{2}]$ modules of the cohomologically relevant $\mathcal{K}$-types.

Let's briefly look at the induced modules that are not cohomological; these are the modules induced from $\chi = (l, d, \underline{m})$ where l is an integer satisfying $2d \equiv l \mod 2$ and $l - 1 \equiv \epsilon(\underline{m}) \equiv 0 \mod 2$. If now $l - 1 + \epsilon(\underline{m}) \geq 2$ then the operator T^{st}_χ is defined over $\mathbb{Q}(i)$, because $\Gamma(1/2)$ appears in the numerator and in the denominator. If $l + \epsilon(\underline{m}) - 1 = 0, -2, -4, \ldots$ then the intertwining operator has a pole. But we can modify the operator and define the normalized operator

$$T^{norm}_\chi = \frac{1}{\Gamma(\frac{z+1-\epsilon(\underline{m})}{2})} T^{st}_\chi,$$

which is holomorphic everywhere, and at the arguments $\chi = (l, d, \underline{m})$ that are not cohomological it is an isomorphism and defined over $\mathbb{Q}(i)$.

8.4 INDUCTION OF HARISH-CHANDRA MODULES

8.4.1 The general context

Let G be a reductive group scheme over $\mathrm{Spec}(\mathbb{Z})$ as in Sect. 8.2.1. We pick a standard parabolic subgroup $P/\mathrm{Spec}(\mathbb{Z})$; let $U_P/\mathrm{Spec}(\mathbb{Z})$ be its unipotent radical and $M = P/U_P$ its Levi quotient. We can also view $M/\mathrm{Spec}(\mathbb{Z})$ to be the Levi subgroup that is stable under the Cartan involution; this means that $M = P \cap P^\Theta$. Then Θ induces a Cartan involution on the semisimple component $M^{(1)}$, it is simply connected. Let $\mathcal{K}^{M,1} \subset M^{(1)}$ be the fixed-point scheme. Let C_M be the connected center of M; it is a split torus. Then we define $\tilde{\mathcal{K}}^M = \mathcal{K}^{M,1} \cdot C_M$. The intersection $P \cap \mathcal{K}^{(1)}$ can be projected down to M and yields a (possibly slightly larger) definite subscheme $\mathcal{K}^M \supset \mathcal{K}^{M,1}$.

Let us assume we have a highest weight module $\mathcal{M}_{\mu,\mathbb{Z}}$ and a $(\mathfrak{m}_{\mathbb{Z}}, \mathcal{K}^M)$ Harish-Chandra module $\mathcal{V}$ over some commutative ring $R \supset \mathbb{Z}$. For instance, R

could be $\mathbb{Z}$ or $\mathbb{Z}[i]$ or $\mathbb{Z}[i, \frac{1}{2}]$. We give a construction of the induced $(\mathfrak{g}_{\mathbb{Z}}, \mathcal{K}^{(1)})$-module $\mathfrak{Ind}_P^G \mathcal{V}$. We are interested in the case that $H^\bullet(\mathfrak{m}_{\mathbb{Z}}, \mathcal{K}^M, \mathcal{V} \otimes \mathcal{M}_{\mu,\mathbb{Z}}) \neq 0$. In this case we compute the cohomology $H^\bullet(\mathfrak{g}_{\mathbb{Z}}, \mathcal{K}^{(1)}, \mathcal{M}_{\lambda,\mathbb{Z}})$ by adapting the method of Delorme.

We start from the module $\mathcal{V} \otimes A(\mathcal{K})$. In the following R_1 will be a *variable* commutative ring containing R. The algebra $A(\mathcal{K})$ is a $\mathcal{K} \times \mathcal{K}$ module; recall that this means that for an element $f \in A(\mathcal{K}) \otimes R_1$ and $x, k_1, k_2 \in \mathcal{K}(R_1)$ we define

$$(R_{(k_1,k_2)} f)(k) = f(k_1^{-1} k k_2).$$

Let $\mathcal{K}^M = P \cap \mathcal{K}$. The projection $\mathcal{K}^M \to M$ is an injective homomorphism; we identify $\mathcal{K}^M$ with its image. This allows us to define the submodule

$$\begin{aligned}&(\mathcal{V} \otimes A(\mathcal{K}))^{\mathcal{K}^M}\\ &\quad = \{\sum v_i \otimes f_i \mid \sum k v_i \otimes f_i = \sum v_i \otimes R_{(k^{-1},e)} f_i,\ \forall k \in \mathcal{K}^M(R_1)\}. \quad (8.48)\end{aligned}$$

We show that this is a $(\mathfrak{g}_{\mathbb{Z}}, \mathcal{K})$-module. The action of $\mathcal{K}$ is by translation from the right on the second factor: For $k \in \mathcal{K}(R_1)$ and any element $v \otimes f \in (\mathcal{V} \otimes A(\mathcal{K}))^{\mathcal{K}^M} \otimes R_1$ we define the translate

$$R_{(e,k)}(v \otimes f) = v \otimes R_{(e,k)} f.$$

Now we have to define the $\mathfrak{g}_{\mathbb{Z}}$ action. We want to define

$$\mathfrak{g}_{\mathbb{Z}} \times ((\mathcal{V} \otimes A(\mathcal{K}))^{\mathcal{K}^M}) \to ((\mathcal{V} \otimes A(\mathcal{K}))^{\mathcal{K}^M}).$$

To do this we discuss again what happens on the R_1-valued points: For $X \in \mathfrak{g}_{\mathbb{Z}}$ and $\sum v_i \otimes f_i \in (\mathcal{V} \otimes A(\mathcal{K}))^{\mathcal{K}^M} \otimes R_1$ and $k \in \mathcal{K}(R_1)$ we have to say what

$$X(\sum v_i \otimes f_i)(k)$$

should be. We work with dual numbers $R_1[\epsilon]$. Then we should have

$$\epsilon X(\sum v_i \otimes f_i)(k) = (\sum v_i \otimes f_i)(k \cdot \exp(\epsilon X)) - (\sum v_i \otimes f_i)(k),$$

except that the first summand on the right-hand side is not yet defined. To define it we consider the parabolic subgroup $k^{-1} P k \subset G \times R_1$ and observe that the linear map

$$\mathrm{Lie}(k^{-1} P k) \oplus \mathfrak{k}_{\mathbb{Z}} \otimes R_1 \to \mathfrak{g}_{\mathbb{Z}} \otimes R_1$$

is surjective. Hence we can write $X = V + U$ where $V \in \mathrm{Lie}(k^{-1}Pk), U \in \mathrm{Lie}(\mathcal{K}) \otimes R_1$. Now we can define

$$(\sum v_i \otimes f_i)(k \cdot \exp(\epsilon X)) = (\sum v_i \otimes f_i)(\exp(\epsilon \mathrm{Ad}(k)(V) k \cdot \exp(\epsilon U)).$$

The expression on the right-hand side is defined. If we recall the definition of $(\mathcal{V} \otimes A(\mathcal{K}))^{\mathcal{K}^M}$ then we see that it is equal to

$$\epsilon(\sum \mathrm{Ad}(k)(V)v_i \otimes f_i(k) + \sum v_i \otimes Uf_i(k)) + \sum v_i \otimes f_i(k).$$

It it also clear that it does not depend on the decomposition of $X = V + U$.

Hence we can define the induced Harish-Chandra module

$$\mathfrak{Ind}_P^G \mathcal{V} = (\mathcal{V} \otimes A(\mathcal{K}))^{\mathcal{K}^M}. \tag{8.49}$$

It is not difficult to show that this satisfies all the conditions (1) to (6) of Sect. 8.2.2. However, condition (2) may require a longish argument. We will discuss an example in the following section and in this example it becomes clear why condition (2) is fulfilled.

8.4.2 The integral version of $\mathbb{D}_\lambda$

We apply this induction process to a special case for the group $\mathrm{GL}_n/\mathbb{Z}$. We want to construct the $\mathbb{Z}$ structure on the modules that are called $\mathbb{D}_\lambda$ in Sect. 3.1. Let $T/\mathbb{Z}$ be the standard split torus and $B \supset T$ the standard Borel subgroup of upper triangular matrices. The parabolic subgroups $P \supset B$ are the standard parabolic subgroups.

Let $\gamma_1, \dots, \gamma_{n-1} \in X^*_{\mathbb{Q}}(T)$ be the dominant fundamental weights, and let δ be the determinant. For this we choose a self-dual highest weight $\lambda = \sum_i^{n-1} a_i\gamma_i + d\delta$, remember that self-dual means $a_i = a_{n-i}$. We use the usual construction to construct the $G/\mathbb{Z}$-module $\mathcal{M}_{\lambda,\mathbb{Z}}$, it is the space of sections $H^0(B\backslash G, \mathcal{L}_{\lambda^-})$ as in (8.14). We use the technique of induction to construct the very specific $(\mathfrak{g}_\mathbb{Z}, \mathcal{K}^{(1)})$ modules $\mathbb{D}_\lambda^\epsilon$ (where $\epsilon = \pm 1$) over $\mathbb{Z}[i, \frac{1}{2}]$ that have nontrivial cohomology in lowest degree b_n

$$H^{b_n}(\mathfrak{g}_\mathbb{Z}, \mathcal{K}^{(1)}, \mathbb{D}_\lambda^\epsilon \otimes \mathcal{M}_{\lambda,\mathbb{Z}}) \xrightarrow{\sim} \mathbb{Z}[i, \frac{1}{2}].$$

(See also Sect. 8.2.4.) We know that there is only a finite set of isomorphism classes of irreducible Harish-Chandra modules over $\mathbb{C}$ that have nontrivial cohomology with coefficients in $\mathcal{M}_\lambda \otimes \mathbb{C}$. If n is even (resp. n is odd) then there are only two (resp. is only one) $(\mathfrak{g}_\mathbb{Z}, \mathcal{K}^{(1)})$-module(s) that are tempered or that can be the infinite component of a cuspidal representation. (See for instance Sect. 3.1, [51], and [67].)

8.4.3 The construction of $\mathbb{D}^{\epsilon}_{\lambda}$

Consider the parabolic subgroup ${}^{\circ}P$ whose simple root system is described by the diagram

$$\circ - \times - \circ - \times - \cdots - \circ(-\times), \tag{8.50}$$

i.e., the set of simple roots $\Pi_{{}^{\circ}M}$ of the semisimple part of the Levi quotient ${}^{\circ}M$ consists of those simple roots that have an odd index. This Levi subgroup can be identified with

$$\prod_{i:\, i\text{odd}} H_{\alpha_i} = \prod \mathrm{GL}_2(\times\mathbb{G}_m); \tag{8.51}$$

i.e., each factor is identified to $\mathrm{GL}_2/\mathrm{Spec}(\mathbb{Z})$, with an extra factor of $\mathbb{G}_m$ if n is odd. We let ${}^{\circ}r$ stand for the number of GL_2 factors in ${}^{\circ}M$. Note that here we have chosen a splitting of the Levi quotient to a Levi subgroup, a splitting that is unique, since we want our Levi subgroup stable under the Cartan involution. Let ${}^{\circ}M^{(1)}$ be the semisimple component, we write as usual ${}^{\circ}M = {}^{\circ}M^{(1)} \cdot C_{{}^{\circ}M}$.

The standard maximal torus is a product $T = \prod_{i:i\text{odd}} T_i(\times\mathbb{G}_m)$ and for each $i = 1, 3, \ldots, m$, where m is the largest odd integer less than n, we have

$$X^*(T_i) \otimes \mathbb{Q} \;=\; \mathbb{Q}\gamma_i^{{}^{\circ}M^{(1)}} \oplus \mathbb{Q}\delta_2^{(i)}.$$

Here $\gamma_i^{{}^{\circ}M^{(1)}} = \frac{\alpha_i}{2}$ and $\delta_2^{(i)}$ is the determinant on that factor. For n odd let $\delta_1^{(n)}$ be the character that sends the last entry t_n to t_n. Let $\bar{B}_i \supset T_i$ be the standard Borel subgroup of upper triangular matrices and let $\bar{B} = \prod_{i:i\text{odd}} \bar{B}_i$ be our Borel subgroup of ${}^{\circ}M$. The $\gamma_i^{{}^{\circ}M^{(1)}}$ are the dominant fundamental weights with respect to the choice of $\bar{B}$.

We return to the conventions in Sect. 8.2 and apply our considerations in Sect. 8.3.2 to the factors H_{α_i}. The Cartan involution induces the Cartan involution on each of the factors H_{α_i}; the group scheme $\prod_i \mathcal{K}_i^{(1)} = T_c$ is a maximal torus in the reductive group $\mathcal{K}^{(1)}$ and its character module is $X^*(T_c \times \mathbb{Z}[i]) = \oplus_i \mathbb{Z}[i]e_i$. The Weyl group W_c of this torus acts on the character module by sending $e_i \mapsto \epsilon_i e_{\sigma(i)}$, where σ is any permutation and $\epsilon_i = \pm 1$ satisfying $\prod \epsilon_i = 1$ if n is even. Let $B_c \supset \mathcal{K}^{(1)} \times \mathbb{Z}[i]$ be the Borel subgroup of $\mathcal{K}^{(1)} \times \mathbb{Z}[i]$ which contains T_c and for which the roots $e_1 - e_3, \ldots, e_{m-2} - e_m, e_{m-2} + e_m$ are the simple positive roots.

We have a very specific Kostant representative $w_{\mathrm{un}} \in W^{{}^{\circ}P}$. The inverse of this permutation is given by

$$w_{\mathrm{un}}^{-1} = \{1 \mapsto 1, 2 \mapsto n, 3 \mapsto 2, 4 \mapsto n-1 \ldots.\}.$$

The length of this element is half the number of roots in the unipotent radical of ${}^\circ P$; i.e., we have

$$l(w_{\mathrm{un}}) = \begin{cases} \frac{1}{4}n(n-2) & \text{if } n \text{ is even} \\ \\ \frac{1}{4}(n-1)^2 & \text{if } n \text{ is odd.} \end{cases} \tag{8.52}$$

Furthermore, the twisted action of w_{un} on λ is given by

$$w_{\mathrm{un}}(\lambda+\rho)-\rho = \sum_{i:i \text{ odd}} b_i \gamma_i^{{}^\circ M^{(1)}} - (2\gamma_2 + 2\gamma_4 + \cdots + 2\gamma_{m-1} + \frac{3}{2}\gamma_{m+1}) + d\delta.$$

Here $\gamma_2, \gamma_4, \ldots$ are the dominant fundamental weights which have an even index and the b_i are the cuspidal parameters

$$b_{2j-1} = \begin{cases} 2a_j + 2a_{j+1} + \cdots + 2a_{\frac{n}{2}-1} + a_{\frac{n}{2}} + n - 2j & \text{if } n \text{ is even} \\ 2a_j + 2a_{j+1} + \cdots + 2a_{\frac{n-1}{2}} + n - 2j & \text{if } n \text{ is odd.} \end{cases}$$

We can express the abelian part in terms of determinants. We have

$$\begin{aligned} &-(2\gamma_2 + 2\gamma_4 + \cdots + 2\gamma_{m-1} + \frac{3}{2}\gamma_{m+1}) + d\delta \\ &= \begin{cases} \frac{n-2}{2}(\sum_1^n e_i) \\ \frac{n-1}{2}(\sum_1^n e_i) \end{cases} -2\Big((e_1+e_2)+(e_1+e_2+e_3+e_4)+\cdots+\begin{cases} e_1+\cdots+e_{n-2} \\ e_1+\cdots+e_{n-1} \end{cases}\Big)+d\delta_n \end{aligned}$$

and this is equal to

$$\begin{aligned} &\begin{cases} \frac{n-2}{2}(\sum_{i:i\ odd} \delta_2^{(i)}) \\ \frac{n-1}{2}((\sum_{i:i\ odd} \delta_2^{(i)}) + \delta_1^{(n)}) \end{cases} \\ &\qquad - 2\Big(\delta_2^{(1)} + (\delta_2^{(1)} + \delta_2^{(3)}) + \cdots + \begin{cases} \delta_2^{(1)} + \cdots + \delta_2^{(n-3)} \\ \delta_2^{(1)} + \cdots + \delta_2^{(n-2)} \end{cases}\Big) + d\delta_n, \end{aligned}$$

where, in the expression inside the bracket $((\delta_2^{(1)} + (\delta_2^{(1)} + \delta_2^{(3)})\ldots)$, the term $\delta_2^{(i)}$ occurs

$$c(i,n) = -\frac{i-1}{2} + \begin{cases} \frac{n-2}{2} \\ \frac{n-1}{2} \end{cases}$$

times. Then we can rewrite the expression for $w_{\mathrm{un}}(\lambda+\rho)-\rho$ as

$$w_{\mathrm{un}} \cdot \lambda = \sum_{i:i \text{ odd}} (b_i \gamma_i^{{}^\circ M^{(1)}} + (-2c(i,n)+d)\delta_2^{(i)}) + \begin{cases} 0 \\ (-\frac{n-1}{2}+d)\delta_n. \end{cases}$$

In simpler terms (for the n even case)

$$w_{\mathrm{un}}\cdot\lambda = b_1\gamma_1^{\circ M^{(1)}} + (-\frac{n-2}{2})\gamma_2 + b_3\gamma_3^{\circ M^{(1)}} + (2-\frac{n-2}{2})\gamma_4 + \dots$$
$$\dots + b_{n-1}\gamma_{n-1}^{\circ M^{(1)}} + (\frac{n-2}{2})\gamma_{n-2}.$$

In this formula the summands

$$\mu_i = b_i\gamma_i^{\circ M^{(1)}} + (-2c(i,n)+d)\delta_2^{(i)} \in X^*(T_i)$$

and $-\frac{n-1}{2}+d\in\mathbb{Z}$. The sum of positive roots in the i-th factor is $2\rho_i=\alpha_i= 2\gamma_i^{\circ M^{(1)}}$. We take the character

$$\mu_i+2\rho_i = (b_i+2)\gamma_i^{\circ M^{(1)}} + (-2c(i,n)+d)\delta_2^{(i)},$$

and apply Sect. 8.3.2 to construct the module $\mathfrak{Ind}_{B_i}^{M_i}(\mu_i+2\rho_i)$ which sits in an exact sequence

$$0\to\mathcal{D}_{\mu_i}\to\mathfrak{Ind}_{B_i}^{M_i}(\mu_i+2\rho_i)\to\mathcal{M}_{\mu_i,\mathbb{Z}}\to 0.$$

Let's put $\mu=w_{\mathrm{un}}\cdot\lambda$, which is a character on the maximal torus T, and we can define the induced module $\mathfrak{Ind}_{{}^\circ B}^{{}^\circ M}(\mu+2^\circ\rho)$. It is clear that this module is a tensor product

$$\mathfrak{Ind}_{{}^\circ B}^{{}^\circ M}(\mu+2^\circ\rho) = \bigotimes_{i:i\mathrm{odd}}\mathfrak{Ind}_{B_i}^{M_i}(\mu_i+2\rho_i)(\otimes\mathbb{Z}(-\frac{n-1}{2}+d)),$$

where the last factor is only there if n is odd. Then this module contains the submodule

$$\mathcal{D}_\mu = \bigotimes_{i:i\mathrm{odd}}\mathcal{D}_{\mu_i}(\otimes\mathbb{Z}(-\frac{n-1}{2}+d)) \hookrightarrow \bigotimes_{i:i\mathrm{odd}}\mathfrak{Ind}_{B_i}^{M_i}(\mu_i+2\rho_i)(\otimes\mathbb{Z}(-\frac{n-1}{2}+d)).$$

We know that $\mathcal{D}_{\mu_i}\otimes\mathbb{Z}[i,\frac{1}{2}]$ decomposes into the two submodules

$$\mathcal{D}_{\mu_i}\otimes\mathbb{Z}[i,\frac{1}{2}] = \mathcal{D}^+_{\mu_i}\otimes\mathbb{Z}[i,\frac{1}{2}] \oplus \mathcal{D}^-_{\mu_i}\otimes\mathbb{Z}[i,\frac{1}{2}].$$

Hence for any choice of signs we define the module

$$\mathcal{D}^{\underline{\epsilon}}_\mu = \bigotimes_{i:i\mathrm{odd}}\mathcal{D}^{\epsilon_i}_{\mu_i}(\otimes\mathbb{Z}(-\frac{n-1}{2}+d))$$

and the induced module

$$\mathbb{D}^{\underline{\epsilon}}_\lambda = \mathfrak{Ind}_{{}^\circ P}^{G}\mathcal{D}^{\underline{\epsilon}}_\mu. \tag{8.53}$$

The module $\mathcal{D}_\mu^{\underline{\epsilon}}$ has as minimal $\mathcal{K}^{\circ M}$ type the character

$$(\underline{\epsilon}, \mu + 2\rho) = \sum_{i:i\text{odd}} \epsilon_i(b_i + 2)e_i - (2\gamma_2 + 2\gamma_4 + \cdots + 2\gamma_{m-1} + \frac{3}{2}\gamma_{m+1}) + d\delta.$$

The $\mathbb{Z}[i, \frac{1}{2}]$ eigenmodule for this character is generated by

$$\mathcal{D}_\mu^{\underline{\epsilon}}(\underline{\epsilon}, \mu + 2\rho) = \mathbb{Z}[i, \frac{1}{2}] \bigotimes_{i:i\text{odd}} \Phi^{(i)}_{d,\epsilon_i(b_i+2)}. \tag{8.54}$$

So it comes with a canonical generator that we denote by

$$\bigotimes_{i:i\text{odd}} \Phi^{(i)}_{d,\epsilon_i(b_i+2)} = \Phi_{\mu,\underline{\epsilon}}. \tag{8.55}$$

The Weyl group W_c contains a subgroup that acts by sign changes on the generators, i.e., $e_i \mapsto \pm e_i$, it acts on the set of characters $(\underline{\epsilon}, \mu)$. We note that W_c acts transitively on this set if n is odd, and if n is even then we see easily that $(\underline{\epsilon}, \mu + 2\rho)$ and $(\underline{\epsilon}', \mu + 2\rho)$ are equivalent under the Weyl group W_c if and only if $\prod_{i:i\text{odd}} \epsilon_i = \prod_{i:i\text{odd}} \epsilon_i'$. Any of our characters $(\underline{\epsilon}, \mu + 2\rho)$ can be conjugated by an element in W_c into a dominant weight with respect to B_c, and an easy computation shows that these dominant weights are

$$\begin{cases} (\underline{1}, \mu + 2\rho) = \sum_{i:i\text{odd}}(b_i + 2)e_i & \text{if } n \text{ odd} \\ (\epsilon, \mu + 2\rho) = \sum_{i:i\text{odd}, i<m}(b_i + 2)e_i + \epsilon(b_m + 2)e_m & \text{if } n \text{ even,} \end{cases}$$

where in the second case ϵ can take either of the values $+1$ or -1. These weights are indeed dominant because $b_{m-2} > b_m$. For $\epsilon = \pm 1$ we define

$$\mathbb{D}_\lambda^\epsilon := \mathbb{D}_\lambda^{(1,1,\ldots,\epsilon)}. \tag{8.56}$$

We have the following proposition.

Proposition 8.5. *The $(\mathfrak{g}_\mathbb{Z}, \mathcal{K}^{(1)})$ modules $\mathbb{D}_\lambda^{\underline{\epsilon}}$ are irreducible. Two such modules are isomorphic if and only if $(\underline{\epsilon}, \mu)$ and $(\underline{\epsilon}', \mu)$ are conjugate under the Weyl group W_c. The module $\mathbb{D}_\lambda^\epsilon$ contains a minimal $\mathcal{K}^{(1)}$ type that has highest weight*

$$\mu_c(\epsilon, \lambda) = \sum_{i:iodd, i<m} (b_i + 2)e_i + \epsilon(b_m + 2)e_m,$$

where $\epsilon = 1$ if n is odd and $\epsilon = \pm 1$ if n is even. This minimal $\mathcal{K}^{(1)}$ type occurs with multiplicity one.

Proof. For the irreducibility we tensor by $\mathbb{C}$ and refer to [51] and [67]. Any element in the Weyl group W_c can be represented by an element $w \in G(\mathbb{Z})$ that normalizes $T_c = \mathcal{K}^{\circ M}$. Then the multiplication from the left by w induces an isomorphism of $(\mathfrak{g}_\mathbb{Z}, \mathcal{K}^{(1)})$ modules:

$$(\mathcal{D}^{\epsilon}_{\underline{\mu}} \otimes A(\mathcal{K}^{(1)}))^{\mathcal{K}^{\circ M}} \xrightarrow{\sim} (\mathcal{D}^{w\epsilon}_{w\underline{\mu}} \otimes A(\mathcal{K}^{(1)}))^{\mathcal{K}^{\circ M}}. \tag{8.57}$$

We prove the assertion concerning the $\mathcal{K}^{(1)}$-types. We have a decomposition of $\mathcal{D}^{\epsilon}_{\mu}$ into T_c-types

$$\mathcal{D}^{\epsilon}_{\mu} = \bigoplus_{k_1 \geq 0, \ldots, k_m \geq 0} \mathbb{Z}[i, \frac{1}{2}]((b_1 + 2 + 2k_1)e_1 + \cdots + \epsilon(b_m + 2 + 2k_m)e_m). \tag{8.58}$$

The character $(b_1 + 2 + 2k_1)e_1 + (b_3 + 2 + 2k_3)e_3 + \cdots + \epsilon(b_m + 2 + 2k_m)e_m$ may not be in the positive chamber (k_m may be too large) but we can conjugate it to $\mu_c(\underline{k})$ in the positive chamber. For this character it is easy to see that

$$\mu_c(\underline{k}) = (b_1 + 2)e_1 + (b_3 + 2)e_3 + \cdots + \epsilon(b_m + 2)e_m + \sum_i m_i \alpha_{i,c},$$

where the $m_i \geq 0$. The highest weight $\mu_c(\epsilon, \lambda) = \mu_c(\underline{0})$. Now we have the classical formula that

$$(\mathbb{Z}[i, \frac{1}{2}](\mu_c(\underline{k})) \otimes A(\mathcal{K}^{(1)}))^{\mathcal{K}^M} \otimes \mathbb{Q} = \bigoplus_{\vartheta} A(\mathcal{K}^{(1)} \otimes \mathbb{Z}[i, \frac{1}{2}])(\mu_c(\underline{k}), \vartheta) \otimes \mathbb{Q},$$

where ϑ runs over the isomorphism classes of irreducible $\mathcal{K}^{(1)} \otimes \mathbb{Z}[i, \frac{1}{2}]$ modules and where

$$A(\mathcal{K}^{(1)} \otimes \mathbb{Z}[i, \frac{1}{2}])(\mu_c(\underline{k})) = \{f | R_{(t^{-1}, e)} f = \mu_c(\underline{k}))f \text{ for all } t \in T_c(R_1)\}.$$

Then it is well known that the multiplicity of ϑ in $A(\mathcal{K}^{(1)} \otimes \mathbb{Z}[i, \frac{1}{2}])(\mu_c(\underline{k}))$ is equal to the multiplicity of $\mu_c(\underline{k})$ in ϑ. We get that the representation $\vartheta_{\mu_c(\epsilon,\lambda)}$ occurs with multiplicity one.

Notice that our argument also implies that for a given ϑ, the number of those $\underline{k}$ for which $\mu_c(\underline{k})$ occurs in ϑ with positive multiplicity is finite. This settles the condition (2) in the definition of Harish-Chandra modules in Sect. 8.2.1 for $\mathfrak{Ind}_P^G \mathcal{D}^{\epsilon}_{\mu}$ but the argument works in the general case too. □

8.4.4 The cohomology $H^\bullet(\mathfrak{g}_\mathbb{Z}, \mathcal{K}^{(1)}, \mathbb{D}^{\epsilon}_{\lambda} \otimes \mathcal{M}_{\lambda, \mathbb{Z}[i, \frac{1}{2}]})$

We define as usual the $(\mathfrak{g}_\mathbb{Z}, \mathcal{K}^{(1)})$-cohomology as the cohomology of the complex

$$\mathrm{Hom}_{\mathcal{K}^{(1)}}(\Lambda^\bullet(\mathfrak{g}_\mathbb{Z}/\mathfrak{k}_\mathbb{Z}), \mathbb{D}^{\epsilon}_{\lambda} \otimes \mathcal{M}_{\lambda,\mathbb{Z}}).$$

If we tensor by $\mathbb{C}$ then we know that $\mathbb{D}^\epsilon_\lambda \otimes \mathbb{C}$ is unitary and since $\mathcal{M}_\lambda \otimes \mathbb{C}$ is dual to its conjugate, it follows that all the differentials in the above complex are trivial; i.e., the complex is equal to its cohomology.

We apply the Delorme method (or Frobenius reciprocity). Let ${}^\circ\mathfrak{m}_{\mathbb{Z}}$ be the Lie algebra of ${}^\circ M$, and let ${}^\circ\mathfrak{m}^{(1)}_{\mathbb{Z}}$ be the Lie algebra of ${}^\circ M^{(1)}$. Let $\mathfrak{u}_{\mathbb{Z}}$ be the Lie algebra of the unipotent radical of ${}^\circ P$ and finally let $\mathfrak{c}_{\mathbb{Z}}$ be the Lie algebra of $C_{{}^\circ M}$. Then

$$\mathfrak{g}_{\mathbb{Z}[\frac{1}{2}]}/\mathfrak{k}_{\mathbb{Z}[\frac{1}{2}]} \;=\; {}^\circ\mathfrak{m}^{(1)}_{\mathbb{Z}[\frac{1}{2}]}/{}^\circ\mathfrak{k}_{\mathbb{Z}[\frac{1}{2}]} \;\oplus\; \mathfrak{c}_{\mathbb{Z}[\frac{1}{2}]} \;\oplus\; \mathfrak{u}_{\mathbb{Z}[\frac{1}{2}]},$$

where now the right-hand side is a $({}^\circ\mathfrak{m}, \mathcal{K}^{{}^\circ M})$–module. The group scheme $\mathcal{K}^{{}^\circ M}$ acts by the adjoint action. It acts trivially on $\mathfrak{c}_{\mathbb{Z}}$, and the adjoint action of $\mathcal{K}^{{}^\circ M}$ on $\mathfrak{u}_{\mathbb{Z}}$ extends to the adjoint action of ${}^\circ M$. (Remember that ${}^\circ M$ is a subgroup of ${}^\circ P$.) We get an isomorphism of complexes

$$\begin{aligned}&\mathrm{Hom}_{\mathcal{K}^{(1)}}(\Lambda^\bullet(\mathfrak{g}_{\mathbb{Z}}/\mathfrak{k}_{\mathbb{Z}}), \mathbb{D}^\epsilon_\lambda \otimes \mathcal{M}_{\lambda,\mathbb{Z}})\\ &\quad = \mathrm{Hom}_{\mathcal{K}^{{}^\circ M}}(\Lambda^\bullet({}^\circ\mathfrak{m}^{(1)}_{\mathbb{Z}}/{}^\circ\mathfrak{k}_{\mathbb{Z}}), \mathcal{D}^\epsilon_\mu \otimes \mathrm{Hom}(\Lambda^\bullet(\mathfrak{u}_{\mathbb{Z}}), \mathcal{M}_{\lambda,\mathbb{Z}}) \otimes \Lambda^\bullet(\mathfrak{c}_{\mathbb{Z}}). \quad (8.59)\end{aligned}$$

In the following we concentrate on the $\Lambda^0(\mathfrak{c}_{\mathbb{Z}})$ component. We claim that the $\mathbb{Z}[i, \frac{1}{2}]$ module

$$\mathrm{Hom}_{\mathcal{K}^{{}^\circ M}}(\Lambda^\bullet({}^\circ\mathfrak{m}^{(1)}_{\mathbb{Z}[i,\frac{1}{2}]}/{}^\circ\mathfrak{k}_{\mathbb{Z}[i,\frac{1}{2}]}), \mathcal{D}^\epsilon_\mu \otimes \mathrm{Hom}(\Lambda^\bullet(\mathfrak{u}_{\mathbb{Z}[i,\frac{1}{2}]}), \mathcal{M}_{\lambda,\mathbb{Z}[i,\frac{1}{2}]}))$$

is free of rank one. To be more precise: we decompose the three $\mathcal{K}^{{}^\circ M}$-modules $\Lambda^\bullet({}^\circ\mathfrak{m}^{(1)}_{\mathbb{Z}[i,\frac{1}{2}]}/{}^\circ\mathfrak{k}_{\mathbb{Z}[i,\frac{1}{2}]}), \mathcal{D}^\epsilon_\mu$, and $\mathrm{Hom}(\Lambda^\bullet(\mathfrak{u}_{\mathbb{Z}[i,\frac{1}{2}]}), \mathcal{M}_{\lambda,\mathbb{Z}[i,\frac{1}{2}]})$ into eigenspaces with respect to characters in $X^*(\mathcal{K}^{{}^\circ M} \otimes \mathbb{Z}[i, \frac{1}{2}])$ and show that there is exactly one triple of characters that contributes nontrivially to the $\mathrm{Hom}_{\mathcal{K}^{{}^\circ M}}$, i.e., that satisfies $\eta_{\mathfrak{m}} = \eta_{\mathcal{D}} + \eta_{\mathfrak{u}}$.

The module $\Lambda^{{}^\circ r}({}^\circ\mathfrak{m}^{(1)}_{\mathbb{Z}[i,\frac{1}{2}]}/{}^\circ\mathfrak{k}_{\mathbb{Z}[i,\frac{1}{2}]})$ (recall that ${}^\circ r$ is the number of GL_2-factors in ${}^\circ M$) contains the submodule

$$\bigoplus_{\underline{\epsilon}} \mathbb{Z}[i, \frac{1}{2}] P_1^{\epsilon_1} \wedge P_3^{\epsilon_3} \wedge \cdots \wedge P_m^{\epsilon_m}$$

and on the individual summand our torus $T_c \cdot C_{{}^\circ M}$ acts by characters

$$\nu(\underline{\epsilon}) = 2(\epsilon_1 e_1 + \cdots + \epsilon_m e_m).C_{{}^\circ M}.$$

We choose for $\underline{\epsilon}$ the value $\underline{\epsilon}_0 = (+, +, \ldots, \epsilon)$, hence $\Lambda^{{}^\circ r}({}^\circ\mathfrak{m}^{(1)}_{\mathbb{Z}[i,\frac{1}{2}]}/{}^\circ\mathfrak{k}_{\mathbb{Z}[i,\frac{1}{2}]})$ contains the direct summand

$$\Lambda^{{}^\circ r}({}^\circ\mathfrak{m}^{(1)}_{\mathbb{Z}[i,\frac{1}{2}]}/{}^\circ\mathfrak{k}_{\mathbb{Z}[i,\frac{1}{2}]})(\nu(\underline{\epsilon}_0)) = \mathbb{Z}[i, \frac{1}{2}](\nu(\underline{\epsilon}_0)).$$

The character $\nu(\underline{\epsilon}_0)$ will be our $\eta_{\mathfrak{m}}$. The action of $C_{^\circ M}$ on $\Lambda^{^\circ r}({}^\circ\mathfrak{m}^{(1)}_{\mathbb{Z}[i,\frac{1}{2}]}/{}^\circ\mathfrak{k}_{\mathbb{Z}[i,\frac{1}{2}]})$ is trivial.

The module $\mathcal{D}^\epsilon_\mu$ contains the submodule $\mathcal{D}^\epsilon_\mu(\mu+2\rho,\underline{\epsilon}_0)$ with multiplicity one, hence

$$\mathcal{D}^\epsilon_\mu(\mu+2\rho,\underline{\epsilon}_0)=\mathbb{Z}[i,\frac{1}{2}]\Phi_{\mu,\epsilon}\ \subset\ \mathcal{D}^\epsilon_\mu.$$

The center $C_{^\circ M}$ acts on $\mathcal{D}^\epsilon_\mu(\mu+2\rho,\underline{\epsilon}_0)$ by the character

$$-\zeta(\mu)\ =\ (2\gamma_2+2\gamma_4+\cdots+2\gamma_{m-1}+\frac{3}{2}\gamma_{m+1})-d\delta. \tag{8.60}$$

Finally, let's investigate the structure of $\mathrm{Hom}(\Lambda^\bullet(\mathfrak{u}_{\mathbb{Z}[i,\frac{1}{2}]}),\mathcal{M}_{\lambda,\mathbb{Z}[i,\frac{1}{2}]})$. Conjugation by the product $\tilde{c}=\prod_{i:i\mathrm{odd}}c_{2,i}$, where the matrices $c_{2,i}$ are as in (8.12), provides an identification

$$\tilde{c}:X^*(\mathcal{K}^{^\circ M}\otimes\mathbb{Z}[i,\frac{1}{2}])\ \to\ X^*(T).$$

Note that $\tilde{c}(e_i)=\gamma_i^{^\circ M^{(1)}}$ and for even indices i we have $\tilde{c}(\gamma_i)=\gamma_i$.

This suggests that we consider $\mathrm{Hom}(\Lambda^\bullet(\mathfrak{u}_{\mathbb{Z}}),\mathcal{M}_{\lambda,\mathbb{Z}})$ as a module for ${}^\circ M$ and we even restrict our attention to the action of the center $C_{^\circ M}$. We have the following proposition (see [46]).

Proposition 8.6. *The character $\zeta(\mu)$ occurs only in degree $l(w_{un})$ and the eigenspace $\mathrm{Hom}(\Lambda^{l(w_{un})}(\mathfrak{u}_{\mathbb{Z}}),\mathcal{M}_{\lambda,\mathbb{Z}})(\zeta(\mu))$ is an irreducible module for ${}^\circ M$ with highest weight $w_{\mathrm{un}}(\lambda+\rho)-\rho$. The homomorphism*

$$\mathrm{Hom}(\Lambda^{l(w_{un})}(\mathfrak{u}_{\mathbb{Z}}),\mathcal{M}_{\lambda,\mathbb{Z}})(\zeta(\mu))\ \to\ H^{l(w_{un})}(\mathfrak{u}_{\mathbb{Z}},\mathcal{M}_{\lambda,\mathbb{Z}})$$

is an isomorphism.

Proof. The Lie algebra $\mathfrak{u}_{\mathbb{Z}}$ has the basis e_β, where $\beta\in\Delta^+\backslash\{\alpha_1,\alpha_3,\ldots,\alpha_m\}$. Let us denote by $e^\vee_\beta$ the dual basis. For any subset $J=\{\beta_1,\beta_2,\ldots,\beta_s\}\subset\Delta^+_{^\circ P}$ we define $e^\vee_J=e^\vee_{\beta_1}\wedge e^\vee_{\beta_2}\wedge\ldots e^\vee_{\beta_s}$. The element $e^\vee_J$ is an eigenvector for the standard maximal torus $T\subset{}^\circ M$; the eigenvalue is the character $\chi_J=-\sum\beta_i$. For any Kostant representative $w\in W^{^\circ P}$ we define the set $\Delta^+(w)=\{\alpha|w^{-1}\alpha<0\}$. Then we know that the restriction of $w_{\mathrm{un}}(\lambda+\rho)-\rho=J_{\Delta^+(w)}$ to $C_{^\circ M}$ is $\zeta(\mu)$. The weight $J_{\Delta^+(w)}$ is the highest weight of an irreducible ${}^\circ M$ submodule $\mathcal{N}$ in $\mathrm{Hom}(\Lambda^{l(w_{\mathrm{un}})}(\mathfrak{u}_{\mathbb{Z}}),\mathcal{M}_{\lambda,\mathbb{Z}})(\zeta(\mu))$ and the weight subspaces in $\mathcal{N}$ are of multiplicity one and of the form $\chi_{J'}$. Now a simple computation shows that a subset $J_1\subset\Delta^+_{^\circ P}$ for which the restriction of χ_{J_1} to $C_{^\circ M}$ is equal to $\zeta(\mu)$ must be one of the $\chi_{J'}$ occurring in $\mathcal{N}$ and hence it follows that $\mathrm{Hom}(\Lambda^{l(w_{\mathrm{un}})}(\mathfrak{u}_{\mathbb{Z}}),\mathcal{M}_{\lambda,\mathbb{Z}})(\zeta(\mu))$ is irreducible. □

This implies that

$$\mathrm{Hom}_{\mathcal{K}^{\circ M}}(\Lambda^\bullet({}^\circ\mathfrak{m}^{(1)}_{\mathbb{Z}[i,\frac{1}{2}]}/{}^\circ\mathfrak{k}_{\mathbb{Z}[i,\frac{1}{2}]}), \mathcal{D}^\epsilon_\mu \otimes \mathrm{Hom}(\Lambda^\bullet(\mathfrak{u}_\mathbb{Z}/\mathfrak{k}_Z), \mathcal{M}_{\lambda,\mathbb{Z}})$$
$$= \mathrm{Hom}_{\mathcal{K}^{\circ M}}(\Lambda^\bullet({}^\circ\mathfrak{m}^{(1)}_{\mathbb{Z}[i,\frac{1}{2}]}/{}^\circ\mathfrak{k}_{\mathbb{Z}[i,\frac{1}{2}]}), \mathcal{D}^\epsilon_\mu \otimes H^{l(w_{\mathrm{un}})}(\mathfrak{u}_\mathbb{Z}, \mathcal{M}_{\lambda,\mathbb{Z}}).$$

If we choose a generator x_λ of the highest weight module $\mathcal{M}_{\lambda,\mathbb{Z}}(\lambda)$ then

$$\xi(w_{\mathrm{un}}\cdot\lambda) \;:=\; e^\vee_{\Delta^+(w_{\mathrm{un}})} \otimes w_{\mathrm{un}} x_\lambda \;\in\; H^{l(w_{\mathrm{un}})}(\mathfrak{u}_\mathbb{Z}, \mathcal{M}_{\lambda,\mathbb{Z}})$$

is a generator of the highest weight module $H^{l(w_{\mathrm{un}})}(\mathfrak{u}_\mathbb{Z}, \mathcal{M}_{\lambda,\mathbb{Z}})(w_{\mathrm{un}}\cdot\lambda)$ which is unique up to a sign. We can modify our Borel subgroup $\bar{B} \subset {}^\circ M$ by flipping into the opposite in some of the factors. Then the highest weight with respect to such a Borel subgroup will be

$$\lambda(w_{\mathrm{un}}, \underline{\epsilon}) = \sum_{i:i \text{ odd}} \epsilon_i b_i \gamma_i^{{}^\circ M^{(1)}} - \zeta(\mu), \tag{8.61}$$

where of course again $\epsilon_i = \pm 1$ and the indices i with $\epsilon_i = -1$ tell us where we flipped the Borel subgroup. To such a weight we have a generating weight vector $\xi(w_{\mathrm{un}}\cdot\lambda, \underline{\epsilon})$. Let us call these weight vectors $\xi(w_{\mathrm{un}}\cdot\lambda, \epsilon)$ the extremal weight vectors.

We replace the split torus T by $\tilde{\mathcal{K}}^{\circ M}$. These two tori have $C_{{}^\circ M}$ in common. Then we see that that we have extremal weight spaces

$$H^{l(w_{\mathrm{un}})}(\mathfrak{u}_{\mathbb{Z}[i,\frac{1}{2}]}, \mathcal{M}_{\lambda,\mathbb{Z}[i,\frac{1}{2}]})(w_{\mathrm{un}}\cdot\lambda)(\tilde{c}^{-1}(\lambda(w_{\mathrm{un}},\underline{\epsilon}))) \;=\; \mathbb{Z}[i,\frac{1}{2}]\tilde{c}^{-1}(\xi(w_{\mathrm{un}}\cdot\lambda,\underline{\epsilon}))$$

and on this weight space the torus $\tilde{\mathcal{K}}^{\circ M}$ acts by the character

$$\tilde{c}^{-1}(\lambda(w_{\mathrm{un}}, -\underline{\epsilon}_0)) = -b_1e_1 - b_3e_3 - \dots \epsilon_m b_m e_m + \zeta(\mu).$$

The module $\mathcal{D}^\epsilon_\mu$ contains the rank one submodule $\mathcal{D}^\epsilon_\mu(\mu_c(\lambda,\epsilon)) = \mathbb{Z}[i,\frac{1}{2}]\Phi_{\mu,\underline{\epsilon_0}}$ on which $\tilde{\mathcal{K}}^{\circ M}$ acts by the character

$$\mu_c(\epsilon,\lambda) \;=\; \sum_{i:i\,\text{odd},\, i<m} (b_i+2)e_i + \epsilon(b_m+2)e_m - \zeta(\mu).$$

Indeed, this is a minimal $\mathcal{K}^{\circ M}$ type (see (8.58)). The sum of these two characters is $\nu(\underline{\epsilon}_0) = 2e_1 + 2e_2 \cdots + \epsilon 2e_m$ and hence we see

$$\mathrm{Hom}_{\mathcal{K}^{\circ M}}(\Lambda^\bullet({}^\circ\mathfrak{m}^{(1)}_{\mathbb{Z}[i,\frac{1}{2}]}/{}^\circ\mathfrak{k}_{\mathbb{Z}[i,\frac{1}{2}]}), \mathcal{D}^\epsilon_\mu \otimes \mathrm{Hom}(\Lambda^\bullet(\mathfrak{u}_{\mathbb{Z}[i,\frac{1}{2}]}), \mathcal{M}_\lambda))$$

is given by

$$\mathrm{Hom}_{\mathcal{K}^{\circ M}}\Big(\Lambda^\bullet({}^\circ\mathfrak{m}^{(1)}_{\mathbb{Z}[i,\frac{1}{2}]}/{}^\circ\mathfrak{k}_{\mathbb{Z}[i,\frac{1}{2}]})(\nu(\underline{\epsilon}_0))\,,\quad \mathcal{D}^\epsilon_\mu(\mu_c(\epsilon,\lambda))\otimes\mathrm{Hom}(\Lambda^\bullet(\mathfrak{u}_{\mathbb{Z}[i,\frac{1}{2}]}),\mathcal{M}_\lambda)(\tilde{c}^{-1}(\lambda(w_{\mathrm{un}},-\underline{\epsilon}_0))\Big)\,.$$

Each of the modules in the argument on the right-hand side is of rank one and we have chosen a generator for each of them. Hence we see

$$\mathrm{Hom}_{\mathcal{K}^{\circ M}}(\Lambda^\bullet({}^\circ\mathfrak{m}^{(1)}_{\mathbb{Z}[i,\frac{1}{2}]}/{}^\circ\mathfrak{k}_{\mathbb{Z}[i,\frac{1}{2}]}),\mathcal{D}^\epsilon_\mu\otimes\mathrm{Hom}(\Lambda^\bullet(\mathfrak{u}_{\mathbb{Z}[i,\frac{1}{2}]}),\mathcal{M}_\lambda) = \mathbb{Z}[i,\frac{1}{2}]\Omega(\lambda,\epsilon), \tag{8.62}$$

where $\Omega(\lambda,\epsilon)$ is the tensor product of the generators and $\epsilon=\pm1$. The generator sits in degree $b_n = {}^\circ r + l(w_{\mathrm{un}})$. (Note that $b_n=[n^2/4]$.)

If n is odd then the choice of $\underline{\epsilon}$ is irrelevant. If n is even we get two irreducible $(\mathfrak{g}_\mathbb{Z},\mathcal{K}^{(1)})$ modules. As we did in the case $G=\mathrm{GL}_2$ we can enlarge the connected group scheme $\mathcal{K}^{(1)}$ to the larger group scheme $\mathcal{K}=\mathcal{K}^{(1)}\ltimes\{\eta\}$, where η is the diagonal matrix which has 1 on the diagonal up to the $(n-1)$-th entry and has -1 in the n-th entry. Then $\mathbb{D}_\lambda=\mathbb{D}^{(+1)}_\lambda\oplus\mathbb{D}^{(-1)}_\lambda$ is an irreducible $(\mathfrak{g}_\mathbb{Z},\mathcal{K})$ module over $\mathbb{Z}[i,\frac{1}{2}]$, and the element η yields an isomorphism between the two summands. Since our weight λ is essentially self-dual, i.e., $a_i=a_{n-i}$, we have the constraint $a_{\frac{n}{2}}\equiv 2d \mod 2$. Then it is clear that

$$\eta\Omega(\lambda,\epsilon) = (-1)^{\frac{a_{\frac{n}{2}}-2d}{2}}\Omega(\lambda,-\epsilon).$$

Once again we consider the elements

$$\omega^{(1)}_\lambda = \Omega(\lambda,+1)+\Omega(\lambda,-1),\quad \omega^{(2)}_\lambda = \Omega(\lambda,+1)-\Omega(\lambda,-1), \tag{8.63}$$

and then these two elements are the generators for the $\pm$ eigenspaces under the action of η. We have the decomposition

$$H^\bullet(\mathfrak{g}_\mathbb{Z},\mathcal{K}^{(1)},\mathbb{D}_\lambda\otimes\mathcal{M}_{\lambda,\mathbb{Z}[\frac{1}{2}]}) = H^\bullet_+(\mathfrak{g}_\mathbb{Z},\mathcal{K}^{(1)},\mathbb{D}_\lambda\otimes\mathcal{M}_{\lambda,\mathbb{Z}[\frac{1}{2}]})\oplus H^\bullet_-(\mathfrak{g}_\mathbb{Z},\mathcal{K}^{(1)},\mathbb{D}_\lambda\otimes\mathcal{M}_{\lambda,\mathbb{Z}[\frac{1}{2}]})$$

and

$$H^\bullet(\mathfrak{g}_\mathbb{Z},\mathcal{K},\mathbb{D}_\lambda\otimes\mathcal{M}_{\lambda,\mathbb{Z}[\frac{1}{2}]}) \xrightarrow{\sim} H^\bullet_+(\mathfrak{g}_\mathbb{Z},\mathcal{K}^{(1)},\mathbb{D}_\lambda\otimes\mathcal{M}_{\lambda,\mathbb{Z}[\frac{1}{2}]}).$$

Note that the cohomology $H^{b_n}(\mathfrak{g}_\mathbb{Z},\mathcal{K}^{(1)},\mathbb{D}_\lambda\otimes\mathcal{M}_{\lambda,\mathbb{Z}[\frac{1}{2}]})$ is the free $\mathbb{Z}[\frac{1}{2}]$-module on the basis $\{\omega^{(1)}_\lambda,\omega^{(2)}_\lambda\}$.

8.4.5 The arithmetic of the intertwining operator

In this section we apply the foregoing considerations to study the arithmetic properties of an intertwining operator between two induced modules.

We start from the group scheme $\mathrm{GL}_N/\mathbb{Z}$; let B be the standard Borel subgroup and consider the standard parabolic subgroups $P \supset B$ (resp., $P' \supset B$) with reductive quotient $M = \mathrm{GL}_n \times \mathrm{GL}_{n'} =: M_1 \times M_2$ (resp., $M' = \mathrm{GL}_{n'} \times \mathrm{GL}_n$). Let U_P (resp., $U_{P'}$) be the unipotent radical. Let $\Pi = \{\alpha_1, \ldots, \alpha_{N-1}\} \subset X^*(T)$ be the set of positive (with respect to B) simple roots. We identify the set of simple roots with the set of indices $\{1, 2, \ldots, N-1\}$. Let us denote by w_N^- the permutation that reverses the order by $i \mapsto i'$, i.e., $i' = N - i$. Then the positive simple roots for M are $\Pi_M = \{\alpha_1, \ldots, \alpha_{n-1}\} \cup \{\alpha_{n+1}, \ldots, \alpha_{N-1}\}$, and accordingly we denote the system of simple roots of M' by $\Pi_{M'} = w_N^-(\Pi_M) = \{\alpha_1, \ldots, \alpha_{n'-1}\} \cup \{\alpha_{n'+1}, \ldots, \alpha_{N-1}\}$. Let γ_n (resp., $\gamma_{n'}$) be the fundamental weight attached to the missing root α_n (resp., $\alpha_{n'}$). If $\Delta^+_{U_P}$ (resp., $\Delta^+_{U_{P'}}$) are the positive roots occurring in these radicals, let ρ_{U_P} (resp., $\rho_{U_{P'}}$) be the half sum over these roots. Then

$$\rho_{U_P} = \frac{N}{2}\gamma_n, \quad \rho_{U_{P'}} = \frac{N}{2}\gamma_{n'}.$$

We choose a highest weight λ for GL_N, and let $\mathcal{M}_\lambda$ be the resulting GL_N-module. Pick a Kostant representative $w \in W^P$, and we write

$$\begin{aligned}\tilde{\mu} &= w(\lambda + \rho) - \rho = \sum_{i=1}^{n-1} a'_i \gamma_i^M + \sum_{i=n+1}^{N-1} a'_i \gamma_i^M + a(w, \lambda)\gamma_n + d \det{}_N \\ &= \sum_{i=1}^{n-1} a'_i \gamma_i^M + d(w, \lambda) \det{}_n + \sum_{i=n+1}^{N-1} a'_i \gamma_i^M + d'(w, \lambda) \det{}_{n'} = \mu_1 + \mu_2. \end{aligned} \tag{8.64}$$

Here μ_1 (resp., μ_2) is a highest weight on M_1 (resp., M_2). In some situations it is more convenient to look at

$$\tilde{\mu} + \rho = w(\lambda + \rho) = \sum_{i=1}^{n-1} b_i \gamma_i^M + \sum_{i=n+1}^{N-1} b_i \gamma_i^M + b(w, \lambda)\gamma_n + d \det{}_N.$$

Then we have the relations $b_i = a'_i + 1$ and $b(w, \lambda) = a(w, \lambda) + \frac{N}{2}$. We define the weight of $\tilde{\mu}$:

$$\mathbf{w}(\tilde{\mu}) = \mathbf{w}(\mu_1) + \mathbf{w}(\mu_2) = \sum_{i=1}^{n-1} b_i + \sum_{i=n+1}^{N-1} b_i.$$

We make certain assumptions on w :

(a) The length $l(w) = 1/2 \dim U_P$- this means that w is balanced.

(b) Both weights μ_1, μ_2 are essentially self-dual.

(c) The weight $\tilde{\mu}$ is in the negative chamber, which means that $a(w, \lambda) \leq -\frac{N}{2}$ or $b(w, \lambda) \leq 0$.

We have the longest Kostant representative $w_P \in W^P$ (resp., $w_Q \in W^Q$) which sends all the roots in $\Delta^+_{U_P}$ (resp., $\Delta^+_{U_{P'}}$) into negative roots. If $s_i \in W$ is the reflection attached to the simple root α_i then we can write any element $w \in W^P$ as a product of reflections $w = s_n s_i s_j \ldots s_k$. We can always complete this product of reflections to get the longest element (it always starts with s_n and stops with $s_{n'}$):

$$s_n s_i s_j \ldots s_k s_{k'} \ldots s_{i'} s_{n'} \;=\; w_P. \tag{8.65}$$

Then $w' = s_{n'} \ldots s_{k'} \in W^Q$ and we get a one-to-one correspondence between W^P and W^Q that is defined by

$$w = w_P w' \text{ or } w' = w_Q w. \tag{8.66}$$

We have $l(w) + l(w') = \dim U_P$, and since w is balanced we see that w' is also balanced. For $w = e$ the identity element we get $w' = w_Q$. (See Lem. 5.6.)

A presentation of w_P as in (8.65) yields a sequence of roots in $\Delta^+_{U_P}$: The first element in this sequence is $\beta_1 = \alpha_n$. Then we find a root $\beta_2 \in \Delta^+_{U_P}$ such that $s_n \beta_2 = \alpha_i$ is a simple root. Then $s_i s_n$ sends the roots β_1 and β_2 into the set of negative roots and we find a root β_3 such that $s_n s_i \beta_3$ is a simple root α_ν and s_ν is the next factor in $w_P = s_n s_i s_\nu \ldots$. To say this in different words: We get an ordering $\{\beta_1, \beta_2, \ldots, \beta_{d_U}\} = \Delta^+_{U_P}$ such that x_k^{-1} conjugates exactly the first k roots $\{\beta_1, \ldots, \beta_k\}$ into negative roots where

$$x_k \;=\; s_n s_i \ldots s_\mu \text{ is the product of the first } k \text{ factors in (8.65).}$$

We can write

$$w_P \;=\; x_1 y_{d_U - 1} \;=\; s_n(s_i \cdots s_{n'}) \;=\; x_2 y_2 \;=\; (s_n s_i)(s_j \cdots s_{n'}) \;=\; x_k y_k$$

and the x_k, y_k are corresponding elements. We also define a function

$$r_w : \{1, 2, \ldots, , d_U\} \;\to\; \{1, 2, \ldots, N-1\} \;=\; \Pi \;=\; \{\alpha_1, \ldots, \alpha_{N-1}\}$$

by the rule

$$x_k \;=\; x_{k-1} s_{r_w(k)} \quad \text{or} \quad x_{k-1} \alpha_{r_w(k)} = \beta_k.$$

For our element w above the corresponding element w' also has length

$\frac{1}{2}\dim_{U_P}$ and we have the corresponding formula

$$\begin{aligned}\tilde{\mu}' &= w'(\lambda+\rho) = \sum_{i=1}^{n'-1} b_{i'}\gamma_{i'}^{M'} + \sum_{i=n'+1}^{N-1} b_{i'}\gamma_{i'}^{M'} + b'(w',\lambda)\gamma_{n'} + d(w',\lambda)\det{}_N \\ &= \sum_{i=n+1}^{N-1} b_{i'}\gamma_{i'}^{M'} + d(w',\lambda)\mathrm{det}_{n'} + \sum_{i=n'+1}^{N-1} b_{i'}\gamma_{i'}^{M'} + d(w',\lambda)\mathrm{det}_n = \mu_1' + \mu_2'.\end{aligned}$$

The formula tells us that the semisimple components $\mu_1'^{(1)} = \mu_2^{(1)}$ and $\mu_2'^{(1)} = \mu_1^{(1)}$. Here we use our assumption that μ_i are essentially self-dual. The coefficients of γ_n and $\gamma_{n'}$ are related by

$$d(w,\lambda) + d(w',\lambda) = -N.$$

The weights μ_1 and μ_2 yield Harish-Chandra modules $\mathbb{D}_{\mu_1}$ (for GL_n) and $\mathbb{D}_{\mu_2}$ for $\mathrm{GL}_{n'}$. (See Sect. 8.4.3.) Hence $\tilde{\mu} = \mu_1 + \mu_2$ provides a Harish-Chandra module $\mathbb{D}_{\tilde{\mu}}$ for M. By the same argument we get a Harish-Chandra module $\mathbb{D}_{\tilde{\mu}'}$ for M'. This $(\mathfrak{m}, \mathcal{K}^M)$ (resp., $(\mathfrak{m}', \mathcal{K}^{M'})$) module has a minimal $\mathcal{K}^M$ type $\mu_c(\epsilon, \tilde{\mu}) = \mu_c(\epsilon_1, \mu_1) + \mu_c(\epsilon_2, \mu_2)$ (resp., $\mathcal{K}^{M'}$ type $\mu_c(\epsilon', \tilde{\mu}')$).

Consider the $(\mathfrak{g}_{\mathbb{Z}}, \mathcal{K}^{(1)})$ modules $\mathfrak{Ind}_P^G\mathbb{D}_{\tilde{\mu}}$ and $\mathfrak{Ind}_Q^G\mathbb{D}_{\tilde{\mu}'}$. Both these modules contain the irreducible module $\vartheta_{\mu_c(\epsilon,\tilde{\mu})}$ as a minimal $\mathcal{K}^{(1)}$ type that appears with multiplicity one. We take the base extensions to $\mathbb{C}$ and twist them by a holomorphic variable. For brevity, let's introduce the notation

$$\mathfrak{Ind}_P^G\mathbb{D}_{\tilde{\mu}} \otimes (z) := \mathfrak{Ind}_P^G\mathbb{D}_{\tilde{\mu}} \otimes \mathbb{C}(|\gamma_n|^z).$$

Then we can write down the standard intertwining operator

$$T^{w_P,\mathrm{st}}(z) : \mathfrak{Ind}_P^G\mathbb{D}_{\tilde{\mu}} \otimes (z) \longrightarrow \mathfrak{Ind}_Q^G\mathbb{D}_{\tilde{\mu}'} \otimes (-z) \tag{8.67}$$

given by the integral

$$\{g \mapsto f(g)\} \mapsto \{g \mapsto \int_{U_P(\mathbb{R})} f(w_P u g) d_\infty u\}$$

where $d_\infty u$ is the Haar measure obtained from the epinglage. This integral converges if $\Re(z) \gg 0$. The action of $\mathcal{K}^{(1)}(\mathbb{R})$ on these modules is independent of z. In particular, it is clear that the above lowest $\mathcal{K}^{(1)}(\mathbb{R})$-type occurs in the deformed modules with multiplicity one. The main theorem of this chapter is the following result.

Theorem 8.7. *(i) This operator extends to a meromorphic operator in the entire z-plane and it is holomorphic at $z = 0$.*

(ii) At $z = 0$ it is an isomorphism and modified by the factor $\frac{1}{\pi^{d_U/2}}$ it is defined over $\mathbb{Q}$; i.e., we get an isomorphism

$$\frac{1}{\pi^{d_U/2}} T^{w_P,\mathrm{st}}(0) : \mathfrak{Ind}_P^G \mathbb{D}_{\tilde{\mu}} \otimes \mathbb{Q} \xrightarrow{\sim} \mathfrak{Ind}_Q^G D_{\tilde{\mu}'} \otimes \mathbb{Q}. \tag{8.68}$$

Proof. The first assertion is over $\mathbb{C}$ and follows from results of Speh [67]. For the second we use the standard strategy and write the operator as a product of operators induced from intertwining operators on GL_2. Here we have to deal with the problem that some of the operators will not be defined because we encounter poles in the backwards operators.

We apply the consideration from Sect. 8.4.2) to M and M', i.e., to the two factors GL_n and $\mathrm{GL}_{n'}$. In particular, we introduce the subgroup scheme ${}^\circ M \subset M$ as the product of the two corresponding groups in the two factors. This also yields the element w^M_{un} in the Weyl group of M. The module $\mathbb{D}_{\tilde{\mu}}$ is induced from a module $\mathcal{D}_{w^M_{un}\cdot\tilde{\mu}+2\rho_{\circ M}}$ and this module is defined as a submodule from $\mathfrak{Ind}_{\bar{B}}^{{}^\circ M} w^M_{\mathrm{un}} \cdot (\tilde{\mu} + 2\rho_{\circ M})$. Hence we get $\mathbb{D}_{\tilde{\mu}} \subset \mathfrak{Ind}_{B_M}^M w^M_{\mathrm{un}} \cdot \tilde{\mu} + 2\rho_{\circ M}$ and finally

$$\mathfrak{Ind}_P^G \mathbb{D}_{\tilde{\mu}} \hookrightarrow \mathfrak{Ind}_B^G\, w^M_{\mathrm{un}} \cdot (\tilde{\mu} + 2\rho_{\circ M}).$$

Now we can try to extend the intertwining operator $T^{w_P,\mathrm{st}}(0)$ to this larger module. This may not be possible. Therefore we apply the usual technique and deform the induced module by a character $|\underline{\gamma}|^{\underline{z}} = \prod_{i=1}^{N-1} |\gamma_i|^{z_i}$ where $\underline{z} \in \mathbb{C}^{N-1}$. We consider the extension

$$\begin{aligned} T^{w_P,\mathrm{st}}(\underline{z}) : \mathfrak{Ind}_B^G w^M_{\mathrm{un}} \cdot (\tilde{\mu} + 2\rho_{\circ M})_{\mathbb{R}} \otimes |\underline{\gamma}|^{\underline{z}} \\ \to \mathfrak{Ind}_B^G w^M_{\mathrm{un}} \cdot (\tilde{\mu}' + 2\rho_{\circ M'})_{\mathbb{R}} \otimes |\underline{\gamma}|^{\underline{z}'}, \end{aligned} \tag{8.69}$$

which is given by the following integral: We write U_P as the product of one-parameter subgroups (note that $\alpha_{n'} = \beta_{d_U}, \alpha_n = \beta_1$):

$$U_P = U_{\alpha_{n'}} \times U_{\beta_{d_U-1}} \times \cdots \times U_{\alpha_n}.$$

Then we may write the operator as

$$\begin{aligned} T^{w_P,\mathrm{st}}(\underline{z})(f)(g) &= \int_{U(\mathbb{R})} f(w_P u g) du \\ &= \int \cdots \int f(w_P u_{\alpha_{n'}} \ldots u_{\alpha_n} g) du_{\alpha_{n'}} \cdots du_{\alpha_n}, \end{aligned}$$

where the measure is the Lebesgue measure which is normalized by the epinglage. This integral converges for $\Re(z_i) \gg 0$ and has a meromorphic extension into the entire complex plane.

We have by definition

$$w_P u_{\alpha_{n'}} u_1 = s_n u_{\alpha_n} y_{d_U - 1} u_1,$$

where $u_1 \in U_1(\mathbb{R}) = U_{\beta_{d_U-1}} \times \cdots U_{\alpha_{i'}}$. Hence our integral becomes

$$T^{w_P,\mathrm{st}}(\underline{z})(f)(g) = \int_{U_1(\mathbb{R})} \int_{U_{\alpha_n}(\mathbb{R})} f(s_n u_{\alpha_n} y_{d_U-1} u_1 g) du_{\alpha_n} du_1. \tag{8.70}$$

The inner integral is an intertwining operator. We write our induced modules now as induced from characters χ on the $\mathbb{R}$ valued points of the torus $T(\mathbb{R})$, and let $\chi = w_{\mathrm{un}} \cdot (\tilde{\mu} + 2\rho_{\circ M})_{\mathbb{R}})$. Then

$$\{g \mapsto f(g)\} \mapsto \{g \mapsto \int_{U_{\alpha_n}(\mathbb{R})} f(s_n u_{\alpha_n} g) du_\alpha\} \tag{8.71}$$

is an intertwining operator

$$T^{\mathrm{st}}(s_n, \chi, \underline{z}) : I_B^G \chi \otimes |\underline{\gamma}|^{\underline{z}} \longrightarrow I_B^G(s_n \cdot (\chi \otimes |\underline{\gamma}|^{\underline{z}})), \tag{8.72}$$

where, of course, $s_n \cdot (\chi \otimes |\underline{\gamma}|^{\underline{z}}) = s_n(\chi \otimes |\underline{\gamma}|^{\underline{z}}) + s_n(|\rho|) - |\rho|$.

This intertwining operator is now induced from an intertwining operator between two SL_2 modules. Let $\tilde{H}_{\alpha_n}$ be the reductive subgroup $H_{\alpha_n} \cdot T$. The group scheme $\tilde{H}_{\alpha_n}$ is then the Levi quotient of a parabolic subgroup P_{α_n}. Let B_{α_n} be the Borel subgroup in $\tilde{H}_{\alpha_n}$. Then the integral in (8.71) also defines an intertwining operator

$$T^{\mathrm{st}}_{\alpha_n}(s_n, \chi, \underline{z}) : I_{B_{\alpha_n}}^{\tilde{H}_{\alpha_n}} \chi \otimes |\underline{\gamma}|^{\underline{z}} \longrightarrow I_{B_{\alpha_n}}^{\tilde{H}_{\alpha_n}}(s_n \cdot (\chi \otimes |\underline{\gamma}|^{\underline{z}})).$$

Our two induced modules can by written as two-step induction

$$I_B^G \chi \otimes |\underline{\gamma}|^{\underline{z}} = I_{P_{\alpha_n}}^G I_{B_{\alpha_n}}^{\tilde{H}_{\alpha_n}}(\chi \otimes |\underline{\gamma}|^{\underline{z}}),$$

$$I_B^G s_n \cdot (\chi \otimes |\underline{\gamma}|^{\underline{z}}) = I_{P_{\alpha_n}}^G I_{B_{\alpha_n}}^{\tilde{H}_{\alpha_n}}(s_n \cdot (\chi \otimes |\underline{\gamma}|^{\underline{z}})),$$

and then our intertwining operator is induced

$$T^{\mathrm{st}}(s_n, \chi, \underline{z}) = I_{P_{\alpha_n}}^G T^{\mathrm{st}}_{\alpha_n}(s_n, \chi, \underline{z}).$$

Hence we can write (8.70) as

$$T^{w_P,\mathrm{st}}(\underline{z})(f)(g) = \int_{U_1(\mathbb{R})} I_{P_{\alpha_n}}^G (T^{\mathrm{st}}_{\alpha_n}(s_n, \chi, \underline{z})(f)(y_{d_U-1} u_1 g) du_1.$$

We iterate this process. We have $y_{d_U-1} = s_i s_j \cdots s_{n'}$, we repeatedly apply the above process to eventually get

$$T^{w_P,\mathrm{st}}(\underline{z})(f) = I^G_{P_{\alpha_{n'}}} T^{\mathrm{st}}_{\beta_{d_U-1}}(s_{n'}, x^{-1}_{d_U-1}\cdot\chi, x^{-1}_{d_U-1}\underline{z})\circ\dots$$
$$\dots\circ I^G_{P_{r_w(k)}} T^{\mathrm{st}}_{r_w(k)}(s_{r_w(k)}, x^{-1}_{k-1}\cdot\chi, x^{-1}_{k-1}(\underline{z}))\circ\dots\circ I^G_{P_{\alpha_n}} T^{\mathrm{st}}_{\alpha_n}(s_n,\chi,\underline{z})(f). \quad (8.73)$$

We understand the GL_2 intertwining operators

$$T^{\mathrm{st}}_{r_w(k)}(s_{r_w(k)}, x^{-1}_{k-1}\underline{z})) : I^{\tilde{H}_{r_w(k)}}_{B_{r_w(k)}}(x^{-1}_{k-1}\cdot\chi\otimes|\underline{\gamma}|^{\underline{z}}) \to I^{\tilde{H}_{r_w(k)}}_{B_{r_w(k)}}(x^{-1}_{k}\cdot\chi\otimes|\underline{\gamma}|^{\underline{z}}).$$

In Sect. 8.3.3 we defined the algebraic intertwining operator $T^{\mathrm{alg}}_{\lambda^-}$ by fixing their value on the lowest $\mathcal{K}$ type. If we put $\chi=\lambda^-_{\mathbb{R}}$ then we can extend these operators to the twisted modules

$$T^{\mathrm{alg}}(s_\alpha,\chi,z) : I^{\mathrm{GL}_2}_{B_\alpha}\chi\otimes|\gamma_\alpha|^z \to I^{\mathrm{GL}_2}_{B_\alpha} s_\alpha\cdot(\chi\otimes|\gamma_\alpha|^z).$$

With respect to the basis Φ_ν the operator $T^{alg}(s_\alpha,\chi,z)$ acts as a diagonal matrix with entries in the rational function field $\mathbb{Q}(\underline{z})$ and we know the factor that compares $T^{\mathrm{alg}}(s_\alpha,\chi,z)$ to $T^{\mathrm{st}}(s_\alpha,\chi,z)$, it is a ratio of Γ-values. To compute $T_{r_w(k)}(s_{r_w(k)}, x^{-1}_{k-1}\cdot\chi)$ we need to know the restriction of $x^{-1}_{k-1}\cdot\chi$ to the torus $T_{r_w(k)}$, recalling that the coroot $\alpha^\vee_{r_w(k)}:\mathbb{G}_m\to T_{r_w(k)}$ provides an identification. Hence the restriction of $x^{-1}_{k-1}\cdot\chi$ to $T_{r_w(k)}$ is a character on $\mathbb{G}_m(\mathbb{R})=\mathbb{R}^\times$ and an easy computation shows that this character is

$$t \mapsto t^{\langle\alpha^\vee_{r_w(k)}, x_{k-1}\chi\rangle}\,|t|^{\langle\alpha^\vee_{r_w(k)}, x^{-1}_{k-1}\rho-\rho\rangle+\langle\alpha^\vee_{r_w(k)},\underline{z}\rangle}.$$

We still can manipulate the exponent. We have $x_{k-1}\alpha^\vee_{r_w(k)} = \beta^\vee_k$. Then the first exponent becomes $\langle\beta^\vee_k,\chi\rangle$ and for the second one we get $\langle\beta^\vee-\alpha^\vee_{r_w(k)},\rho\rangle = h(\beta_k)$, where for $\beta=\alpha_\nu+\dots+\alpha_{\nu+h}$ we put $h(\beta)=h$. Hence our character is

$$t \mapsto t^{\langle\beta^\vee_k,\chi\rangle}|t|^{h(\beta_k)+\langle\alpha^\vee_{r_w(k)},\underline{z}\rangle}.$$

Then we put $\epsilon_w(k,\chi)=0$ if $\langle\beta^\vee_k,\chi\rangle\equiv 0 \mod 2$ and $\epsilon_w(k,\chi)=1$ else. Then we get from our formulae in Sect. 8.3.4

$$T^{\mathrm{st}}_{r_w(k)}(s_{r_w(k)}, x^{-1}_{k-1}\cdot\chi, x^{-1}_{k-1}\underline{z})$$
$$= \frac{\Gamma(\frac{\langle\beta^\vee_k,\chi\rangle+\epsilon_w(k,\chi)+h(\beta_k)-1+\langle\alpha^\vee_{r_w(k)},\underline{z}\rangle}{2})}{\Gamma(\frac{\langle\beta^\vee_k,\chi\rangle+\epsilon_w(k,\chi)+h(\beta_k)+\langle\alpha^\vee_{r_w(k)},\underline{z}\rangle}{2})}\Gamma(\frac{1}{2})M_k(\underline{z}) \quad (8.74)$$

where $M_k(\underline{z})$ is a diagonal matrix with entries in the field of rational functions

$\mathbb{Q}(\underline{z})$. It may have a pole at the hyperplane $\langle \alpha^\vee_{r_w(k)}, \underline{z}\rangle = 0$ but the ratio

$$M_k^*(\underline{z}) = \frac{M_k(\underline{z})}{\Gamma(\frac{\langle \beta_k^\vee, \chi\rangle + \epsilon_w(k,\chi) + h(\beta_k) + \langle \alpha^\vee_{r_w(k)}, \underline{z}\rangle}{2})}$$

is holomorphic on this hyperplane and hence can be evaluated at $\underline{z} = 0$.

By definition the number $\langle \beta_k^\vee, \chi\rangle + \epsilon_w(k,\chi)$ is even and hence this operator is holomorphic at $\underline{z} = 0$ if $h(\beta_k)$ is even. In this case the character restricted to the torus $T_{r_w(k)}$ is cohomological. We find

$$T^{\mathrm{st}}_{r_w(k)}(s_{r_w(k)}, x_{k-1}^{-1}\cdot\chi, x_{k-1}^{-1}\underline{z})|_{\underline{z}=0} = \Gamma(\frac{\langle \beta_k^\vee, \chi\rangle + \epsilon_w(k,\chi) + h(\beta_k) - 1}{2})\Gamma(\frac{1}{2})M_k^*(0), \quad (8.75)$$

where $M_k^*(0)$ is a matrix with rational entries and the factor in front is $\boldsymbol{\pi}$ times a rational number. This tells us that

$$T^{\mathrm{st}}_{r_w(k)}(s_{r_w(k)}, x_{k-1}^{-1}\cdot\chi)|_{\underline{z}=0} = \boldsymbol{\pi}\times T^{alg}_{r_w(k)}(s_{r_w(k)}, x_{k-1}^{-1}\cdot\chi)q_k, \quad (8.76)$$

where $q_k \in \bar{\mathbb{Q}}^\times$.

If $h(\beta_k)$ is odd then the hyperplane $\langle \alpha^\vee_{r_w(k)}, \underline{z}\rangle = 0$ may be a first-order pole, that happens exactly when

$$\langle \beta_k^\vee, \chi_{alg}\rangle + \epsilon_w(k,\chi) + h(\beta_k) - 1 = 0, -2, -4, \ldots.$$

We put $m_k = 1$ if $h(\beta_k)$ is odd and we encounter a pole and $m_k = 0$ else. Then we manipulate the right-hand side in (8.74) and change it to

$$\langle \alpha_{r_w(k)}, \underline{z}\rangle^{m_k} \frac{\Gamma(\frac{\langle \beta_k^\vee, \chi_{alg}\rangle + \epsilon_w(k,\chi) + h(\beta_k) - 1 + \langle \alpha^\vee_{r_w(k)}, \underline{z}\rangle}{2})}{\Gamma(\frac{\langle \beta_k^\vee, \chi_{alg}\rangle + \epsilon_w(k,\chi) + h(\beta_k) + \langle \alpha^\vee_{r_w(k)}, \underline{z}\rangle}{2})}\Gamma(\frac{1}{2})\frac{M_k(\underline{z})}{\langle \alpha_{r_w(k)}, \underline{z}\rangle^{m_k}}.$$

The last factor to the right is still a diagonal matrix with entries in the field $\mathbb{Q}(\underline{z})$. The expression in values of the Γ-function can be evaluated at $\underline{z} = 0$. The result is a rational number, and the two contributions of $\sqrt{\boldsymbol{\pi}}$ cancel.

We return to our factorization of the intertwining operator $T^{w_P,\mathrm{st}}(\underline{z})$. It is an intertwining operator between two Harish-Chandra modules with a $\mathbb{Q}$-structure. They have a decomposition into $\mathcal{K}^{(1)}$ types (which are of course the base-change to $\mathbb{C}$ of $\mathbb{Q}$ vector spaces). We consider the restriction to a $\mathcal{K}^{(1)}$ type ϑ. Then our product decomposition yields

$$(\prod_k \frac{\langle \alpha_{r_w(k)}, \underline{z}\rangle^{m_k}\Gamma(\frac{\langle \beta_k^\vee, \chi\rangle + \epsilon_w(k,\chi) + h(\beta_k) - 1 + \langle \alpha^\vee_{r_w(k)}, \underline{z}\rangle}{2})}{\Gamma(\frac{\langle \beta_k^\vee, \chi\rangle + \epsilon_w(k,\chi) + h(\beta_k) + \langle \alpha^\vee_{r_w(k)}, \underline{z}\rangle}{2})}\Gamma(\frac{1}{2}))M(\vartheta, \underline{z}),$$

where $M(\vartheta, \underline{z})$ is in

$$\mathrm{Hom}_{\mathcal{K}^{(1)}}(\mathfrak{Ind}_B^G w_{un}^M \cdot (\mu + 2\rho_{\circ M})(\vartheta), \mathfrak{Ind}_B^G w_{\mathrm{un}}^{M'} \cdot (\mu' + 2\rho_{\circ M'})(\vartheta)) \otimes \mathbb{Q}(\underline{z}).$$

The factor in front can be evaluated at $\underline{z} = 0$. Each factor contributes by a nonzero rational number or $\boldsymbol{\pi}$ times a nonzero rational number. We get a factor $\boldsymbol{\pi}$ in the cases where $h(\beta_k)$ is even and this happens $d_U/2$ number of times. So we would be finished with the proof if we evaluate $M(\vartheta, \underline{z})$ at $\underline{z} = 0$ and observe that this is a matrix with entries rational numbers. But we do not know whether $M(\vartheta, \underline{z})$ can be evaluated at zero; we have moved the poles in the Γ-factors into $M(\vartheta, \underline{z})$, and the extended intertwining operator in (8.69) may not be regular at $\underline{z} = 0$.

We are interested only in the restriction of the operator to $\mathfrak{Ind}_P^G \mathbb{D}_{\tilde{\mu}} \otimes \mathbb{C}[|\gamma_n|^z]$ (see (8.67)); i.e., we restrict it to the line $z|\gamma_n| \subset \mathbb{C}^{N-1}$. We notice that this line is not contained in any of the hyperplanes $\langle \alpha^\vee_{r_w(k)}, \underline{z} \rangle = k \in \mathbb{Z}$. Hence we see that the operator $M(\vartheta, z)$ is a meromorphic function in the variable z.

The modules $\mathfrak{Ind}_P^G \mathbb{D}_{\tilde{\mu}}$ and $\mathfrak{Ind}_Q^G \tilde{\mu}'$ contain the special irreducible $\mathcal{K}^{(1)}$-module $\mathcal{X}[\tilde{\mu}]$ with highest weight $\mu_c(\epsilon, \tilde{\mu})$ with multiplicity one. This $\mathcal{K}^{(1)}$ module occurs with higher multiplicity t in $\mathfrak{Ind}_B^G w_{\mathrm{un}}^M \cdot (\tilde{\mu} + 2\rho_{\circ M}) \otimes \mathbb{C}|\gamma_n|^z$. Restriction to this $\mathcal{K}^{(1)}$ type yields a diagram

$$\begin{array}{ccc}
\mathcal{X}[\tilde{\mu}] \otimes (z) & \xrightarrow{T^{w_P, \mathrm{st}}(z)} & \mathcal{X}[\tilde{\mu}] \otimes (-z) \\
\downarrow & & \downarrow \\
(\mathcal{X}[\tilde{\mu}])^t \otimes (z) & \xrightarrow{T^{w_P, \mathrm{st}}(z)} & (\mathcal{X}[\tilde{\mu}])^t \otimes (-z),
\end{array}$$

where the down arrows are the inclusion by the first coordinate. Then the matrix $M(\vartheta_{\mu_c(\epsilon,\tilde{\mu})}, z)$ will be a $t \times t$ matrix with entries $C_{l,m}(\vartheta_{\mu_c(\epsilon,\tilde{\mu})}, z) \in \mathbb{Q}(z)$. We look at the first row, which tells us what happens to the first coordinate under $T^{w_P, \mathrm{st}}(z)$. This first row is $(C_{1,1}(\vartheta_{\mu_c(\epsilon,\tilde{\mu})}, z), 0 \ldots, 0)$. The rational function $(C_{1,1}(\vartheta_{\mu_c(\epsilon,\tilde{\mu})}, z) \in \mathbb{Q}(z)$ is regular at $z = 0$. Therefore we can evaluate the first row at $z = 0$. The result will be $(C_{1,1}(\vartheta_{\mu_c(\epsilon,\tilde{\mu})}, 0), 0, \ldots, 0)$ with $C_{1,1}(\vartheta_{\mu_c(\epsilon,\tilde{\mu})}, 0) \in \mathbb{Q}^\times$. □

It is clear that the operator $\frac{1}{\pi^{\frac{d_U}{2}}} T^{w_P, \mathrm{st}}(0)$ induces an isomorphism in cohomology

$$T^{w_P, \bullet} : H^\bullet(\mathfrak{g}_\mathbb{Z}, \mathcal{K}, \mathfrak{Ind}_P^G \mathbb{D}_{\tilde{\mu}} \otimes \mathcal{M}_\lambda) \otimes \mathbb{Q} \to H^\bullet(\mathfrak{g}_\mathbb{Z}, \mathcal{K}, \mathfrak{Ind}_Q^G \mathbb{D}_{\tilde{\mu}'} \otimes \mathcal{M}_\lambda) \otimes \mathbb{Q}. \quad (8.77)$$

We can compute these cohomology groups using Delorme's method. Let $\mathcal{K}^{(1),M} \subset$

M be the connected maximal definite group scheme. Then

$$H^\bullet(\mathfrak{g}_\mathbb{Z}, \mathcal{K}, \mathfrak{Ind}_P^G \mathbb{D}_{\tilde\mu} \otimes \mathcal{M}_\lambda) \hookrightarrow H^\bullet_+(\mathfrak{m}, \mathcal{K}^{(1),M}, \mathbb{D}_{\tilde\mu} \otimes \mathcal{M}_{w_{\mathrm{un}}\cdot\lambda}) \oplus H^\bullet_-(\mathfrak{m}, \mathcal{K}^{(1),M}, \mathbb{D}_{\tilde\mu} \otimes \mathcal{M}_{w_{\mathrm{un}}\cdot\lambda})$$

and the same holds for the Q part:

$$H^\bullet(\mathfrak{g}_\mathbb{Z}, \mathcal{K}, \mathfrak{Ind}_Q^G \mathbb{D}_{\tilde\mu'} \otimes \mathcal{M}'_\lambda) \hookrightarrow H^\bullet_+(\mathfrak{m}, \mathcal{K}^{(1),M'}, \mathbb{D}_{\tilde\mu'} \otimes \mathcal{M}_{w_{\mathrm{un}}\cdot\lambda}) \oplus H^\bullet_-(\mathfrak{m}, \mathcal{K}^{(1),M'}, \mathbb{D}_{\tilde\mu'} \otimes \mathcal{M}_{w_{\mathrm{un}}\cdot\lambda}),$$

and the inclusion is always an isomorphism to the + component. Now we remember that $M = M_1 \times M_2$ and one of the factors is a GL_n with n even, and the other is $\mathrm{GL}_{n'}$. We will consider the situation when n' is odd (leaving the very similar details when n' is even to the reader). Then in the lowest degree the cohomology $H^\bullet_+(\mathfrak{m}, \mathcal{K}^{(1),M}, \mathbb{D}_\mu \otimes \mathcal{M}_{w_{\mathrm{un}}\cdot\lambda})$ is generated by an element $\omega_{\mu_1}^{(e)} \otimes \omega_{\mu_2}$ where $e = 1$ or 2. The cohomology $H^\bullet_+(\mathfrak{m}, \mathcal{K}^{(1),M'}, \mathbb{D}_{\tilde\mu'} \otimes \mathcal{M}_{w'_{\mathrm{un}}\cdot\lambda})$ is generated by $\omega_{\mu'_2} \otimes \omega_{\mu'_1}^{(e')}$ where $1' = 2, 2' = 1$. (This introduces the relative period.) Then we get

$$T^{w_P, b_n + b_{n'}}(\omega_{\mu_1}^{(e)} \otimes \omega_{\mu_2}) = C_{1,1}(\vartheta_{\mu_c(\epsilon,\tilde\mu)}, 0)\omega_{\mu'_2} \otimes \omega_{\mu'_1}^{(e')}.$$

This rationality result is used in Thm. 7.25 to prove a rationality result for ratios of critical values of Rankin–Selberg L-functions at consecutive critical arguments. We recall that we can attach a local L-function $L_\infty^{\mathrm{Coh}}(\mathbb{D}_{\tilde\mu}, s)$ to the Harish-Chandra module $\mathbb{D}_{\tilde\mu}$. Then we can rewrite the above formula into

$$T^{w_P,\mathrm{st}}(0)(\omega_{\mu_1}^{(e)} \otimes \omega_{\mu_2}) = c_\infty(\tilde\mu) \frac{L_\infty^{\mathrm{Coh}}(\mathbb{D}_{\tilde\mu}, \mathbf{w}(\tilde\mu) + b(w,\lambda))}{L_\infty^{\mathrm{Coh}}(\mathbb{D}_{\tilde\mu}, \mathbf{w}(\tilde\mu) + b(w,\lambda) + 1)} (\omega_{\mu'_2} \otimes \omega_{\mu'_1}^{(e')}). \quad (8.78)$$

The local L-function can be expressed in terms of products of the functions $\Gamma_\mathbb{C}(z) = 2(2\pi)^{-s}\Gamma(s)$, and using this expression we find

$$\frac{L_\infty^{\mathrm{Coh}}(\mathbb{D}_{\tilde\mu}, \mathbf{w}(\tilde\mu) + b(w,\lambda))}{L_\infty^{\mathrm{Coh}}(\mathbb{D}_{\tilde\mu}, \mathbf{w}(\tilde\mu) + b(w,\lambda) + 1)} = \frac{\pi^{d_U/2}}{\prod N_i(w,\tilde\mu)}, \quad (8.79)$$

where $N_i(w,\tilde\mu)$ are certain integers. The combinatorial lemma implies that under the given conditions the L-values are critical and hence $N_i(w,\tilde\mu) \neq 0$. Our rationality result is then equivalent to the assertion that $c_\infty(\tilde\mu) \in \mathbb{Q}^\times$. It enters in the proof of the main theorem of this monograph, Thm. 7.21.

8.4.6 An intriguing question

It is certainly possible to prove the necessary rationality result at infinity with somewhat lesser effort. In Sect. 5.2, the relative period is defined after we make some choices of basis vectors in various vector spaces (most of the time one-dimensional). Then the computation of the intertwining operator comes down to seeing its effect on these basis elements and this computation can be carried out quite directly, which is carried out in Chap. 9.

We develop the concept of Harish-Chandra modules over $\mathbb{Z}$ because this gives us some motivation for the choice of these basis elements. But we can get more profit out of it. We have seen that the cohomology modules

$$H^\bullet(\mathfrak{g}_\mathbb{Z}, \mathcal{K}, \mathfrak{Ind}_P^G \mathbb{D}_{\tilde{\mu}} \otimes \mathcal{M}_\lambda) \otimes \mathbb{Z}[\frac{1}{2}], \quad H^\bullet(\mathfrak{g}_\mathbb{Z}, \mathcal{K}, \mathfrak{Ind}_Q^G \mathbb{D}_{\mu'} \otimes \mathcal{M}_\lambda) \otimes \mathbb{Z}[\frac{1}{2}] \tag{8.80}$$

in lowest degree are free of rank one and the intertwining multiplied by $1/\pi^{d_U/2}$ induces an isomorphism if we tensor by $\mathbb{Q}$. But we may also consider the slightly modified operator

$$\tilde{T}^{w_P}(\tilde{\mu}) = \frac{L_\infty^{\mathrm{Coh}}(\mathbb{D}_{\tilde{\mu}}, \mathbf{w}(\tilde{\mu}) + b(w,\lambda) + 1)}{L_\infty^{\mathrm{Coh}}(\mathbb{D}_{\tilde{\mu}}, \mathbf{w}(\tilde{\mu}) + b(w,\lambda))}) T^{w_P,\mathrm{st}}(0), \tag{8.81}$$

which also induces an isomorphism between the two modules after we tensor them by $\mathbb{Q}$. We ask the

Question 8.82. Is the modified operator

$$\tilde{T}^{w_P}(\tilde{\mu}) : H^\bullet(\mathfrak{g}_\mathbb{Z}, \mathcal{K}, \mathfrak{Ind}_P^G \mathbb{D}_{\tilde{\mu}} \otimes \mathcal{M}_\lambda) \otimes \mathbb{Z}[\frac{1}{2}] \;\to\; H^\bullet(\mathfrak{g}_\mathbb{Z}, \mathcal{K}, \mathfrak{Ind}_Q^G \mathbb{D}_{\tilde{\mu}'} \otimes \mathcal{M}_\lambda) \otimes \mathbb{Z}[\frac{1}{2}]$$

an isomorphism?

This is of course equivalent to the assertion $c_\infty(\tilde{\mu}) \in \mathbb{Z}[\frac{1}{2}]^\times$. At the time of writing this chapter, the only nontrivial case where we knew that this is true was the case $N = 3$ (see [74]). In the meanwhile, this question has been answered by Weselmann in Chap. 9, and a similar question is also discussed in Harder [32].

8.4.7 Refining the relative periods

In Thm. 7.21, the rationality result involves a certain relative period $\Omega(\tilde{\sigma}_f)$. This period is basically a nonzero complex number that is defined modulo $E^\times$, where E is a number field over which $\tilde{\sigma}_f$ is defined. We will show here that we can make this choice of periods more precise so that they are essentially defined modulo the units $\mathcal{O}_E^\times$. This allows us to speak of the prime factorization of the ratios of critical values (divided by the period) and this is of great arithmetic interest.

Assume n even and $G = \mathrm{GL}_n/\mathbb{Z}$, for simplicity we assume $F = \mathbb{Q}$. We consider the inner cohomology $H^\bullet_!(\mathcal{S}^G_{K_f}, \mathcal{M}_{\lambda,\mathbb{Z}})$. This is a finitely generated $\mathbb{Z}$ module and denote $H^\bullet_{!,\,\mathrm{int}}(\mathcal{S}^G_{K_f}, \mathcal{M}_{\lambda,\mathbb{Z}})$ for its quotient by torsion. We have an action of the integral Hecke algebra on these cohomology groups (see [30]). If we extend $\mathbb{Q}$ to finite extension $E/\mathbb{Q}$ and tensor by the ring of integers $\mathcal{O}_E$ then we get a decomposition up to isogeny

$$H^\bullet_{!,\,\mathrm{int}}(\mathcal{S}^G_{K_f}, \mathcal{M}_{\lambda,\mathcal{O}_E}) \supset \bigoplus_{\pi_f \in \mathrm{Coh}(G,\lambda,K_f)} H^\bullet_{!,\,\mathrm{int}}(\mathcal{S}^G_{K_f}, \mathcal{M}_{\lambda,\mathcal{O}_E})(\pi_f). \tag{8.83}$$

We have the action of $\pi_0(\mathrm{GL}_n(\mathbb{R}))$ on these cohomology groups giving

$$H^\bullet_{!,\,\mathrm{int}}(\mathcal{S}^G_{K_f}, \mathcal{M}_{\lambda,\mathcal{O}_E[\frac{1}{2}]})(\pi_f)(+) \oplus H^\bullet_{!,\,\mathrm{int}}(\mathcal{S}^G_{K_f}, \mathcal{M}_{\lambda,\mathcal{O}_E[\frac{1}{2}]})(\pi_f)(-).$$

If π_f is strongly inner and if we tensor by $\mathbb{Q}$ the two summands become isomorphic $\mathcal{H}^G_{K_f}$ modules. If S is a finite set of primes containing the primes where K_f is ramified then $\mathcal{H}^{G,S}_{K_f} = \prod_{p \notin S} \mathcal{H}^G_{K_p}$ is a central subalgebra of $\mathcal{H}^G_{K_f}$. We say that π_f is weakly split by E if the restriction of π_f to $\mathcal{H}^{G,S}_{K_f}$ is a homomorphism $\psi^S(\pi_f) : \mathcal{H}^{G,S}_{K_f} \to \mathcal{O}_E$, i.e., the eigenvalues of the Hecke operators at places outside S lie in $\mathcal{O}_E$. We define

$$\begin{aligned} &H^\bullet_{!,\,\mathrm{int}}(\mathcal{S}^G_{K_f}, \mathcal{M}_{\lambda,\mathcal{O}_E[\frac{1}{2}]})(\psi(\pi_f), \epsilon) \\ &\quad = \{\xi \in H^\bullet_{!,\,\mathrm{int}}(\mathcal{S}^G_{K_f}, \mathcal{M}_{\lambda,\mathcal{O}_E[\frac{1}{2}]})(\epsilon) \mid h\xi = \psi(\pi_f)(h)\xi \text{ for all } h \in \mathcal{H}^{G,S}_{K_f}\}. \end{aligned}$$

The isomorphism type π_f is uniquely determined by $\psi(\pi_f)$ (strong multiplicity one for GL_n), and it is absolutely irreducible. More precisely, we have

$$H^\bullet_{!,\,\mathrm{int}}(\mathcal{S}^G_{K_f}, \mathcal{M}_{\lambda,\mathcal{O}_E[\frac{1}{2}]})(\pi_f, \epsilon) = H^\bullet_{!,\,\mathrm{int}}(\mathcal{S}^G_{K_f}, \mathcal{M}_{\lambda,\mathcal{O}_E[\frac{1}{2}]})(\psi(\pi_f), \epsilon). \tag{8.84}$$

If we define $E(\pi_f) \subset E$ to be the subfield of E generated by the values $\psi(\pi_f)(h)$ (it is independent of the choice of S) then

$$\begin{aligned} &H^\bullet_{!,\,\mathrm{int}}(\mathcal{S}^G_{K_f}, \mathcal{M}_{\lambda,\mathcal{O}_E[\frac{1}{2}]})(\pi_f)(\epsilon) \\ &\qquad = H^\bullet_{!,\,\mathrm{int}}(\mathcal{S}^G_{K_f}, \mathcal{M}_{\lambda,\mathcal{O}_{E(\pi_f)}[\frac{1}{2}]})(\pi_f)(\epsilon) \otimes_{\mathcal{O}_{E(\pi_f)}[\frac{1}{2}]} \mathcal{O}_E. \end{aligned}$$

If π_f is strongly inner then the cohomology modules in lowest degree

$$H^{b_n}_{!!}(\mathcal{S}^G_{K_f}, \mathcal{M}_{\lambda,\mathcal{O}_E})(\pi_f, \epsilon)$$

are absolutely irreducible $\mathcal{H}^G_{K_f}$ modules. (This also means that the homomorphism $\mathcal{H}^G_{K_f} \to \mathrm{End}_E(H^{b_n}(\mathcal{S}^G_{K_f}, \mathcal{M}_{\lambda,E})(\pi_f, \epsilon))$ is surjective.) The module of homomorphisms

$$\mathcal{T}^{\mathrm{alg}}(\pi_f,\epsilon)$$
$$= \mathrm{Hom}_{\mathcal{H}^G_{K_f}}(H^{b_n}_{!!,\,\mathrm{int}}(\mathcal{S}^G_{K_f},\mathcal{M}_{\lambda,\mathcal{O}_E})(\pi_f,\epsilon),\ H^{b_n}_{!!,\,\mathrm{int}}(\mathcal{S}^G_{K_f},\mathcal{M}_{\lambda,\mathcal{O}_E})(\pi_f,-\epsilon))$$

is a finitely generated, torsion-free $\mathcal{O}_E$-module of rank one. We consider it as an invertible sheaf for the Zariski topology on $\mathrm{Spec}(\mathcal{O}_E)$. For any open subset $U \subset \mathrm{Spec}(\mathcal{O}_E)$ we use the usual notation $\mathcal{T}^{\mathrm{alg}}(\pi_f,\epsilon)(U)$ for the module of sections over U; this is a module for $\mathcal{O}(U)$. It is clear that we find a covering of $\mathrm{Spec}(\mathcal{O}_E)$ by two open sets U_1, U_2 such that

$$\mathcal{T}^{\mathrm{alg}}(\pi_f,\epsilon)(U_i) = \mathcal{O}_E(U_i)T_i^{\mathrm{alg}}(\pi_f,\epsilon)$$

where $T_i^{\mathrm{alg}}(\pi_f,\epsilon) \in \mathcal{T}^{\mathrm{alg}}(\pi_f,\epsilon)$. These homomorphisms are unique up to an element in $\mathcal{O}_E(U_i)^\times$. For a given level K_f we can find a finite Galois extension $E/\mathbb{Q}$ such that all π_f that occur in $H^\bullet(\mathcal{S}^G_{K_f},\mathcal{M}_{\lambda,E})$ are weakly split, or what amounts to the same, absolutely irreducible. We have the action of the Galois group $\mathrm{Gal}(E/\mathbb{Q})$ on the set of isomorphism classes $\mathrm{Coh}(G,\lambda,K_f)$. For a given π_f we choose a covering $\mathrm{Spec}(\mathcal{O}_{E(\pi_f)})$ by open subsets U_1, U_2 and generators $T_i^{\mathrm{alg}}(\pi_f,\epsilon)(U_i)$. Let us call such a choice a *local trivialization* of $\mathcal{T}^{\mathrm{alg}}(\pi_f,\epsilon)$. Then it is clear that we choose our trivialization such that it is invariant under the action of the Galois group: To get this we choose a π_f in an orbit and our local trivialization $\{T_i^{\mathrm{alg}}(\pi_f,\epsilon)\}_{i=1,2}$ over $\mathcal{O}_{E(\pi_f)}$ as above. For $\tau \in \mathrm{Gal}(E/\mathbb{Q})$ we define

$$T_i^{\mathrm{alg}}(\tau(\pi_f),\epsilon) = \tau(T_i^{\mathrm{alg}}(\pi_f,\epsilon))$$

and then this system of local trivializations

$$\{T_i^{\mathrm{alg}}(\tau(\pi_f),\epsilon)\}_{\pi_f\in\mathrm{Coh}_{!!}(G,\lambda,K_f),\epsilon,i}$$

is defined over $\mathbb{Q}$, i.e., invariant under the Galois group $\mathrm{Gal}(E/\mathbb{Q})$.

We return to the transcendental level. We assume $\pi_f \in \mathrm{Coh}_{!!}(G,\lambda,K_f)$ and choose a model space for π_f, say $H_{\pi_f} = H^{b_n}_{!!}(\mathcal{S}^G_{K_f},\mathcal{M}_{\lambda,\mathcal{O}_E})(\pi_f,+)$. For any $\iota : E \to \mathbb{C}$ we get an inclusion

$$\Phi(\lambda,\pi_f,\iota) : \mathbb{D}_\lambda \otimes H_{\pi_f} \otimes_{E,\iota} \mathbb{C} \hookrightarrow \mathcal{A}(G(\mathbb{Q})\backslash G(\mathbb{A})),$$

which is Hecke equivariant and therefore unique up to a scalar. This map provides an isomorphism for the cohomology

$$\Lambda^\bullet(\Phi(\lambda,\pi,\iota)) : H^\bullet(\mathfrak{g}_\mathbb{Z}/\mathfrak{k}_\mathbb{Z},\mathbb{D}_\lambda\otimes\mathcal{M}_\lambda)\otimes H_{\pi_f}\otimes_{E,\iota}\mathbb{C} \xrightarrow{\sim} H^\bullet(\mathcal{S}^G_{K_f},\mathcal{M}_{\lambda,\mathbb{C}})(\pi_f),$$

which respects the action of $\pi_0(\mathrm{GL}_n(\mathbb{R}))$. We have seen in Sect. 8.4.4 that we have canonical generators for the + and − eigenspaces (see (8.63)):

$$H^{b_n}(\mathfrak{g}_{\mathbb{Z}}/\mathfrak{k}_{\mathbb{Z}}, \mathbb{D}_\lambda \otimes \mathcal{M}_{\lambda,\mathbb{C}}) = \mathbb{C}\omega_\lambda^{(+)} \oplus \mathbb{C}\omega_\lambda^{(-)}.$$

The choice of the generators was motivated by integrality considerations and in this sense they are canonical. But using the explicit description we could just write them down in an ad hoc manner. The actual choice is not so important; what really matters is that they are *entangled*, which means that once we choose $\omega_\lambda^{(+)}$, the choice of $\omega_\lambda^{(-)}$ is forced upon us; see Sect. 5.2.2.

For $\epsilon = \pm$ we have the two isomorphisms

$$\Psi(\lambda, \pi_f, \epsilon, \iota) : H_{\pi_f} \otimes_{E,\iota} \mathbb{C} \longrightarrow H^{b_n}(\mathcal{S}^G_{K_f}, \mathcal{M}_{\lambda,E})(\pi_f, \epsilon) \otimes_{E,\iota} \mathbb{C},$$

which are given by the composition

$$\psi_f \mapsto \omega_\lambda^{(\epsilon)} \times \psi_f \mapsto \Lambda^{b_n}(\Phi(\lambda, \pi_f, \iota)(\omega_\lambda^{(\epsilon)} \times \psi_f).$$

We get a composition $T^{\text{trans}}(\pi_f, \iota, \epsilon) = \Psi(\lambda, \pi_f, \epsilon, \iota)^{-1} \circ \Psi(\lambda, \pi_f, -\epsilon, \iota)$ which is an isomorphism

$$\begin{aligned} T^{\text{trans}}(\pi_f, \iota, \epsilon) : H^{b_n}(\mathcal{S}^G_{K_f}, \mathcal{M}_{\lambda,E})(\pi_f, \epsilon) \otimes_{E,\iota} \mathbb{C} & \\ \to H^{b_n}(\mathcal{S}^G_{K_f}, \mathcal{M}_{\lambda,E})(\pi_f, -\epsilon) \otimes_{E,\iota} \mathbb{C}. & \end{aligned}$$

This isomorphism does not depend on the choice of the embedding $\Phi(\lambda, \pi_f, \iota)$. For $i = 1, 2$ we define the periods by comparing the two isomorphisms between the $\pm$ eigenspaces (see Def. 5.3):

$$\Omega_i(\pi_f, \iota, \epsilon) T^{\text{trans}}(\pi_f, \iota, \epsilon) = T_i^{\text{alg}}(\pi_f, \epsilon) \otimes_{E,\iota} 1. \tag{8.85}$$

These refined relative periods $\Omega_i(\pi_f, \iota, \epsilon)$ are complex numbers that are well-defined modulo $\iota(\mathcal{O}_{(E(\pi_f)}(U_i))^\times$, and the ratio $\Omega_1(\pi_f, \iota, \epsilon)/\Omega_2(\pi_f, \iota, \epsilon)$ is an element in $\iota(\mathcal{O}_{E(\pi_f)})(U_1 \cap U_2)^\times$.

If we now work with these refined periods, then the assertion

$$\frac{1}{\Omega^{\varepsilon'}({}^\iota\sigma_f)} \frac{L^{\text{Coh}}(\iota, \sigma \times \sigma'^{\vee}, \mathsf{m}_0)}{L^{\text{Coh}}(\iota, \sigma \times \sigma'^{\vee}, 1 + \mathsf{m}_0)} \in \iota(E)$$

in Thm. 7.21 remains unchanged; however, now it makes sense to ask for the decomposition of these numbers into prime ideals. We have evidence that this decomposition into prime factors has some influence on the structure of cohomology of arithmetic groups. The prime factors should be related to denominators of Eisenstein classes but this relationship could be spoiled by primes dividing the factor $c_\infty(\tilde{\mu})$ above; see the next chapter by Uwe Weselmann. This subtle issue, also discussed in [30], definitely warrants a deeper study.

Chapter Nine

The Archimedean Intertwining Operator for GL_N

By Uwe Weselmann

In this chapter we generalize an identity conjectured by Harder [29] and proved by Zagier [74] from the case of $\mathrm{GL}_3(\mathbb{R})$-representations to the case of general $\mathrm{GL}_N(\mathbb{R})$-representations, which will be useful in applying the results of Harder and Raghuram on quotients of special values of L-functions as in Chap. 7 of this monograph. The notations in this chapter are mostly self-contained and are broadly compatible with the notations of the rest of the monograph.

Let $N = n + n'$ with positive integers n, n'. Let $P \subset G = \mathrm{GL}_N$ be the standard (i.e., including the Borel $B = B_N$ of upper triangular matrices) parabolic subgroup with Levi quotient $M = \mathrm{GL}_n \times \mathrm{GL}_{n'}$ and let $Q \subset \mathrm{GL}_N$ be the standard parabolic subgroup with Levi quotient $\mathrm{GL}_{n'} \times \mathrm{GL}_n$. Then Q is the conjugate of the opposite parabolic P^- by an element w in the normalizer of the standard diagonal torus T, i.e., by a representative of a Weyl-group element. The Harder–Zagier identity deals with intertwining operators between parabollically induced representations $\mathrm{Ind}_P^G(\pi_1 \otimes \pi_2)$ and $\mathrm{Ind}_Q^G(\pi_2 \otimes \pi_1)$ for representations π_1 of $\mathrm{GL}_n(\mathbb{R})$ and π_2 of $\mathrm{GL}_{n'}(\mathbb{R})$. Integration over the unipotent radical of the opposite parabolic P^- leads to the analytically defined standard intertwining operator T_w. In the case that π_1 and π_2 are themselves induced from characters of the corresponding Borel subgroups B_n and $B_{n'}$, or are subrepresentations of such representations, we deal with induced representations of the Borel B_N or their subrepresentations.

Let K be the maximal compact subgroup O_N of $\mathrm{GL}_N(\mathbb{R})$. Using the Iwasawa decomposition $G(\mathbb{R}) = B(\mathbb{R})\cdot K$, we may identify the representation space $\mathrm{Ind}_B^G\chi$ as a K-module with the induced representation $\mathrm{Ind}_\Sigma^K\chi$, where $\Sigma = B(\mathbb{R}) \cap K$ is an elementary 2-group. Using the Peter–Weyl theorem, the Harish-Chandra module (in the following lines of the introduction to this chapter we will not distinguish between a representation and its Harish-Chandra module) of the latter space may be written in the form

$$\bigoplus_{\vartheta \in \hat{K}} (V_\vartheta^*)^\chi \otimes V_\vartheta,$$

where K acts via the second factor. Subrepresentations of a principal series representation (i.e., induced from the Borel subgroup B) may now be described by a family of subspaces of the first factors. Intertwining operators between different principal series representations may be described by a family of linear maps between the corresponding $(V_\vartheta^*)^\chi$ spaces. Now, we may assume that the Weyl-element w lies in K. Then one may ask whether there is a relation between the effect of the analytically defined intertwining operator T_w and the action of w on V_ϑ^*. The main result of this chapter, Thm. 9.13, states that these actions on certain vectors v differ by a scalar, which is a quotient of two consecutive archimedean L-values. One has to consider subrepresentations of $\mathrm{Ind}_B^G\chi$, which correspond to induced representations from discrete series representations of $\mathrm{GL}_2(\mathbb{R})$ factors, and v has to be an averaged highest weight vector in a minimal K-type of this subrepresentation as defined in Sect. 9.4.

The proof of the main theorem, described in Sect. 9.5, consists of several reduction steps to the case GL_3, which is already known by the work of Harder and Zagier. But in Sect. 9.5.7 we give an independent proof of this special case, which is of a more representation-theoretic nature.

In Sect. 9.6 we apply the main theorem to the cohomology of Harish-Chandra modules. In fact the notion of an *averaged highest weight vector* corresponds to those K-submodules of the induced representations, which generate nontrivial cohomology classes. We then connect our results to the work of Harder and Raghuram as in Chap. 7. This involves some more technical considerations, since our identification of the cohomology of the unipotent radicals a priori differs from their identification: On the level of Lie algebras we use the canonical embeddings having the same image: $\mathfrak{n} \hookrightarrow \mathfrak{g}/\mathfrak{k} \hookleftarrow \mathfrak{n}_-$, which leads to an O_N-equivariant identification via transposition. But Harder and Raghuram use the fact that the middle exterior powers of $\mathfrak{n}$ and $\mathfrak{n}^-$ may be compared as a module for the Levi subgroup M via the trace pairing between $\mathfrak{n}$ and $\mathfrak{n}^-$ and via the $\wedge$ product into the highest exterior power. Then one has to show that these two identifications give the same for the vectors that are involved in the construction of cohomology.

9.1 THE GENERAL SETTING

9.1.1 Some group theoretic data

Let $G/\mathbb{R}$ be a split connected reductive group. Fix a Borel subgroup B of G and a maximal compact subgroup $K = K_\infty$ of $G(\mathbb{R})$. We remark that every other pair (B', K') is conjugate by a single element $g \in G(\mathbb{R})$ to the given pair: It is well known that all Borel subgroups and all maximal compact subgroups are conjugates of each other; i.e., there exist $g_B, g_K \in G(\mathbb{R})$ such that $B' = g_B \cdot B \cdot g_B^{-1}$ and $K' = g_K \cdot K \cdot g_K^{-1}$. By the Iwasawa decomposition there exists $b \in B(\mathbb{R})$ and $k \in K$ with $g_B^{-1} \cdot g_K = b \cdot k$. Put $g = g_B \cdot b = g_K \cdot k^{-1}$. Then we have $B' = gBg^{-1}$ and $K' = gKg^{-1}$. Observe that g is unique up to right multiplication by an element of $\Sigma = B(\mathbb{R}) \cap K$, since $B(\mathbb{R})$ and K are their own normalizers inside $G(\mathbb{R})$.

Now we introduce some additional data that depend only on the pair (B, K). Let U denote the unipotent radical of B. Let $\Theta : G \to G$ be the Cartan involution, such that K is the set of its fixed points. Then it is easy to see that $T = B \cap \Theta(B)$ is a maximal split torus. This means that B and the Borel subgroup $\Theta(B)$ are in opposition and is equivalent to the statement that the Θ-stable algebraic subgroup $V := U \cap \Theta(U)$ is trivial. (We get an action of Θ on the Lie algebra $\mathrm{Lie}(V)$, and if $V \neq \{1\}$, then we would get a nontrivial eigenvector v for the action of Θ on $\mathrm{Lie}(V)$. Now the Cartan involution Θ has the property, that $\beta_\Theta : (x, y) \mapsto \beta(x, \Theta(y))$ is a negative definite quadratic form on $\mathrm{Lie}(G^{\mathrm{der}})$, where β denotes the Killing form and the derived group G^{der} contains U and thus V. But since $ad(v)$ is nilpotent, we have $\beta(v, v) = 0$ and thus $\beta_\Theta(v, v) = \beta(v, \pm v) = 0$ in contradiction to the definiteness of β_Θ.)

Let Φ be the subset of roots inside the character group $X^*(T)$. The unipotent radical $U \subset B$ determines the set of positive roots $\Phi^+ \subset \Phi$. Let $\Delta \subset \Phi^+$ be the subset of simple roots. The Cartan involution Θ acts as $t \mapsto t^{-1}$ on T, since its set of fixed points has to be a compact subgroup of $T(\mathbb{R})$. For each $\alpha \in \Phi$ let U_α be the corresponding unipotent one-parameter subgroup. Then we have $\Theta(U_\alpha) = U_{-\alpha}$. Let M_α be the reductive group generated by U_α and $U_{-\alpha}$. It is isomorphic to either SL_2 or to PSL_2.

Let K_α be the connected component of the identity of $M_\alpha(\mathbb{R}) \cap K$. It is a compact group, which is isomorphic to the rotation group S^1. If $I \subset \Phi$ is a subset of mutually orthogonal roots, then K_α and K_β commute for $\alpha, \beta \in I$. Thus $K_I = \prod_{\alpha \in I} K_\alpha$ is a compact torus in K.

For simple $\alpha \in \Delta$ we get a parabolic subgroup $P_\alpha = B \cdot M_\alpha$ of semisimple rank 1, such that the root space of $-\alpha$ is contained in the Lie algebra of P_α. Then K_α is the connected component of the identity of $P_\alpha(\mathbb{R}) \cap K$.

We have $\Sigma = T(\mathbb{R}) \cap K \cong \{\pm 1\}^r$, if T is of rank r. If N_T denotes the normalizer of the torus T, then we get a finite group $N_c = N_T(\mathbb{R}) \cap K$, such that the quotient N_c/Σ is isomorphic to the Weyl group $W = N_T/T = N_T(\mathbb{R})/T(\mathbb{R})$. For $\alpha \in \Delta$ let $w_\alpha \in N_c \cap P_\alpha(\mathbb{R})$ be an element that represents the reflection at the simple root α.

For $\alpha \in \Phi$ let ι_α be an isomorphism between the additive group $\mathbb{G}_a$ and the root subgroup U_α. Then the element $\phi(x) = \Theta(\iota_\alpha(x)) \cdot \iota_\alpha(x) \cdot \Theta(\iota_\alpha(x))$ is an element of N_c for certain values of x. For example, here is an illustrative identity to be construed as a calculation in M_α:

$$\begin{pmatrix} 1 & 0 \\ -x & 1 \end{pmatrix} \begin{pmatrix} 1 & x \\ 0 & 1 \end{pmatrix} \begin{pmatrix} 1 & 0 \\ -x & 1 \end{pmatrix} = \begin{pmatrix} 1-x^2 & x \\ x^3 - 2x & 1-x^2 \end{pmatrix} = \begin{pmatrix} 0 & \pm 1 \\ \mp 1 & 0 \end{pmatrix} \ if\ x = \pm 1.$$

For $\alpha \in \Delta$ we may fix ι_α in such a way that $\phi(1)$ lies in N_c. This ι_α is unique up to sign. The canonical measure on $\mathbb{R}$ induces via ι_α a measure on $U_\alpha(\mathbb{R})$, which is now unique.

The triple $(B, T, \{\iota_\alpha\}_{\alpha\in\Delta})$ is a splitting of G, and it is unique up to the action of Σ (roughly speaking up to signs), once the pair (B, K) is fixed.

9.1.2 The induced representations

The group $T(\mathbb{R})$ is canonically the direct product

$$T(\mathbb{R}) \;=\; T(\mathbb{R})^\circ \times \Sigma$$

of its connected component $T(\mathbb{R})^\circ$ of the identity and the group $\Sigma = T(\mathbb{R}) \cap K = T(\mathbb{R})[2]$ of its 2-torsion points. We may write $t \in T(\mathbb{R})$ in the form $t = (|t|, \operatorname{sign}(t)) \in T(\mathbb{R})^\circ \times \Sigma$. The group X of continuous characters of $T(\mathbb{R})$ may be described in the form $X = (X^*(T) \otimes \mathbb{C}) \oplus \hat{\Sigma}$, where $\hat{\Sigma} = \operatorname{Hom}(\Sigma, \{\pm 1\})$, and $\eta \otimes s \in X^*(T) \otimes \mathbb{C}$ corresponds to the map $t \mapsto \eta(|t|)^s \in \mathbb{C}^*$. Write $\chi = (\chi^\circ, \epsilon_\chi)$.

Denote by $\delta \in X$ the square root of the modulus character. We have $\delta = ((\sum_{\alpha\in\Phi^+} \alpha) \otimes \frac{1}{2}, 1)$; i.e., it corresponds to the half sum of all positive roots as a character on $T(\mathbb{R})^\circ$ and is trivial on Σ.

For $\chi \in X$, let I_χ be the Harish-Chandra module attached to the unitarily induced representation $\operatorname{Ind}_B^G \chi$. This is the space of K-finite vectors inside the space of functions

$$\{f \in \mathcal{C}^\infty(G(\mathbb{R}), \mathbb{C}) \mid f(utg) = (\chi + \delta)(t) \cdot f(g) \;\; \forall g \in G(\mathbb{R}), t \in T(\mathbb{R}), u \in U(\mathbb{R})\}$$

on which the group action is by right translations.

9.2 THE INTERTWINING OPERATORS

9.2.1 Definition of the intertwining operators

Let $w \in N(T)(\mathbb{R})$ be an element in the normalizer of the torus representing some element in the Weyl group $W = N(T)/T$. Let $U^- = \Theta(U)$ be the unipotent radical of the Borel $B^- = \Theta(B)$, which is in opposition to B. Let $U_w = w \cdot U^- \cdot w^{-1} \cap U$ and $w(\chi) : t \mapsto \chi(w^{-1}tw)$. Then we consider the intertwining operator $T_w : I_\chi \to I_{w(\chi)}$, which converges for χ in some open subset of $X^*(T) \otimes \mathbb{C}$, and is given by

$$(T_w f)(g) = \int_{(wU(\mathbb{R})w^{-1}\cap U(\mathbb{R}))\backslash U(\mathbb{R})} f(w^{-1}ug)du = \int_{U_w(\mathbb{R})} f(w^{-1}ug)du.$$

We have the following well-known lemma of factorizing such an intertwining operator.

Lemma 9.1. *In the case* $l(w_1w_2) = l(w_1) + l(w_2)$ *we have a diffeomorphism given by multiplication*

$$U_{w_1w_2} \cong w_1U_{w_2}w_1^{-1} \times U_{w_1},$$

which implies the relation

$$T_{w_1w_2} = T_{w_1} \circ T_{w_2}. \tag{9.1}$$

Proof. Working with the Lie algebras, we have

$$\mathrm{Lie}(U \cap wUw^{-1}) = \sum_{\alpha\in\Phi^+,\, w^{-1}\alpha>0} \mathfrak{u}_\alpha,$$

where $\mathfrak{u}_\alpha = \mathrm{Lie}(U_\alpha)$. Hence,

$$\mathrm{Lie}((U \cap wUw^{-1})\backslash U) = \sum_{\alpha\in\Phi^+,\, w^{-1}\alpha<0} \mathfrak{u}_\alpha = \mathrm{Lie}\left(U \cap wU^-w^{-1}\right) = \mathrm{Lie}(U_w).$$

To see the decomposition of $U_{w_1w_2} = w_1U_{w_2}w_1^{-1} \times U_{w_1}$, when $l(w_1w_2) = l(w_1) + l(w_2)$, we can work at the level of Lie algebras. For $\alpha \in \Phi^+$ such that $(w_1w_2)^{-1}\alpha < 0$, consider the two disjoint and exhaustive cases:

1. If $w_1^{-1}\alpha > 0$ then $\mathfrak{u}_{w_1^{-1}\alpha} \subset \mathrm{Lie}(U_{w_2})$, whence $\mathfrak{u}_\alpha \subset w_1\mathrm{Lie}(U_{w_2})w_1^{-1}$.
2. If $w_1^{-1}\alpha < 0$ then $\mathfrak{u}_\alpha \subset \mathrm{Lie}(U_{w_1})$.

The factorization of $T_{w_1w_2}$ is clear. □

9.2.2 The Peter–Weyl isomorphism

The Iwasawa decomposition implies that the functions in I_χ are uniquely determined by their restrictions to K. Therefore the space I_χ is isomorphic as a K-module to the space of K-finite functions on K, which transform under left translations by elements of $\Sigma = K \cap B(\mathbb{R})$ with the character ϵ. In view of the Peter–Weyl theorem we get a canonical isomorphism of K-modules

$$i_{PW} : I_\chi \xleftarrow{\sim} \bigoplus_{\vartheta\in\hat{K}} (V_\vartheta^*)^\epsilon \otimes V_\vartheta,$$

where ϑ runs over equivalence classes of irreducible representations of K, and V_ϑ is some representation space inside the class ϑ, and V_ϑ^* is its dual. A matrix coefficient $v^* \otimes v$ corresponds to the function

$$i_{PW}(v^* \otimes v) \;=\; f_{v^*\otimes v}: \quad G \ni utk \;\mapsto\; (\chi+\delta)(t)\cdot v^*(kv),$$

which is well defined if $v^* \in (V_\vartheta^*)^\epsilon = \{v^* \in V_\vartheta^* \mid \sigma v^* = \epsilon(\sigma)\cdot v^* \quad \text{for all } \sigma \in \Sigma\}$.

9.2.3 Calculations in SL_2

For a simple root $\alpha \in \Delta$ we fix a surjective map $\iota_\alpha : \mathrm{SL}_2 \to M_\alpha = \langle U_\alpha, U_{-\alpha}\rangle$ in such a way that the standard diagonal torus maps into T and such that $\begin{pmatrix} 1 & u \\ 0 & 1 \end{pmatrix}$ maps to $\iota_\alpha(u)$ in the notations of Sect. 9.1.1. We put

$$w_\alpha \;=\; \iota_\alpha\begin{pmatrix} 0 & -1 \\ 1 & 0 \end{pmatrix}, \quad T_\alpha \;=\; T_{w_\alpha},$$
$$R(\phi) \;=\; \begin{pmatrix} \cos(\phi) & -\sin(\phi) \\ \sin(\phi) & \cos(\phi) \end{pmatrix}, \; R_\alpha(\phi) \;=\; \iota_\alpha(R(\phi)).$$

Thus $w_\alpha = R_\alpha(\pi/2)$. Now it is easy to see that the subgroup $\{R_\alpha(\phi) | \phi \in \mathbb{R}\}$ is the connected component K_α of the identity of $M_\alpha(\mathbb{R}) \cap K$.

We have $U_{w_\alpha} = \iota_\alpha((\begin{smallmatrix} 1 & * \\ 0 & 1 \end{smallmatrix}))$. Now suppose that $v^* \in V_\vartheta^*$ transforms under $R_\alpha(\phi)$ with the eigenvalue $e^{i\phi n}$ for some $n \in \mathbb{Z}$ and all $\phi \in \mathbb{R}$, i.e.,

$$v^*(R_\alpha(\phi)v) = e^{-i\phi n} v^*(v), \quad \forall v \in V_\vartheta.$$

The Iwasawa decomposition

$$\begin{pmatrix} 0 & -1 \\ 1 & 0 \end{pmatrix}\cdot\begin{pmatrix} 1 & u \\ 0 & 1 \end{pmatrix} \;=\; \begin{pmatrix} \tau & \mu \\ 0 & \tau^{-1} \end{pmatrix}\cdot\begin{pmatrix} x & -y \\ y & x \end{pmatrix}$$

with $x = -\mu = \frac{u}{\sqrt{1+u^2}}, \quad y = \tau = \frac{1}{\sqrt{1+u^2}}$ implies

$$\begin{aligned}
T_\alpha f_{v^*\otimes v}(k) \;&=\; \int_{U_{w_\alpha}} f_{v^*\otimes v}\,(w_\alpha \cdot u \cdot k)\, du \\
&=\; \int_{\mathbb{R}} \tau^{z+1} \cdot v^*\left(\iota_\alpha\begin{pmatrix} x & -y \\ y & x \end{pmatrix} kv\right) du \\
&=\; \int_{\mathbb{R}} \sqrt{1+u^2}^{\,-z-1} \cdot \left(\frac{u+i}{\sqrt{1+u^2}}\right)^{-n} \cdot v^*(kv)du \\
&=\; c(z,n)\cdot f_{v^*\otimes v}(k),
\end{aligned}$$

where $z = \langle \chi^\circ, \check{\alpha} \rangle \in \mathbb{C}$ and

$$c(z,n) = \int_{-\infty}^{\infty} \sqrt{1+u^2}^{-z-1} \cdot \left(\frac{u-i}{\sqrt{1+u^2}} \right)^n du.$$

Here $\check{\alpha} \in X_*(T) = \mathrm{Hom}(\mathbb{G}_m, T)$ denotes the dual root of α.

9.2.4 The factors $\gamma_n(z)$

Making the substitution $u = \tan\theta$ we get immediately

$$\begin{aligned} c(z,n) &= \int_{-\pi/2}^{\pi/2} (\cos\theta)^{z+1} \cdot (\sin\theta - i\cos\theta)^n \cdot (\cos\theta)^{-2} d\theta \\ &= (-i)^n \cdot \gamma_n(z), \qquad \text{where} \\ \gamma_n(z) &= \int_{-\pi/2}^{\pi/2} e^{i\theta n} \cdot (\cos\theta)^{z-1} d\theta, \end{aligned} \tag{9.2}$$

which coincides with the formula in Zagier [74, (8)]. The formula $\cos\theta = \frac{1}{2} \cdot (e^{i\theta} + e^{-i\theta})$ implies the relation

$$\gamma_n(z+1) = \frac{1}{2} \cdot (\gamma_{n-1}(z) + \gamma_{n+1}(z)). \tag{9.3}$$

The substitution $\theta \mapsto -\theta$ gives the symmetry property

$$\gamma_{-n}(z) = \gamma_n(z), \qquad \text{which implies together with (9.3):} \tag{9.4}$$

$$\gamma_1(z) = \gamma_0(z+1). \tag{9.5}$$

The derivative of the function $f(\theta) = -2i \cdot e^{i\theta n} \cdot (\cos\theta)^z$ is

$$\begin{aligned} &(2n\cos\theta + 2iz\sin\theta) \cdot e^{i\theta n}(\cos\theta)^{z-1} \\ &\quad = (n+z) \cdot e^{i\theta(n+1)}(\cos\theta)^{z-1} - (z-n) \cdot e^{i\theta(n-1)}(\cos\theta)^{z-1}. \end{aligned}$$

Integration gives in view of $f(-\frac{\pi}{2}) = f(\frac{\pi}{2}) = 0$ for $\Re z > 0$:

$$(z+n) \cdot \gamma_{n+1}(z) = (z-n) \cdot \gamma_{n-1}(z). \tag{9.6}$$

This formula can also be obtained by comparing the formulas for the action of $P_\pm$ with the interpretation of $\gamma_n(z)$ as describing the intertwining operator as in Sect. 8.3.4. If we use (9.3) twice and then (9.6) we get

$$\begin{aligned}\gamma_n(z+2) &= \frac{1}{4}\cdot(\gamma_{n-2}(z)+2\gamma_n(z)+\gamma_{n+2}(z)) \\ &= \frac{1}{4}\cdot\left(\frac{z+n-1}{z-n+1}+2+\frac{z-n-1}{z+n+1}\right)\cdot\gamma_n(z) \\ &= \frac{z(z+1)}{(z+1-n)(z+1+n)}\cdot\gamma_n(z).\end{aligned} \tag{9.7}$$

This equation can be used to continue the γ_n, which are a priori defined only for $\Re(z)>0$, to meromorphic functions on the whole complex plane, satisfying all the relations above.

We define

$$\Gamma_{\mathbb{R}}(z) = \Gamma\left(\frac{z}{2}\right)\cdot\pi^{-z/2}, \qquad \Gamma_{\mathbb{C}}(z) = 2\cdot\Gamma(z)\cdot(2\pi)^{-z}.$$

Then the doubling formula for the Γ-function is equivalent to

$$\Gamma_{\mathbb{C}}(z) = \Gamma_{\mathbb{R}}(z)\cdot\Gamma_{\mathbb{R}}(z+1). \tag{9.8}$$

We compute

$$\begin{aligned}\Gamma_{\mathbb{R}}(z+1)\cdot\gamma_0(z) &= \pi^{-(z+1)/2}\cdot\int_0^\infty\int_{-\pi/2}^{\pi/2} e^{-t}\cdot t^{\frac{z-1}{2}}\cdot(\cos\theta)^{z-1}\,d\theta\,dt \\ &= \pi^{-(z+1)/2}\cdot\int_0^\infty\int_{-\pi/2}^{\pi/2} e^{-r^2}\cdot r^{z-1}\cdot(\cos\theta)^{z-1}\,d\theta\,2rdr \\ &= \pi^{-(z+1)/2}\cdot\int_0^\infty\int_{-\infty}^{\infty} e^{-x^2-y^2}\cdot x^{z-1}\,dy2dx \\ &= \pi^{-(z+1)/2}\cdot\int_{-\infty}^{\infty} e^{-y^2}dy\cdot\int_0^\infty e^{-t}\cdot t^{\frac{z}{2}-1}\,dt \\ &= \pi^{-z/2}\cdot\Gamma\left(\frac{z}{2}\right) = \Gamma_{\mathbb{R}}(z),\end{aligned}$$

where we made the substitutions: $t=r^2$ in the second, $x=r\cdot\cos\theta, y=r\cdot\sin\theta$ in the third, and $x^2=t$ in the fourth line. Thus we have

$$\gamma_0(z) = \frac{\Gamma_{\mathbb{R}}(z)}{\Gamma_{\mathbb{R}}(z+1)}, \tag{9.9}$$

$$\gamma_0(z)\cdot\gamma_0(z+1) = \frac{\Gamma_{\mathbb{R}}(z)}{\Gamma_{\mathbb{R}}(z+2)} = \frac{2\pi}{z} = \frac{\Gamma_{\mathbb{C}}(z)}{\Gamma_{\mathbb{C}}(z+1)}. \tag{9.10}$$

As a consequence we get

$$\gamma_n(z) = \frac{\pi}{2^{z-1}} \cdot \frac{\Gamma(z)}{\Gamma\left(\frac{z+1+n}{2}\right) \cdot \Gamma\left(\frac{z+1-n}{2}\right)} = \frac{\Gamma_{\mathbb{R}}(z) \cdot \Gamma_{\mathbb{R}}(z+1)}{\Gamma_{\mathbb{R}}(z+1+n) \cdot \Gamma_{\mathbb{R}}(z+1-n)}. \tag{9.11}$$

The second equation follows from the definition of $\Gamma_{\mathbb{R}}, \Gamma_{\mathbb{C}}$ and the formula (9.8). That $\gamma_n(z)$ coincides with the right-hand side has just been proved for $n = 0$, and by (9.5) it follows for $n = 1$. Using induction on $|n|$ we may deduce the claim from the recursive relation (9.6) and from the functional equation of the Γ-function. □

9.2.5 The meromorphic intertwining operators

If V is a finite-dimensional K-module and α a simple root, we may decompose V into eigenspaces with respect to the action of the subgroup $\iota_\alpha(\mathrm{SO}_2)$:

$$V = \bigoplus_{n \in \mathbb{Z}} V_n, \quad \text{where } R_\alpha(\phi)v = e^{i\phi n} \cdot v \text{ for all } v \in V_n.$$

Now we can define $\tilde{T}_\alpha = \tilde{T}_\alpha(z) : V_\vartheta^* \to V_\vartheta^*$ to be the meromorphic endomorphism, which is multiplication by $c(z,n) = (-i)^n \cdot \gamma_n(z)$ on each eigenspace $(V_\vartheta^*)_n$. Then we have the commutativity relation

$$i_{PW} \circ \left(\tilde{T}_\alpha \otimes id_{V_\vartheta}\right) = T_\alpha \circ i_{PW} \qquad \text{on } (V_\vartheta^*)^\epsilon \otimes V_\vartheta. \tag{9.12}$$

We proved this relation for z in the domain of convergence of T_α, but it may be used to define the operator T_α for arbitrary z, depending meromorphically on z. Then the relation (9.1) may be used to define $\tilde{T}_w$ for arbitrary $w \in S_n$. The identity theorem for meromorphic functions tells us that this leads to well-defined meromorphic intertwining operators, and they satisfy (9.1) in full generality.

Lemma 9.2. *The following relation holds:*

$$\tilde{T}_\alpha(\sigma \cdot v^*) = w_\alpha(\sigma) \cdot \tilde{T}_\alpha(v^*) \qquad \textit{for } \sigma \in \Sigma,\ v \in V_\vartheta^*.$$

Proof. To prove this we may assume $v^* \in (V_\vartheta^*)_n$ for some $n \in \mathbb{Z}$. Since $\sigma \in T(\mathbb{R}) \cap K$ normalizes the connected component K_α of $M_\alpha(\mathbb{R}) \cap K$ and since it acts as continuous involution by conjugation, we either have $\sigma \cdot R_\alpha(\phi) \cdot \sigma^{-1} = R_\alpha(\phi)$ or $\sigma \cdot R_\alpha(\phi) \cdot \sigma^{-1} = R_\alpha(-\phi)$ for all ϕ. In the first case we get $\sigma \cdot v^* \in (V_\vartheta^*)_n$ and $w_\alpha(\sigma) = R_\alpha(\pi/2) \cdot \sigma \cdot R_\alpha(\pi/2)^{-1} = \sigma$ and thus the claim is immediate, since $\tilde{T}_\alpha$ acts as a scalar. In the second case we get $\sigma \cdot v^* \in (V_\vartheta^*)_{-n}$ and $w_\alpha(\sigma) = R_\alpha(\pi/2) \cdot \sigma \cdot R_\alpha(-\pi/2) = R_\alpha(\pi/2) \cdot R_\alpha(-(-\pi/2)) \cdot \sigma = R_\alpha(\pi) \cdot \sigma$. The claim now reads $c(z,-n) \cdot \sigma \cdot v^* = R_\alpha(\pi) \cdot \sigma \cdot c(z,n) \cdot v^*$ and is a consequence of the relation

$c(z,-n) = (-1)^n \cdot c(z,n)$ and of $R_\alpha(\boldsymbol{\pi})\tilde{v}^* = (-1)^n \cdot \tilde{v}^*$ for $\tilde{v}^* \in (V_\vartheta^*)_{-n}$. □

Corollary 9.3. *The intertwining operator $\tilde{T}_w$ maps $(V_\vartheta^*)^\epsilon$ to $(V_\vartheta^*)^{w\epsilon}$.*

Here $w\epsilon$ denotes the character $\sigma \mapsto \epsilon(w^{-1}\sigma)$. For a simple reflection $w = w_\alpha$ this is a consequence of the lemma. The general case may be deduced from this by writing an arbitrary w as a product of simple reflections.

Let's summarize the above discussion in the following diagram:

$$\begin{array}{ccc}
(V_\vartheta^*)^\epsilon \otimes V_\vartheta & \xrightarrow{\ i_{PW}\ } & I_\chi \\
{\scriptstyle \tilde{T}_w \otimes 1_{V_\vartheta}} \downarrow & & \downarrow {\scriptstyle T_w} \\
(V_\vartheta^*)^{w(\epsilon)} \otimes V_\vartheta & \xrightarrow{\ i_{PW}\ } & I_{w(\chi)}
\end{array}$$

9.2.6 Discrete series constructions

Let us assume that

$$z \;=\; \langle \chi^\circ, \check{\alpha} \rangle \;=\; D - 1 \geq 0, \quad \text{with a positive integer } D \geq 1,$$

and that

$$\epsilon_\chi \circ \check{\alpha}(-1) = (-1)^D.$$

The action of $R_\alpha(\pi) = \check{\alpha}(-1)$ shows that $(V_\theta^*)^\epsilon \subset \bigoplus_{n \equiv D \mod 2} (V_\theta^*)_n$. In (9.11) the first Γ-factor in the denominator has a pole for $n = -D - 2b$ for any integer $b \geq 0$, while the second Γ-factor in the denominator has a pole for $n = D + 2a$ for any integer $a \geq 0$.

If $D \geq 2$, then $\Gamma(z)$ is finite, and we get $c(z,n) = \gamma_n(z) = 0$ for $|n| \geq D$. Furthermore all Γ-factors of (9.11) are finite and nonzero for $|n| < D$. The intertwining operator T_α is holomorphic, but has a nontrivial kernel

$$\Delta_\alpha(I_\chi) = \ker(T_\alpha : I_\chi \to I_{w_\alpha \chi}).$$

In the case $D = 1$ we get by analytic continuation of the previous formulas (9.4), (9.5) and (9.11): $\gamma_{-1}(0) = \gamma_1(0) = \gamma_0(1) = \frac{\boldsymbol{\pi} \cdot \Gamma(1)}{\Gamma(1)^2} = \boldsymbol{\pi}$. Furthermore by (9.6) we get $\gamma_{1+2k}(0) = \gamma_{-1-2k}(0) = (-1)^k \cdot \boldsymbol{\pi}$ and thus T_α is an isomorphism in this case.

If one defines the corresponding kernels

$$\Delta_\alpha\left((V_\theta^*)^\epsilon\right) = \ker\left(\tilde{T}_\alpha : (V_\theta^*)^\epsilon \to (V_\theta^*)^{w_\alpha \epsilon}\right)$$
$$= (V_\theta^*)^\epsilon \cap \bigoplus_{n\equiv D \pmod 2,\ |n|\geq D} (V_\theta^*)_n$$

then one gets an isomorphism

$$i_{PW} : \Delta_\alpha(I_\chi) \xleftarrow{\sim} \bigoplus_{\vartheta\in\hat{K}} \Delta_\alpha\left((V_\vartheta^*)^\epsilon\right) \otimes V_\vartheta.$$

Definition 9.4. *Let $J \subset \Delta$ denote a set of simple roots.*

(a) An element $w \in W$ is called J-admissible, iff $w(J) \subset \Delta$.
(b) A character $\chi = (\chi^\circ, \epsilon_\chi) \in X$ is called J-admissible, iff $\langle \chi^\circ, \check{\alpha}\rangle$ is an integer of the form $D_\alpha - 1$ with $D_\alpha \geq 1$ and if $\epsilon_\chi \circ \check{\alpha}(-1) = (-1)^{D_\alpha}$ for all $\alpha \in J$.

If we assume that χ is J-admissible for a set of simple roots $J \subset \Delta$, then we may define the modules

$$\Delta_J(I_\chi) = \bigcap_{\alpha\in J} \Delta_\alpha(I_\chi) \qquad \text{and } \Delta_J\left((V_\theta^*)^\epsilon\right) = \bigcap_{\alpha\in J} \Delta_\alpha\left((V_\theta^*)^\epsilon\right).$$

We get a similar isomorphism i_{PW} from $\bigoplus_{\vartheta\in\hat{K}} \Delta_J\left((V_\vartheta^*)^\epsilon\right) \otimes V_\vartheta$ to $\Delta_J(I_\chi)$.

9.2.7 The case GL_n

In the case of the group $G = \mathrm{GL}_n$ let B be the Borel subgroup of upper triangular matrices. The standard orthogonal group $K = \mathrm{O}_n \subset G(\mathbb{R})$ is the fixed point set of the Cartan-involution $\Theta : g \mapsto {}^t g^{-1}$. Then $T \subset G = \mathrm{GL}_n(\mathbb{R})$ is the diagonal torus, $U \subset G$ (resp., U^-) is the subgroup of all unipotent upper (resp., lower) triangular matrices. The characters of $T(\mathbb{R})$ are parametrized by $\chi = (s;\epsilon) \in X = \mathbb{C}^n \times (Z/2\mathbb{Z})^n$ in such a way that we have for $s = (s_1, \ldots, s_n) \in \mathbb{C}^n$, $\epsilon = (\epsilon_1, \ldots, \epsilon_n) \in (Z/2\mathbb{Z})^n$ and $t = \mathrm{diag}(t_1, \ldots, t_n)$:

$$\chi(t) = \prod_{j=1}^{n} |t_j|^{s_j} \cdot (\mathrm{sign}(t_j))^{\epsilon_j}.$$

Using this description we may write $\delta = (\frac{n-1}{2}, \frac{n-3}{2}, \ldots, \frac{1-n}{2}; 0, \ldots, 0) \in X$.

For $1 \leq j \leq n-1$ let $\alpha_j = e_j - e_{j+1} \in \Delta$ be the corresponding simple root. The corresponding embedding $\iota_j = \iota_{\alpha_j} : \mathrm{SL}_2 \to G$ may be extended to an embedding

$$\iota_j: \quad \mathrm{GL}_2(\mathbb{R}) \hookrightarrow G = \mathrm{GL}_n(\mathbb{R}), \qquad A \mapsto \begin{pmatrix} 1_{j-1} & 0 & 0 \\ 0 & A & 0 \\ 0 & 0 & 1_{n-j-1} \end{pmatrix},$$

where 1_ν denotes the unit matrix of size ν.

We have $z = \langle \chi^\circ, \check{\alpha}_j \rangle = s_j - s_{j+1}$. The condition $\epsilon_\chi \circ \check{\alpha}_j(-1) = (-1)^D$ now reads $\epsilon_j + \epsilon_{j+1} \equiv D \mod 2$. Here $\epsilon_\chi : \Sigma \to \{\pm 1\}$ denotes the character, which is described by the n-tuple $\epsilon \in (\mathbb{Z}/2\mathbb{Z})^n$.

9.3 J-ADMISSIBLE PERMUTATIONS

From now on we restrict our considerations to the case of $G = \mathrm{GL}_n$.

9.3.1 Transpositions and orthogonal subsets

We may identify the set of simple roots Δ with the set $\{1, 2, \ldots, n-1\}$ in such a way that we identify α_j with j. Thus we may write $w_j = w_{\alpha_j} \in W = S_n$ and this is the transposition which interchanges j with $j+1$. The action on the simple roots may be described by the formula:

$$w_j \alpha_i \quad = \quad \begin{cases} -\alpha_i & \text{if } i = j, \\ \alpha_i + \alpha_j & \text{if } i = j \pm 1, \\ \alpha_i & \text{else.} \end{cases} \tag{9.13}$$

Now let $J \subset \Delta = \{1, \ldots, n-1\}$ be a subset of mutually orthogonal simple roots. This means $|j - j'| \geq 2$ for all $j, j' \in J$ with $j \neq j'$. We call such subsets orthogonal subsets of Δ.

An element $w \in W$ is J-admissible iff $w(j+1) = w(j) + 1$ for all $j \in J$. We get the following list of examples of J-admissible elements in the Weyl group:

1. $w = w_j$ is J-admissible iff $j-1, j$ and $j+1$ are not elements of J.
2. If $j \in J$, but $j + 2 \notin J$ then $w = w_j \cdot w_{j+1}$ is J-admissible (mapping α_j to α_{j+1}).
3. If $j+1 \in J$, $j - 1 \notin J$, then $w = w_{j+1} \cdot w_j$ is J-admissible (mapping α_{j+1} to α_j).
4. If $j, j+2 \in J$, then $w = w_{j+1} \cdot w_j \cdot w_{j+2} \cdot w_{j+1} = w_{j+1} \cdot w_{j+2} \cdot w_j \cdot w_{j+1}$ is J-admissible; it interchanges j with $j+2$ and $j+1$ with $j+3$ and therefore interchanges the simple roots α_j and α_{j+2}.

We call the J-admissible permutations w of the above form *elementary J-admissible permutations.*

9.3.2 S_J and $S^>_{J,w}$

Now let $w \in S_n$ be a J-admissible permutation. Let S_J be the set (of cardinality $n - \#J$), whose elements are all sets of the form $\{j, j+1\}$ for $j \in J$ and all sets of the form $\{j\}$ where $j, j-1 \notin J$ and $j \leq n$. The elements of S_J (which are itself sets) have an obvious natural order induced by the order of their elements. Put

$$S^>_{J,w} = \{(a,b) \in S_J^2 \mid a < b, w(a) > w(b)\}, \quad \text{and} \quad l_J(w) = \#(S^>_{J,w}).$$

We call $l_J(w)$ the J-length of w. Thus w is elementary if and only if its J-length is 1. Furthermore, w induces a bijection from S_J to $S_{w(J)}$.

Lemma 9.5. *Let $w, \tilde{w} \in S_n$, and w be J-admissible.*

(a) $\tilde{w} \cdot w$ is J-admissible if and only if $\tilde{w}$ is $w(J)$-admissible.
(b) If $\tilde{w}$ is $w(J)$-admissible and if $l_J(\tilde{w} \cdot w) = l_{w(J)}(\tilde{w}) + l_J(w)$ then we have a disjoint decomposition

$$S^>_{J,\tilde{w}w} = S^>_{J,w} \,\dot{\cup}\, w^{-1}(S^>_{w(J),\tilde{w}}). \tag{9.14}$$

Proof. (a) is straightforward. For (b), we have a disjoint decomposition of $S^>_{J,\tilde{w}w}$ into the subsets:

$$\begin{aligned} S^{>>}_{J,\tilde{w}w,w} &= \{(a,b) \in S_J^2 \mid a < b, w(a) > w(b), \tilde{w}w(a) > \tilde{w}w(b)\} \qquad \text{and} \\ S^{<>}_{J,\tilde{w}w,w} &= \{(a,b) \in S_J^2 \mid a < b, w(a) < w(b), \tilde{w}w(a) > \tilde{w}w(b)\}. \end{aligned}$$

Now $S^{>>}_{J,\tilde{w}w,w}$ is a subset of $S^>_{J,w}$, which has cardinality $l_J(w)$, and $S^{<>}_{J,\tilde{w}w,w}$ is a subset of $w^{-1}(S^>_{w(J),\tilde{w}})$, which has cardinality $l_{w(J)}(\tilde{w})$. If $l_J(\tilde{w} \cdot w) = l_{w(J)}(\tilde{w}) + l_J(w)$ holds, then both inclusions must be equalities of sets and the claim follows. □

Lemma 9.6. *Each J-admissible permutation $w \in S_n = W(\mathrm{GL}_n)$ is the product of elementary permutations: $w = w_k \cdot \ldots \cdot w_1$ where w_i is an elementary J_i-admissible permutation with $J_i = w_{i-1} \cdot \ldots \cdot w_1(J)$.*

Proof. The lemma is a modification of the fact that S_n is generated by simple reflections; i.e. each permutation is a product of transpositions that interchange subsequent elements. The proof is by induction on $l_J(w)$, the case $l_J(w) = 0$ being trivial: So let us assume $S^>_{J,w} \neq \emptyset$. Then there exists a pair of subsequent elements $(a_0, b_0) \in S^>_{J,w}$; i.e. there is no $c \in S_J$ with $a_0 < c < b_0$. Let w_1

be the elementary J-admissible permutation that exchanges a_0 and b_0. Then $S^{>}_{J,w_1} = S^{>>}_{J,w,w_1} = \{(a_0, b_0)\}$ and $\tilde{w} = w \cdot w_1^{-1}$ is $w_1(J)$-admissible and we have $S^{<>}_{J,w,w_1} = w_1^{-1}(S_{w_1(J),\tilde{w}})$, since every (a, b) which is in the right-hand side, but not in the left-hand side has to satisfy $a > b, w_1(a) < w_1(b)$ and $w(a) > w(b)$. The first two conditions are satisfied only for $a = b_0$ and $b = a_0$, but then the third condition is violated. From this we conclude $l_{w_1(J)}(\tilde{w}) = \#(S^{<>}_{J,w,w_1}) = l_J(w) - 1$, and the claim is now an immediate consequence of the induction assumption. □

9.4 REPRESENTATIONS AND L-FUNCTIONS

9.4.1 Irreducible Representations of $W_{\mathbb{R}}$

Let $W_{\mathbb{R}} = \mathbb{C}^* \cup \mathbb{C}^*\sigma$ be the Weil group of $\mathbb{R}$ with the multiplication rules $\sigma \cdot z = \bar{z} \cdot \sigma$ for $z \in \mathbb{C}^*$ and $\sigma^2 = -1 \in \mathbb{C}^*$. Recall the classification of irreducible continuous complex representations of $W_{\mathbb{R}}$ and their (Artin)-L-functions:

The group of continuous homomorphisms from $\mathbb{C}^*$ to $\mathbb{C}^*$ is isomorphic to

$$\Xi = \{(s, t) \in \mathbb{C}^2 \mid s - t \in \mathbb{Z}\}$$

in such a way that (s, t) corresponds to the homomorphism

$$z \mapsto z^{(s,t)} := z^s \cdot \bar{z}^t := z^{s-t} \cdot (z\bar{z})^t.$$

In order that (s, t) extends to homomorphism $\rho : W_{\mathbb{R}} \to \mathbb{C}^*$ we must have $\epsilon = \rho(\sigma) \in \{\pm 1\}$ and $\rho(\bar{z}) = \rho(z)$. The latter condition is equivalent to $s = t$; i.e. $\rho : z \mapsto (z\bar{z})^s$ for arbitrary $s \in \mathbb{C}$. We denote this one-dimensional representation by ρ_s^ϵ. We put

$$L(\rho_s^{+1}, z) = \Gamma_{\mathbb{R}}(z + s) \quad \text{and} \quad L(\rho_s^{-1}, z) = \Gamma_{\mathbb{R}}(z + s + 1).$$

Since the irreducible continuous representations of $\mathbb{C}^*$ (which has index 2 in $W_{\mathbb{R}}$) are all one-dimensional each irreducible representation of $W_{\mathbb{R}}$ is either one-dimensional or induced from a one-dimensional representation of $\mathbb{C}^*$. Define $\rho_{(s,t)} = \mathrm{Ind}_{\mathbb{C}^*}^{W_{\mathbb{R}}}(s, t)$ for $(s, t) \in \Xi$. Then $\rho_{(s,t)}$ is irreducible iff $s \neq t$. In view of the isomorphism $\rho_{(s,t)} \sim \rho_{(t,s)}$ we have to include only those $\rho_{(s,t)}$ in the list of irreducible representations for which $s - t$ is a positive integer. We have a partial order on $\mathbb{C}$ if we define $s \geq t$ for $s - t \in \mathbb{R},\ s - t \geq 0$. We put

$$L(\rho_{(s,t)}, z) = \Gamma_{\mathbb{C}}(z + s) \qquad \text{if } s \geq t.$$

We remark that this equation is valid also in the reducible case in view of the decomposition $\rho_{(s,s)} = \rho_s^{+1} \oplus \rho_s^{-1}$ and (9.8). Of course the L-function of a reducible semisimple representation of $W_{\mathbb{R}}$ is the product of the L-functions of its irreducible components.

We get the following relations between representations:

$$\begin{aligned} \rho_{(s,t)} \otimes \rho_u^{\epsilon} &\cong \rho_{(s+u,t+u)}, \quad \text{especially,} \quad \rho_{(s,t)} \otimes \rho_0^{\epsilon} \cong \rho_{(s,t)} \\ \rho_s^{\epsilon_1} \otimes \rho_t^{\epsilon_2} &\cong \rho_{s+t}^{\epsilon_1\epsilon_2} \\ \rho_{(s,t)} \otimes \rho_{(u,v)} &\cong \rho_{(s+u,t+v)} \oplus \rho_{(s+v,t+u)} \\ (\rho_{(s,t)})^{\vee} &\cong \rho_{(-t,-s)} \qquad \text{and } (\rho_s^{\epsilon})^{\vee} \cong \rho_{-s}^{\epsilon}, \end{aligned}$$

where $\rho^{\vee}$ denotes the contragredient of ρ.

9.4.2 The L-functions $L(\chi, J, w, z)$

Let $\chi = (s, \eta_\chi)$ be a J-admissible character, where $s = (s_1, \ldots, s_n) \in \mathbb{C}^n$ and $\eta_\chi = (\eta_1, \ldots, \eta_n) \in (\mathbb{Z}/2\mathbb{Z})^n$. Put $\epsilon_i = (-1)^{\eta_i}$. For $a \in S_J$ we define an irreducible representation $\rho(a, \chi)$ of $W_{\mathbb{R}}$ in the following way: If $a = \{j\}$, then $\rho(a, \chi) = \rho_{s_j}^{\epsilon_j}$. If $a = \{j, j+1\}$, then $\rho(a, \chi) = \rho_{(s_j, s_{j+1})}$. Then we define the L-function

$$L(\chi, J, w, z) \quad := \quad \prod_{(a,b) \in S_{J,w}^{>}} L(\rho(a,\chi) \otimes \rho(b,\chi)^{\vee}, z).$$

Lemma 9.7. *Let $\tilde{w}w$ and w be J-admissible permutations and let χ be a J-admissible character. Assume $l_J(\tilde{w} \cdot w) = l_{w(J)}(\tilde{w}) + l_J(w)$. Then:*

$$L(\chi, J, \tilde{w}w, z) \quad = \quad L(\chi, J, w, z) \cdot L(w(\chi), w(J), \tilde{w}, z).$$

Proof. By Lem. 9.5 we may decompose $L(\chi, J, \tilde{w}w, z)$ into the product

$$\prod_{(a,b) \in S_{J,w}^{>}} L(\rho(a,\chi) \otimes \rho(b,\chi)^{\vee}, z) \cdot \prod_{(a,b) \in w^{-1}(S_{w(J),\tilde{w}}^{>})} L(\rho(a,\chi) \otimes \rho(b,\chi)^{\vee}, z).$$

Now we observe $\rho(a, \chi) = \rho(w(a), w(\chi))$. Replacing (a, b) in the last product by $(\tilde{a}, \tilde{b}) = (w(a), w(b)) \in S_{w(J),\tilde{w}}^{>}$ we thus get as second factor

$$\prod_{(\tilde{a},\tilde{b}) \in S_{w(J),\tilde{w}}^{>}} L(\rho(\tilde{a}, w(\chi)) \otimes \rho(\tilde{b}, w(\chi))^{\vee}, z) \quad = \quad L(w(\chi), w(J), \tilde{w}, z)$$

and the formula is proved. □

The following observation is crucial for the reduction of the main theorem from the case $n = 4$ to the case $n = 3$.

Lemma 9.8. *Let $n = 4$, $J = \{1, 3\}$, $\chi = (s, \eta_\chi)$ be an J-admissible character and $w = w_2w_1w_3w_2$ the elementary J-admissible permutation. Then we have*

$$L(\chi, J, w, z) \quad = \quad L(\chi, \tilde{J}, w, z),$$

where we may put

1. $\tilde{J} = \{1\}$, *if* $s_1 - s_2 \geq s_3 - s_4$, *and*
2. $\tilde{J} = \{3\}$, *if* $s_1 - s_2 \leq s_3 - s_4$.

Proof. We use the fact that $L(\rho_{(s,t)}, z) = \Gamma_{\mathbb{C}}(z + s)$ depends on s but not on t if $s \geq t$. The left-hand side is

$$L(\rho_{(s_1,s_2)} \otimes (\rho_{(s_3,s_4)})^\vee, z) \;=\; L(\rho_{(s_1-s_4,s_2-s_3)}, z) \cdot L(\rho_{(s_1-s_3,s_2-s_4)}, z).$$

Since χ is J-admissible we have $s_1 \geq s_2$ and $s_3 \geq s_4$, so that $s_1 - s_4 \geq s_2 - s_3$. Now assume $s_1 - s_2 \geq s_3 - s_4$ and put $\tilde{J} = \{1\}$. Then the product of L-functions is (observe $s_1 - s_3 \geq s_2 - s_4$):

$$\Gamma_{\mathbb{C}}(z + s_1 - s_4) \cdot \Gamma_{\mathbb{C}}(z + s_1 - s_3) \;=\; L(\rho_{(s_1-s_4,s_2-s_4)}, z) \cdot L(\rho_{(s_1-s_3,s_2-s_3)}, z),$$

which coincides with the right-hand side. The case $s_1 - s_2 \leq s_3 - s_4$ is completely analogous (observe that in case of equality both relations hold and $\rho_{(s_1-s_3,s_2-s_4)}$ decomposes into two one-dimensional representations). □

Definition 9.9. *Let $J \subset \Delta = \{1, \ldots, n-1\}$ be an orthogonal subset.*

(a) The compact torus associated to J is the torus $T_J = \prod_{\alpha \in J} K_\alpha$.
(b) J is called maximal orthogonal if and only if $\#J = \left[\frac{n}{2}\right]$.

We remark that J is maximal if and only if T_J is a maximal torus in $K^\circ = \mathrm{SO}_n$. For a J-admissible $w \in N_T(\mathbb{R}) \cap K$ we have $w(T_J) = T_{w(J)}$. In the case of even $n = 2m$ the set $J = \{1, 3, 5, \ldots, 2m - 1\}$ is the only maximal orthogonal subset. If $n = 2m + 1$ is odd, then the maximal orthogonal subsets are of the form $J_{j_0} = \{1, 3, \ldots, 2j_0 - 1, 2j_0 + 2, 2j_0 + 4, \ldots, 2m\}$ for some $0 \leq j_0 \leq m$. We write $J = \{\alpha_1, \ldots, \alpha_m\}$ in both cases.

9.4.3 Averaged highest weight vectors

Now let the character χ be admissible with respect to the maximal orthogonal subset J. This means that the differences $D_j = \langle \chi^\circ, \check{\alpha}_j \rangle + 1 = s_j - s_{j+1} + 1$ are positive integers for $j \in J$. We get a character of the maximal torus T_J:

$$\chi_T : \prod_{\alpha \in J} R_\alpha(\phi_\alpha) \quad \mapsto \prod_{\alpha \in J} e^{i\phi_\alpha \cdot D_\alpha}.$$

The character is positive with respect to some Borel subalgebra $\mathfrak{b} \subset \mathfrak{k}_\mathbb{C} = \mathrm{Lie}(K) \otimes \mathbb{C}$, which contains the Lie algebra $\mathfrak{t}$ of the torus T_J.

We want to make the construction of $\mathfrak{b}$ explicit: For $1 \leq j, k \leq n$ let E_{jk} denote the elementary matrix, which has a 1 in the j-th row and in the k-th column and zeros elsewhere. Let $K_{jk} = E_{jk} - E_{kj}$ for $j \neq k$. Then we have $K_{kj} = -K_{jk}$ and these matrices form a basis of $\mathfrak{k} = Lie(O_n)$. For $j \in J$ we write $t_j = i \cdot K_{j,j+1}$. These elements form a basis of $\mathfrak{t}$. Let $t_i^\vee$ denote the dual basis.

For $j, k \in J$ with $j \neq k$ and $\epsilon_1, \epsilon_2 \in \{\pm 1\}$ we introduce the elements

$$K_{j,k}^{\epsilon_1,\epsilon_2} = K_{jk} - i\epsilon_1 \cdot K_{j,k+1} - i\epsilon_2 \cdot K_{j+1,k} - \epsilon_1\epsilon_2 \cdot K_{j+1,k+1}.$$

Then we have $[t_j, K_{j,k}^{\epsilon_1,\epsilon_2}] = \epsilon_2 \cdot K_{j,k}^{\epsilon_1,\epsilon_2}$ and $[t_k, K_{j,k}^{\epsilon_1,\epsilon_2}] = \epsilon_1 \cdot K_{j,k}^{\epsilon_1,\epsilon_2}$. Since $K_{j,k}^{\epsilon_1,\epsilon_2}$ commutes with all other t_l this means that $K_{j,k}^{\epsilon_1,\epsilon_2}$ generates the root space of the root $\epsilon_2 \cdot t_j^\vee + \epsilon_1 \cdot t_k^\vee$. We have $K_{k,j}^{\epsilon_1,\epsilon_2} = -K_{j,k}^{\epsilon_2,\epsilon_1}$.

In the case of even n all roots are of the form $\pm t_j^\vee \pm t_k^\vee$.

If $n = 2m+1$ is odd and if $J = J_{j_0} = \{1, 3, \ldots, 2j_0-1, 2j_0+2, 2j_0+4, \ldots, 2m\}$ then there are additional matrices which generate the missing root spaces:

$$K_j^\epsilon = K_{j,2j_0+1} - i\epsilon \cdot K_{j+1,2j_0+1} \qquad \text{for } j \in J, \ \epsilon = \pm 1.$$

They satisfy $[t_j, K_j^\epsilon] = \epsilon \cdot K_j^\epsilon$ and $[t_k, K_j^\epsilon] = 0$ for $k \neq j$. Thus K_j^ϵ spans the root space of the root $\epsilon \cdot t_j^\vee$. From the commutation relation $[K_{j,k}, K_{k,l}] = K_{j,l}$ we deduce the relations for $j \neq k \neq l \neq j$:

$$\begin{aligned}
[K_{j,k}^{\epsilon_1,\epsilon_2}, K_{k,l}^{\epsilon_3,\epsilon_4}] &= (1 - \epsilon_1\epsilon_4) \cdot K_{j,l}^{\epsilon_3,\epsilon_2}, \\
[K_{j,k}^{\epsilon_1,\epsilon_2}, K_k^\epsilon] &= (1 - \epsilon\epsilon_1) \cdot K_j^{\epsilon_2}, \\
[K_{j,k}^{\epsilon_1,\epsilon_2}, K_{k,j}^{\epsilon_3,\epsilon_4}] &= (1 - \epsilon_1\epsilon_4)(\epsilon_2 - \epsilon_3) \cdot t_j - (1 - \epsilon_2\epsilon_3)(\epsilon_4 - \epsilon_1) \cdot t_k, \\
[K_j^{\epsilon_1}, K_k^{\epsilon_2}] &= -K_{j,k}^{\epsilon_2,\epsilon_1}.
\end{aligned}$$

The differential of the character χ_T is $d\chi_T = \sum_{j\in J} D_j \cdot t_j^\vee$. We introduce some total order $\prec$ on J in such a way that we have $j \prec k$ if $D_j < D_k$. Now let $\mathfrak{b}$ be the subspace generated by t_j for all $j \in J$, by $K_{j,k}^{1,1}$ for all $j, k \in J$ with $j \neq k$, by $K_{j,k}^{1,-1}$ for all j, k with $j \prec k$ and finally in the case of odd n by K_j^1 for all $j \in J$.

Then it is easy to see using the commutation relations, that $\mathfrak{b}$ is a Borel subalgebra of $\mathfrak{k}_\mathbb{C}$ and that $d\chi_T$ is in the positive chamber with respect to this algebra.

An irreducible representation V^0 of the connected compact group $K^0 = \mathrm{SO}_n(\mathbb{R})$ is uniquely described by its highest weight χ inside this positive chamber. Put $\Sigma^0 = \Sigma \cap \mathrm{SO}_n(\mathbb{R})$. For odd n put $\sigma_{\mathrm{out}} = -id$, for even n let $\sigma_{\mathrm{out}} \in \Sigma$

be the unique nontrivial element that stabiles the Borel algebra $\mathfrak{b}$ (and thus the positive chamber). Then $\sigma_{out} \in \Sigma - \Sigma^0$.

An irreducible representation V of $K = \mathrm{O}_n(\mathbb{R})$ is either induced from a highest weight representation of K^0 or an extension of such a representation to K. For odd n it is always an extension. For even n it may be properly induced and then the restriction of V to K^0 decomposes into two highest weight representations: $V = V^0 \oplus \sigma_{\mathrm{out}}(V^0)$. If V^0 has highest weight χ, then $\sigma_{\mathrm{out}}(V^0)$ has highest weight $\sigma_{\mathrm{out}}(\chi) \neq \chi$. We call such induced representation a highest weight representation of K (with highest weight χ or $\sigma_{\mathrm{out}}(\chi)$).

Definition 9.10. *Let $J \subset \Delta$ be a maximal orthogonal set and χ be an admissible character with respect to J.*

(a) An irreducible representation V of $K = \mathrm{O}_n(\mathbb{R})$ or of $K = \mathrm{SO}_n(\mathbb{R})$ is called a (J, χ)-minimal K-type if and only if it has χ_T as a highest weight with respect to some Borel subalgebra $\mathfrak{b} \subset \mathfrak{k}_{\mathbb{C}}$ as earlier.
(b) A nontrivial vector $v \in V$, with V as in (a), is called an (ϵ-)averaged highest weight vector if and only if it is of the form $v = \sum_{\sigma \in \Sigma} \epsilon(\sigma) \cdot \sigma v_0$, where v_0 is a highest weight vector of weight χ_T, and where Σ is the Σ^0 from above in case $K = \mathrm{SO}_n(\mathbb{R})$.

Remark: If $\tilde{v} = \sum_{\sigma \in \Sigma^0} \epsilon(\sigma) \cdot \sigma v_0 \in V^0$ is an (ϵ-)averaged highest weight vector with respect to $\mathrm{SO}_n(\mathbb{R})$, then $v = \tilde{v} + \tilde{v}_1$ with $\tilde{v}_1 = \epsilon(\sigma_{\mathrm{out}}) \cdot \sigma_{\mathrm{out}}(\tilde{v})$ is an (ϵ-)averaged highest weight vector with respect to $\mathrm{O}_n(\mathbb{R})$, if it is nontrivial. Observe that we have $\tilde{v}_1 = \pm\tilde{v}$ in the extension case. In the induced case we may write $v = (\tilde{v}, \tilde{v}_1) \in V^0 \oplus \sigma_{\mathrm{out}}(V^0)$. Observe that the operators $\tilde{T}_w$ respect the deomposition $V = V^0 \oplus \sigma_{\mathrm{out}}(V^0)$.

Lemma 9.11. *If v is an averaged highest weight vector in a (J, χ)-minimal K-type V where $K = \mathrm{O}_n(\mathbb{R})$, then $\Delta_J(V^\epsilon)$ is one-dimensional and generated by v.*

Proof. At first we prove that $\Delta_J(V)$ considered as a T_J module contains the character χ_T and all its Σ-transforms with multiplicity 1 and no other characters: Let $m = \mathrm{rank}(K) = \dim(T_J)$ and write each character of T_J as an element of $\mathbb{Z}^m$ with respect to the basis t_i introduced earlier. In case $n = 2m$ we assume that the ordering of the indices is such that $\alpha_1 = e_1 + e_2$ and $\alpha_i = e_i - e_{i-1}$ for $2 \leq i \leq m$ form the set of simple roots of $\mathfrak{b}$. In the case $n = 2m + 1$ let $\alpha_1 = e_1$ and $\alpha_i = e_i - e_{i-1}$ for $i \geq 2$ be the set of simple roots. Let $\phi = (n_1, n_2, \ldots, n_m) \in \mathbb{Z}^m$ be a character of T_J appearing in $\Delta_J(V)$. Since we have an action of Σ on $\Delta_J(V)$, all characters in a Σ-orbit appear with the same multiplicity. We may thus assume that ϕ satisfies $n_2 \geq 0, \ldots, n_m \geq 0$ and that the sign of n_1 is chosen in such a way that ϕ appears in $\Delta(V_0)$, where V_0 is the SO_n-subrepresentation of V generated by the highest weight vector v_0. Let $\chi_T = (k_1, k_2, \ldots, k_m)$ be the highest weight. Since V_0 is a quotient of the

Verma-module generated by v_0, we have $\phi = \chi_T - \sum_{i=1}^{m} \mu_i \cdot \alpha_i$, where $\mu_i \geq 0$ are integers and the α_i are the simple roots. Thus we have in the case $n = 2m$: $n_1 = k_1 - \mu_1 + \mu_2$, $n_2 = k_2 - \mu_1 - \mu_2 + \mu_3$, $n_i = k_i - \mu_i + \mu_{i+1}$ for $3 \leq i \leq m-1$ and finally $n_m = k_m - \mu_m$. In the case $n = 2m+1$ the second equation is replaced by $n_2 = k_2 - \mu_2 + \mu_3$. But by the definition of $\Delta_J(V)$ we must have $n_i \geq k_i$ for $i \geq 2$, since we assumed $n_i \geq 0$. Since the μ_i are all nonnegative this implies successively $\mu_m = 0, \mu_{m-1} = 0, \ldots, \mu_3 = 0, \mu_2 = \mu_1 = 0$. Thus $\phi = \chi_T$. Since χ_T has multiplicity one in the Verma module, it has multiplicity one in $\Delta_J(V)$ and the claim is proved.

As a consequence we get, that each Σ-eigenspace in $\Delta_J(V)$ is one-dimensional. But these are the spaces $\Delta_J(V^\epsilon)$. Since the averaged highest weight vectors are nontrivial vectors in these spaces, the lemma is proved. □

Lemma 9.12. *If v is an ϵ-averaged highest weight vector in a (J, χ)-minimal K-type V and if $w \in N_T(\mathbb{R}) \cap K$ represents a J-admissible permutation, then w^{-1} represents a $w(J)$-admissible permutation, V is also a $(w(J), w\chi)$-minimal K-type, and $w(v)$ is a $w(\epsilon)$-averaged highest weight vector in V.*

Proof. If we consider J to be a set of orthogonal simple roots, then $w(J)$ is again an orthogonal set of simple roots, since w represents a J-admissible permutation. But from $J = w^{-1}(w(J))$ we conclude that w^{-1} represents a $w(J)$-admissible permutation. If v_0 is a highest weight vector of weight χ_{T_J} with respect to the Borel subalgebra $\mathfrak{b}$, then $w(v_0)$ is a highest weight vector of weight $w(\chi_{T_J}) = w(\chi)_{T_{w(J)}}$ with respect to the Borel subalgebra $w(\mathfrak{b})$. Thus V is a $(w(J), w\chi)$-minimal K-type. If $v = \sum_{\sigma\in\Sigma} \epsilon(\sigma) \cdot \sigma v_0$, then

$$w(v) = \sum_{\sigma\in\Sigma} \epsilon(\sigma) \cdot w(\sigma v_0) = \sum_{\sigma\in\Sigma} \epsilon(w^{-1}\sigma w) \cdot \sigma(w(v_0)) = \sum_{\sigma\in\Sigma} w(\epsilon)(\sigma) \cdot \sigma(w(v_0)),$$

and the last claim is proved. □

9.5 THE MAIN THEOREM ON ARCHIMEDEAN INTERTWINING OPERATOR

Now we can state the main theorem:

Theorem 9.13. *Let $G = \mathrm{GL}_n$, $J \subset \{1, 2, \ldots, n-1\}$ be a maximal orthogonal subset, χ be a J-admissible character and let $w \in N_T(\mathbb{R}) \cap K$ represent a J-admissible permutation. Let V be a (J, χ)-minimal K-type, where $K = \mathrm{O}_n(\mathbb{R})$*

or $K = \mathrm{SO}_n(\mathbb{R})$, *and* $v \in V$ *be an averaged highest weight vector. Then we have*

$$\tilde{T}_w v \quad = \quad \frac{L(\chi, J, w, 0)}{L(\chi, J, w, 1)} \cdot w(v).$$

9.5.1 Plan of proof

It is easy to see that the claim remains unchanged if we replace w by $w\sigma$ for some $\sigma \in \Sigma$. Furthermore, it is an easy consequence of the considerations in Sect. 9.4.3, that the statements for $\mathrm{O}_n(\mathbb{R})$ and for $\mathrm{SO}_n(\mathbb{R})$ are equivalent. In the following we will deal with the $\mathrm{O}_n(\mathbb{R})$ case unless otherwise stated. Also, the theorem is trivial if $w \in \Sigma$ represents the trivial permutation. Therefore the case $n = 2$ is trivial, since then $J = \{1\}$ and the identity is the only J-admissible permutation. We will reduce the proof of the general case ($n > 3$) in several steps to the case $n = 3$: After the first reduction step we may assume, that w represents an elementary J-admissible permutation. The next step is to reduce to the cases $3 \le n \le 4$, and the last step will be the reduction of the case $n = 4$ to the case $n = 3$. The first proof in the case $n = 3$ is due to Zagier [74], but we will present an independent argument of a more algebraic and representation theoretic nature.

Let's add a remark here. It should be noted that the theorem should be true in slightly more generality, if we assume J to be orthogonal, but not maximal orthogonal. J-admissible characters and J-admissible permutations are still defined. But at the moment we do not have a general concept of a (J, χ)-minimal K-type V and of an averaged highest weight vector $v \in V$.

As an example of the validity of the formula let us consider the case $n = 2$, $J = \emptyset$ and $w = \begin{pmatrix} 0 & 1 \\ -1 & 0 \end{pmatrix}$. Then every character $\chi = (\chi_1, \chi_2)$ with data $s_1, s_2 \in \mathbb{C}$ and $\eta_1, \eta_2 \in \mathbb{Z}/2\mathbb{Z}$ is J-admissible. Put $s = s_1 - s_2$ and $\epsilon = (-1)^{\eta_1 + \eta_2}$.

We have $S_J = \{\{1\}, \{2\}\}$ and $S^{>}_{J,w} = \{(\{1\}, \{2\})\}$. Thus

$$L(\chi, J, w, z) \;=\; L(\rho(\chi_1) \otimes \rho(\chi_2)^{\vee}, z) \;=\; L(\rho_s^{\epsilon}, z)$$

and consequently $L(\chi, J, w, z) = \Gamma_{\mathbb{R}}(z + s)$ if $\epsilon = +1$ and $L(\chi, J, w, z) = \Gamma_{\mathbb{R}}(z + s + 1)$ if $\epsilon = -1$. Thus we get

$$\frac{L(\chi, J, w, 0)}{L(\chi, J, w, 1)} = \gamma_0(s) \quad \text{for } \epsilon = +1 \text{ by (9.9) and}$$

$$\frac{L(\chi, J, w, 0)}{L(\chi, J, w, 1)} = \gamma_0(s + 1) = \gamma_1(s) \quad \text{for } \epsilon = -1.$$

Now in the case $\epsilon = +1$ the restriction of the induced representation I_χ to $\mathrm{O}_2(\mathbb{R})$ contains the trivial one-dimensional representation as a subrepresentation. This should be a (J,χ)-minimal K-type. Each generating vector should be an averaged highest weight vector v. Then $\tilde{T}_w$ acts by multiplication with $c(s,0)=\gamma_0(s)$ on v, and v is invariant under the action of $w\in \mathrm{SO}_2(\mathbb{R})$. Thus the formula is valid in this case.

In the case $\epsilon=-1$ we have an $\mathrm{O}_2(\mathbb{R})$-submodule in I_χ of the form $V=V_1\oplus V_{-1}$, where $V_1=\mathbb{C}\cdot v_1$ and $V_{-1}=\mathbb{C}\cdot v_{-1}$ with $v_{-1}=\delta v_1$ for $\delta=\left(\begin{smallmatrix}1&0\\0&-1\end{smallmatrix}\right)$. Then $v=v_1+\epsilon\cdot v_{-1}$ should be considered as an averaged highest weight vector in the (J,χ)-minimal K-type V. We have $w(v)=(-i)\cdot v_1+\epsilon\cdot(-i)^{-1}\cdot v_{-1}$ and

$$\tilde{T}_w v \;=\; c(1,s)\cdot v_1+\epsilon\cdot c(-1,s)\cdot v_{-1} \;=\; (-i)\cdot\gamma_1(s)\cdot v_1+\epsilon\cdot(-i)^{-1}\cdot\gamma_{-1}(s)\cdot v_{-1}.$$

But this is $\gamma_1(s)\cdot w(v)$, since we have $\gamma_{-1}(s)=\gamma_1(s)$. Thus the formula is still valid in the case $\epsilon=-1$.

9.5.2 Reduction to elementary permutations

Now we reduce to the case that w is an elementary J-admissible permutation: By Lem. 9.6 we may write w as a product of elementary permutations. Assume $w=w_1\cdot w_2$ where $l_J(w)=l_{w_2(J)}(w_1)+l_J(w_2)$. As in the proof of Lem. 9.12 one can see that w_1 represents a $w_2(J)$-admissible permutation and that V is a $(w_2(J),w_2(\chi))$-minimal K-type with $w_2(v)$ as an averaged highest weight vector. Then it is immediate from Lem. 9.7 and the relation (9.1) that the claim for w_1 and for w_2 implies the claim for the product w. In order to apply (9.1), we leave it as an exercise to the reader that the relation for the l_J length implies the analogous relation for the length: $l(w)=l(w_1)+l(w_2)$.

9.5.3 Reduction to $n\le 4$

Next we reduce the claim to the case of the linear groups GL_3 and GL_4: For $2\le k<n$ let $\tilde{G}=\mathrm{GL}_k$. For $1\le l\le n-k+1$ let $\iota=\iota_l:\tilde{G}\hookrightarrow G=\mathrm{GL}_n$ be the embedding $A\mapsto diag(1_{l-1},A,1_{n-k-l+1})$. For $k=2$ this coincides with our earlier notation ι_l, if we consider l to be a simple root. Write $\tilde{K}=\mathrm{O}_k(\mathbb{R})\subset\mathrm{GL}_k(\mathbb{R})$. Then we have $\iota(\tilde{K})\subset K$.

Let $\tilde{J}\subset\tilde{\Delta}=\{1,\dots,k-1\}$ be a maximal orthogonal subset. Consider the embedding $\iota_l:\tilde{\Delta}\hookrightarrow\Delta, j\mapsto j+l-1$. Put $I_l=\{l,\dots,l+k-1\}=\iota_l(\tilde{\Delta})\cup\{l+k-1\}\subset\{1,\dots,n\}$. We say that $\tilde{J}$ is *adapted* to J and ι_l if and only if we have $\iota_l(\tilde{J})=J\cap I_l$ and if furthermore for all $j\in J$: $j\in I_l$ if and only if $j+1\in I_l$. This is equivalent to the facts that I_l is a union of members of the set S_J introduced in 9.3.2 and that $\tilde{J}$ is the inverse image of J under ι_l.

If this is the case, the intersection $\iota_l(\tilde{K})\cap T_J$ is the image under ι_l of a maximal torus $T_{\tilde{J}}$ in $\tilde{K}$. Let $\tilde{\chi}_T$ be the composition of the restriction of χ_T to $\iota_l(T_{\tilde{J}})$ with ι_l. On the diagonal torus $\tilde{T}=\iota_l^{-1}(\iota_l(\tilde{G})\cap T)$ in $\tilde{G}$ we consider the character $\tilde{\chi}=\chi\circ\iota_l$.

Lemma 9.14. *If $v \in V$ is an averaged highest weight vector in the (J, χ)-minimal K-type V and if $\tilde{J}$ is adapted to J and ι_l, then v generates an irreducible representation $\tilde{V} \subset V$ of $\tilde{K}$, which is a $(\tilde{J}, \tilde{\chi})$-minimal $\tilde{K}$-type and in which v is an averaged highest weight vector.*

Proof. Write $v = \sum_{\sigma \in \Sigma} \epsilon(\sigma) \cdot \sigma v_0$ as in Def. 9.10 where v_0 is a highest weight vector of weight χ_T with respect to the Borel subalgebra $\mathfrak{b}$, which is constructed as in 9.4.3. This means especially $uv_0 = 0$ for all u in the nilpotent radical $\mathfrak{u}$ of $\mathfrak{b}$. We have an obvious direct product decomposition $\Sigma = \tilde{\Sigma} \times \Sigma'$, where Σ' consists of all $\sigma = (\epsilon_1, \ldots, \epsilon_n) \in \Sigma$ with $\epsilon_i = 1$ for $i \in I_l$ and where $\tilde{\Sigma}$ consists of all $\sigma = (\epsilon_1, \ldots, \epsilon_n) \in \Sigma$ with $\epsilon_i = 1$ for $i \notin I_l$. Now it is easy to see that the intersection $\tilde{\mathfrak{b}}$ of $\mathfrak{b}$ with the complexified Lie algebra $\tilde{\mathfrak{k}}_{\mathbb{C}}$ of $\tilde{K}$ is a Borel subalgebra of $\tilde{\mathfrak{k}}_{\mathbb{C}}$: Since $\tilde{J}$ is adapted to J and ι_l, the algebra $\tilde{\mathfrak{b}}$ is the span of all those generators of $\mathfrak{b}$, for which the indices j, k and $j_0 + 1$ all lie in I_l.

Thus v_0 is annihilated by the unipotent radical $\tilde{\mathfrak{u}}$ of $\tilde{\mathfrak{b}}$. Since the action of Σ' commutes with the action of $\tilde{\mathfrak{b}}$, the same holds for the vector $\tilde{v}_0 = \sum_{\sigma' \in \Sigma'} \epsilon(\sigma') \cdot \sigma' v_0$. We remark $v = \sum_{\tilde{\sigma} \in \tilde{\Sigma}} \epsilon(\tilde{\sigma}) \cdot \tilde{\sigma} \tilde{v}_0$. Let W be the $\tilde{K}$ subrepresentation of V, which is generated by $\tilde{v}_0$, and let $W^0 \subset W$ be the $\tilde{K}^0 \simeq \mathrm{SO}_k(\mathbb{R})$-subrepresentation generated by $\tilde{v}_0$. Since $\tilde{v}_0$ is annihilated by $\tilde{\mathfrak{u}}$ and since $\tilde{v}_0$ has weight $\tilde{\chi}_T$ as a $T_{\tilde{J}}$ module, W^0 is a quotient of the Verma module. Since it is finite-dimensional, it has to be the irreducible quotient. Thus W^0 is an irreducible finite-dimensional $\tilde{K}^0$ module with highest weight $\tilde{\chi}$, and W is an irreducible finite-dimensional $\tilde{K}$ module with highest weight $\tilde{\chi}$. It is thus generated by each of its nontrivial elements, especially by v and thus is just the $\tilde{V}$ of the lemma. By construction v is an averaged highest weight vector in $\tilde{V}$.

Now $\tilde{V}$ is a $(\tilde{J}, \tilde{\chi})$-minimal $\tilde{K}$-type: This is an immediate consequence of the definitions, if we remark, that the character $\tilde{\chi}_{\tilde{K}}$, constructed from the character $\tilde{\chi}$ on the split torus $\tilde{T}$, is the same as the character $\tilde{\chi}_T$. Now the proof is complete. □

9.5.4 Proof of the reduction to $n \leq 4$

The reduction of the general case to the cases $n = 3, 4$ is now straightforward: Let $G = \mathrm{GL}_n$ for arbitrary $n \geq 4$, and let $w \in N_T(\mathbb{R}) \cap K$ represent an elementary J-admissible permutation $\underline{w}$. The case $\underline{w} = \underline{w}_j = (j, j+1)$ cannot occur, since J is maximal orthogonal. In the case $\underline{w} = \underline{w}_j \cdot \underline{w}_{j+1} = (j, j+1, j+2)$ we put $k = 3$ and $\tilde{J} = \{1\}$; in the case $\underline{w} = \underline{w}_{j+1} \cdot \underline{w}_j = (j, j+2, j+1)$ we put $k = 3$ and $\tilde{J} = \{2\}$; in the case $\underline{w} = \underline{w}_{j+1} \cdot \underline{w}_j \cdot \underline{w}_{j+2} \cdot \underline{w}_{j+1} = (j, j+2) \cdot (j+1, j+3)$ we put $k = 4$ and $\tilde{J} = \{1, 3\}$. For $\tilde{G} = \mathrm{GL}_k$ consider the embedding $\iota_j : \tilde{G} \hookrightarrow G$. Let $\tilde{K} = \mathrm{O}_k(\mathbb{R})$. Then there exists $\tilde{w} \in N_{\tilde{T}}(\mathbb{R}) \cap \tilde{K}$, which represents a $\tilde{J}$-admissible permutation $\underline{\tilde{w}}$, such that we may assume $w = \iota_j(\tilde{w})$ (perhaps after replacing w by some $w\sigma$). Furthermore, $\tilde{J}$ is adapted to J and ι_l. We assume $k < n$; i.e., ι_j is a proper embedding. Of course this is an additional condition only for $n = 4$.

Let χ be a J-admissible character. On the diagonal torus $\tilde{T}$ in $\tilde{G}$ we consider the character $\tilde{\chi} = \chi \circ \iota_l$. Let V be a (J,χ)-minimal K-type and $v \in V$ be an averaged highest weight vector. By lemma 9.14 the vector v generates an irreducible subrepresentation $\tilde{V} \subset V$, which is a $(\tilde{J},\tilde{\chi})$-minimal $\iota_l(\tilde{K})$-type and in which v is an averaged highest weight vector. By transport of structure via ι_l we can consider $\tilde{V}$ to be a $\tilde{K}$-module. By the definition of $\tilde{T}_w v$ we can calculate this entirely in the module $\tilde{V}$, which means that we have $\tilde{T}_w v = \tilde{T}_{\tilde{w}} v$. Since we furthermore have $L(\chi, J, w, s) = L(\tilde{\chi}, \tilde{J}, \tilde{w}, s)$ and $\tilde{w}(v) = w(v)$, the identity $\tilde{T}_w v = \frac{L(\chi,J,w,0)}{L(\chi,J,w,1)} \cdot w(v)$ is now reduced to the case of the group $\tilde{G} = \mathrm{GL}_k$ for $k = 3$ or $k = 4$.

9.5.5 Reduction of the case $n = 4$ to the case $n = 3$

For $n = 4$ the only maximal orthogonal subset is $J = \{1, 3\}$. We have two embeddings of $\tilde{G} = \mathrm{GL}_3$ into $G = \mathrm{GL}_4$: ι_1 into the upper left corner and ι_2 into the lower right corner.

Let $w \in N_T(\mathbb{R}) \cap K$ represent an elementary J-admissible permutation $\underline{w}$, such that the k of the preceding section equals 4. Thus w is not in the image of either ι_1 or ι_2, so that we cannot apply the preceding reduction step. Then $\underline{w} = \underline{w}_2 \cdot \underline{w}_1 \cdot \underline{w}_3 \cdot \underline{w}_2 = (13)(24) = \underline{w}_2 \cdot \underline{w}_3 \cdot \underline{w}_1 \cdot \underline{w}_2$. Let χ be a J-admissible character. Thus we have positive integers $D_1 = s_1 - s_2 + 1$ and $D_3 = s_3 - s_4 + 1$.

The main point of the reduction Lem. 9.14 was the observation, that the intersection $\tilde{\mathfrak{b}}$ of the Borel subalgebra $\mathfrak{b}$ with the image of the Lie algebra $\tilde{\mathfrak{k}}_{\mathbb{C}}$ was a Borel subalgebra of the latter. The reason for this was, that $\tilde{J}$ was adapted to J and ι_l and thus one could construct a basis of $\tilde{\mathfrak{b}}$ as a subset of a basis of $\mathfrak{b}$. In this simple form the argument does not work for the reduction step for both embeddings, but it still works in a modified form for one of them, if we choose this embedding in dependence of the order between D_1 and D_3. As generators for the nilradical of $\tilde{\mathfrak{b}}$ we can take linear combinations of nilpotent basis elements of $\mathfrak{b}$.

Case A: $D_1 \geq D_3$. Then we can put $\tilde{J} = \{1\}$, $l = 1$ and $j_0 = 1$. Furthermore, let $w' = w_2 \cdot w_3$ and $w'' = w_1 \cdot w_2$.

Case B: $D_1 \leq D_3$. Then we can put $\tilde{J} = \{3\}$, $l = 2$ and $j_0 = 0$. Furthermore, let $w' = w_2 \cdot w_1$ and $w'' = w_3 \cdot w_2$.

In both cases we put $\tilde{K} = \iota_l(\mathrm{O}_3)$. We always have $w = w' \cdot w''$, where w'' and w both are $\tilde{J}$-admissible; furthermore, $w''(\tilde{J}) = \{2\}$.

Lemma 9.15. *If $v \in V$ is an averaged highest weight vector in the (J,χ)-minimal K-type V and if $\tilde{J}$ and l are as defined earlier, then v generates an irreducible representation $\tilde{V} \subset V$ of $\tilde{K}$, which is a $(\tilde{J},\tilde{\chi})$-minimal $\tilde{K}$-type and in which v is an averaged highest weight vector.*

Proof. The proof of the first reduction lemma applies with the modification, that we can construct the generators of the nilradical of $\tilde{\mathfrak{b}}$ as linear combinations of basis elements of $\mathfrak{b}$:

In case A we can construct the Borel subalgebra $\mathfrak{b}$ with respect to the order relation $3 \prec 1$ and get the nilpotent generators $K_{3,1}^{1,1} = -K_{1,3}^{1,1}$ and $K_{3,1}^{1,-1} = -K_{1,3}^{-1,1}$. Since the highest weight vector v_0 is annihilated by both of them, it is annihilated by the linear combination $\frac{1}{2} \cdot \left(K_{1,3}^{1,1} + K_{1,3}^{-1,1}\right) = K_1^1$, where $K_1^1 = K_{1,3} - i \cdot K_{2,3}$ is constructed with respect to $\tilde{J} = \{1\}$ and $2j_0 + 1 = 3$. Thus v_0 is a highest weight vector with respect to $\tilde{\mathfrak{b}} = \mathfrak{b} \cap \iota_1(\tilde{\mathfrak{k}})$.

In case B we construct $\mathfrak{b}$ with respect to the relation $1 \prec 3$ and get the nilpotent generators $K_{1,3}^{1,1}$ and $K_{1,3}^{1,-1}$. Since the highest weight vector v_0 is annihilated by both of them, it is annihilated by the linear combination $\frac{1}{2} \cdot \left(K_{1,3}^{1,1} - K_{1,3}^{1,-1}\right) = i \cdot \iota_2(K_2^1)$, where $K_2^1 = K_{2,1} - i \cdot K_{3,1}$ is constructed with respect to $2j_0 + 1 = 1$. Thus v_0 is a highest weight vector with respect to $\tilde{\mathfrak{b}} = \mathfrak{b} \cap \iota_2(\tilde{\mathfrak{k}})$.

The rest of the argument is the same as in the proof of Lem. 9.14. □

9.5.6 End of reduction to $n = 3$

Now we can finish the reduction by collecting the results of several lemmas. At first we remark that we have $L(\chi, J, w, z) = L(\chi, \tilde{J}, w, z)$ in both cases by Lem. 9.8. By Lem. 9.7 we have

$$L(\chi, \tilde{J}, w, z) = L(\chi, \tilde{J}, w'', z) \cdot L(w''(\chi), w''(\tilde{J}), w', z),$$

since $w = w' \cdot w''$ and w'' represent $\tilde{J}$-admissible permutations. Thus we have to prove:

$$\tilde{T}_{w'}\tilde{T}_{w''}v = \frac{L(\chi, \tilde{J}, w'', 0)}{L(\chi, \tilde{J}, w'', 1)} \cdot \frac{L(w''(\chi), w''(\tilde{J}), w', 0)}{L(w''(\chi), w''(\tilde{J}), w', 1)} \cdot w'(w''(v)).$$

If we assume the theorem to be true in case $k = 3$ then we get

$$\tilde{T}_{w''}v = \frac{L(\chi, \tilde{J}, w'', 0)}{L(\chi, \tilde{J}, w'', 1)} \cdot w''(v),$$

since $w'' \in \tilde{K}$ and the assumptions of the theorem are fulfilled by Lem. 9.15.

Recall that we assumed that χ is a J-admissible character, $w \in N_T(\mathbb{R}) \cap K$ represents a J-admissible permutation, V is a (J, χ)-minimal K-type and $v \in V$ is an averaged highest weight vector. By Lem. 9.12 we get, that $w^{-1} = (w'')^{-1} \cdot (w')^{-1}$ represents a $w(J) = J$-admissible permutation, $w(\chi)$ is a $w(J) = J$-admissible character, V is a $(J, w(\chi))$-minimal K-type and $w(v)$ is an averaged highest weight vector. Observe that in $w(\chi)$ the roles of D_1 and D_3

are interchanged. Therefore let $\tilde{K}' = \iota_{l'}(O_3)$ with $l' = 3 - l$ be the *opposite* small compact subgroup, and let $\tilde{J}' = J - \tilde{J}$ be the opposite subset. Then $(w')^{-1} \in \tilde{K}'$ represents a $\tilde{J}'$-admissible permutation, and we can again apply Lem. 9.15 to this opposite situation and get that $\widetilde{w(v)}$ generates an irreducible representation $\tilde{V}' \subset V$ of $\tilde{K}'$, which is a $(\tilde{J}', \widetilde{w(\chi)})$-minimal $\tilde{K}'$-type and in which $w(v)$ is an averaged highest weight vector. But now we can apply Lem. 9.12 again with respect to the $\tilde{K}'$-modules and get that $w''(v) = (w')^{-1}(w(v))$ is an averaged highest weight vector in $\tilde{V}'$, that w' represents a $(w')^{-1}(w(\tilde{J})) = w''(\tilde{J})$-admissible permutation and that $\tilde{V}'$ is a minimal $(((w')^{-1}(\tilde{J}'), (w')^{-1}(\widetilde{w(\chi)})) = (w''(\tilde{J}), w''(\tilde{\chi}))$-type. Then we can apply the case $k = 3$ of the theorem and get

$$\tilde{T}_{w'}(w''(v)) \quad = \quad \frac{L(w''(\chi), w''(\tilde{J}), w', 0)}{L(w''(\chi), w''(\tilde{J}), w', 1)} \cdot w'(w''v)).$$

From these two equations the claim in the case $n = 4$ of the theorem follows immediately. □

9.5.7 The case GL_3

Now we will present an algebraic and representation theoretic argument that settles the $n = 3$ case (which was first proved by Zagier [74]). Here we will deal with the SO_3 version of the theorem. (In the following, for brevity, we write SO_3 for the group of its $\mathbb{R}$-points.)

9.5.7.1 Representations of SU_2 *and* SO_3

Let D be a positive integer and $V = V_\vartheta^* = \mathrm{Sym}^{2D}$ be the $(2D + 1)$-dimensional irreducible representation of SU_2. We may identify Sym^{2D} with the space of homogeneous polynomials of degree $2D$ in the variables X and Y in such a way that

$$g = \begin{pmatrix} a & b \\ c & d \end{pmatrix} \in \mathrm{GL}_2(\mathbb{C}) \text{ acts via } X \mapsto aX + cY,\ Y \mapsto bX + dY.$$

Since we deal with homogeneous polynomials of even degree, $-Id \in \mathrm{SU}_2$ acts trivially and the action of SU_2 induces an action of the quotient group $\mathrm{SO}_3 = \mathrm{SU}_2/\{\pm 1\}$.

In the case $D = 1$ we consider the following base of Sym^2, the space of quadratic polynomials with complex coefficients:

$$g_1 \ = \ -i(X^2 - Y^2), \qquad g_2 \ = \ X^2 + Y^2, \qquad g_3 \ = \ 2i \cdot XY.$$

With respect to this base the representation of U_2 on Sym^2 is described by a homomorphism $\pi : U_2 \to \mathrm{GL}_3$. It satisfies

$$\pi\begin{pmatrix} e^{i\alpha} & 0 \\ 0 & e^{-i\alpha}\end{pmatrix} = \begin{pmatrix} R(2\alpha) & 0 \\ 0 & 1\end{pmatrix}, \qquad \pi\begin{pmatrix} \cos\alpha & i\sin\alpha \\ i\sin\alpha & \cos\alpha\end{pmatrix} = \begin{pmatrix} 1 & 0 \\ 0 & R(2\alpha)\end{pmatrix}.$$

Since SU_2 is generated by the matrices of the left-hand side and since the matrices on the right-hand side are in SO_3 we see that π is a homomorphism from SU_2 to SO_3. In fact it is the double cover. In the following we will consider Sym^{2D} as a representation of SO_3 in such a way that $\gamma \in \mathrm{SO}_3$ acts by the standard action on Sym^{2D} of some inverse image of γ under π.

On the standard basis $e_b = X^b \cdot Y^{2D-b}$ $(b = 0, 1, \ldots, 2D)$ of V we get a diagonal action of $R_1(\alpha)$:

$$R_1(\alpha) = \begin{pmatrix} R(\alpha) & 0 \\ 0 & 1\end{pmatrix} : \quad e_b \mapsto e^{i\alpha(b-D)} \cdot e_b.$$

Similarly we get a diagonal action of $R_2(\alpha)$ on the basis

$$f_b \;=\; 2^{-D}\cdot(X+Y)^b\cdot(X-Y)^{2D-b} \;=\; Ae_b, \quad \text{where } A = \frac{1}{\sqrt{2}}\cdot\begin{pmatrix} 1 & 1 \\ 1 & -1\end{pmatrix}:$$

$$R_2(\alpha) = \begin{pmatrix} 1 & 0 \\ 0 & R(\alpha)\end{pmatrix} : \quad f_b \mapsto e^{i\alpha(b-D)} \cdot f_b.$$

On V we have a positive definite and U_2 invariant sesquilinear pairing, which is given on the standard basis by the formula

$$\langle \beta \cdot e_b, \gamma \cdot e_c\rangle \;=\; \beta \cdot \bar{\gamma} \cdot \delta_{b,c} \cdot \binom{2D}{b}^{-1}.$$

Write

$$Ae_b = f_b = \sum_{c=0}^{2D} C_{b,c}^{2D} \cdot e_c$$

with coefficients $C_{b,c}^{2D}$. Since the matrix A is symmetric of order 2, it is orthogonal (thus unitary) and we get

$$C_{c,b}^{2D} \cdot \binom{2D}{b}^{-1} \;=\; \langle e_b, Ae_c\rangle \;=\; \langle A^2 e_b, Ae_c\rangle \;=\; \langle Ae_b, e_c\rangle \;=\; C_{b,c}^{2D} \cdot \binom{2D}{c}^{-1}.$$

Thus we have the symmetry property (compare [74, (6)]):

$$C_{c,b}^{2D} \cdot \binom{2D}{c} \;=\; C_{b,c}^{2D} \cdot \binom{2D}{b}. \tag{9.15}$$

9.5.7.2 *The case $J = \{1\}$*

Now we want to prove the theorem in the case of $n = 3$ and the maximal orthogonal subset $J = \{1\}$. So let χ be a J-admissible character, described by a triple of complex numbers (s_1, s_2, s_3) and a triple $(\epsilon_1, \epsilon_2, \epsilon_3) \in (\mathbb{Z}/2\mathbb{Z})^3$. Since χ is J-admissible we have $s_1 - s_2 = D - 1$ for some $D \geq 1$ and $\epsilon_1 + \epsilon_2 \equiv D \mod 2$. Thus $s_2 = s_1 - D + 1$. We put $z = s_1 - s_3$ and thus get $s_2 - s_3 = z - D + 1$.

The only nontrivial J-admissible permutation is the cycle $(1, 2, 3) \in S_3$. It is represented by

$$w = \begin{pmatrix} 0 & 0 & 1 \\ 1 & 0 & 0 \\ 0 & 1 & 0 \end{pmatrix} = w_1 \cdot w_2 = R_1\left(\frac{\pi}{2}\right) \cdot R_2\left(\frac{\pi}{2}\right) = \pi(\tilde{A}),$$

where

$$\tilde{A} = \frac{1+i}{2} \cdot \begin{pmatrix} 1 & i \\ 1 & -i \end{pmatrix} = \begin{pmatrix} e^{i\frac{\pi}{4}} & 0 \\ 0 & e^{-i\frac{\pi}{4}} \end{pmatrix} \cdot \begin{pmatrix} \cos(\frac{\pi}{4}) & i\sin(\frac{\pi}{4}) \\ i\sin(\frac{\pi}{4}) & \cos(\frac{\pi}{4}) \end{pmatrix}.$$

Then we have $S^{>}_{J,w} = \{(\{1, 2\}, \{3\})\}$. Thus

$$L(\chi, J, w, s) = L(\rho_{(s_1,s_2)} \otimes (\rho_{s_3}^{\epsilon_3})^\vee, s) = L(\rho_{(s_1-s_3,s_2-s_3)}, s) = \Gamma_{\mathbb{C}}(z+s).$$

Consequently, we get

$$\frac{L(\chi, J, w, 0)}{L(\chi, J, w, 1)} = \frac{\Gamma_{\mathbb{C}}(z)}{\Gamma_{\mathbb{C}}(z+1)} = \frac{2\pi}{z} = \gamma_0(z) \cdot \gamma_0(z+1) \qquad (9.16)$$

by (9.9).

9.5.7.3 *The averaged highest weight vectors*

It is clear that $e_{2D} = X^{2D}$ is a highest weight vector for the compact torus K_1. Now we determine the averaged highest weight vectors: We have

$$\pi\begin{pmatrix} 0 & i \\ i & 0 \end{pmatrix} = \begin{pmatrix} 1 & 0 & 0 \\ 0 & -1 & 0 \\ 0 & 0 & -1 \end{pmatrix}, \quad \pi\begin{pmatrix} i & 0 \\ 0 & -i \end{pmatrix} = \begin{pmatrix} -1 & 0 & 0 \\ 0 & -1 & 0 \\ 0 & 0 & 1 \end{pmatrix}.$$

In order that X^{2D} has a nontrivial ϵ average $P_\epsilon(X^{2D}) = \sum_{\sigma \in \Sigma^0} \epsilon(\sigma) \cdot \sigma(X^{2D})$, it is necessary and sufficient that we have $D \equiv \epsilon_1 + \epsilon_2 \mod 2$. Let us assume that this is the case. Then we get the following averaged highest weight vector in V^ϵ:

$$X^{2D} + (-1)^{\epsilon_2+\epsilon_3} \begin{pmatrix} 0 & i \\ i & 0 \end{pmatrix} X^{2D} = X^{2D} + (-1)^{D+\epsilon_2+\epsilon_3} \cdot Y^{2D} = X^{2D} + (-1)^{\epsilon_1+\epsilon_3} \cdot Y^{2D}.$$

Similarly, if the congruence $D \equiv \epsilon_1 + \epsilon_2 \mod 2$ is satisfied, then $(X+Y)^{2D} + (-1)^{D+\epsilon_1+\epsilon_3} \cdot (X-Y)^{2D}$ is an averaged highest weight vector for the torus $K_2 = \{R_2(\alpha)\}$ in $V^{\epsilon'}$, where $\epsilon' = (\epsilon_3, \epsilon_1, \epsilon_2)$.

9.5.7.4 *The representation theoretic argument*

If we put $\epsilon = (-1)^{\epsilon_1+\epsilon_3}$ we can reformulate the claim of the theorem in the form

$$\tilde{T}_1(s_1 - s_3)\tilde{T}_2(s_2 - s_3)(X^{2D} + \epsilon \cdot Y^{2D}) = \gamma_0(z) \cdot \gamma_0(z+1) \cdot \tilde{A}(X^{2D} + \epsilon \cdot Y^{2D}),$$

or more explicitly:

$$\begin{aligned} &\tilde{T}_1(z)\tilde{T}_2(z-D+1)(X^{2D} + \epsilon \cdot Y^{2D}) \\ &= \gamma_0(z) \cdot \gamma_0(z+1) \cdot 2^{-D} \cdot i^D \cdot \left((X+Y)^{2D} + (-1)^D \cdot \epsilon \cdot (X-Y)^{2D}\right). \end{aligned} \tag{9.17}$$

By Lem. 9.11 the averaged highest weight vector $X^{2D} + \epsilon \cdot Y^{2D}$ spans the one-dimensional space $\Delta_J(V^\epsilon)$. Since this space is defined representation theoretically, the meromorphic intertwining operator $\tilde{T}_w$ maps it into the corresponding space that is generated by the averaged highest weight vector on the right-hand side.

Thus it is clear that the formula is valid up to a factor. To prove the formula it is thus sufficient that the coefficients of some fixed monomial coincide on both sides of the formula. In case $\epsilon = 1$ we compare the coefficients of the monomial $X^D Y^D$, in case $\epsilon = -1$ we use $X^{D-1}Y^{D+1}$.

9.5.7.5 *Beginning the computation*

Now we can compute:

$$X^{2D} = \left(\frac{X+Y}{2} + \frac{X-Y}{2}\right)^{2D} = 2^{-D} \cdot \sum_{b=0}^{2D} \binom{2D}{b} \cdot f_b.$$

Since the intertwining operator $\tilde{T}_2$ acts diagonally with respect to the f-basis, we get that $T_2(z-D+1)(X^{2D})$ is

$$2^{-D} \cdot \sum_{b=0}^{2D} \binom{2D}{b} \cdot (-i)^{D-b} \cdot \gamma_{D-b}(z-D+1) \cdot \sum_{c=0}^{2D} C_{b,c}^{2D} \cdot e_c$$

$$= 2^{-D} \cdot \sum_{c=0}^{2D} \sum_{b=0}^{2D} \binom{2D}{c} \cdot (-i)^{D-b} \cdot \gamma_{D-b}(z-D+1) \cdot C_{c,b}^{2D} \cdot e_c,$$

where we used the definition in 9.2.5 in the first line, and (9.15) in the second. We apply $\tilde{T}_1(z)$ to both sides and get

$$\tilde{T}_1(z)\tilde{T}_2(z-D+1)(X^{2D})$$
$$= 2^{-D} \cdot \sum_{c=0}^{2D} \sum_{b=0}^{2D} (-i)^{2D-b-c} \cdot \binom{2D}{c} \cdot C_{c,b}^{2D} \cdot \gamma_{D-b}(z-D+1) \cdot \gamma_{D-c}(z) \cdot e_c.$$

If we replace X^{2D} by Y^{2D} on the left-hand side, we have to introduce an additional factor $(-1)^b$ in each summand on the right-hand sides of the formulas above.

9.5.7.6 Computation with the + sign

Thus we get $\tilde{T}_1(z)\tilde{T}_2(z-D+1)(X^{2D}+Y^{2D})$ is

$$2^{-D+1} \cdot \sum_{c=0}^{2D} \sum_{\beta=0}^{D} (-i)^{2D-2\beta-c} \cdot \binom{2D}{c} \cdot C_{c,2\beta}^{2D} \cdot \gamma_{D-2\beta}(z-D+1) \cdot \gamma_{D-c}(z) \cdot e_c.$$

The formula $(X+Y)^D \cdot (X-Y)^D = (X^2-Y^2)^D = \sum_{\beta=0}^{D} (-1)^{D-\beta} \binom{D}{\beta} X^{2\beta} \cdot Y^{2D-2\beta}$ implies $C_{D,b}^{2D} = 0$ for odd b and $C_{D,2\beta}^{2D} = 2^{-D} \cdot (-1)^{D-\beta} \cdot \binom{D}{\beta}$.

Now it is an immediate consequence of (9.3) using induction and the Pascal triangle that we have

$$\sum_{\beta=0}^{D} (-i)^{-2\beta} \cdot C_{D,2\beta}^{2D} \cdot \gamma_{D-2\beta}(z-D+1)$$
$$= 2^{-D} \cdot (-1)^D \cdot \sum_{\beta=0}^{D} \binom{D}{\beta} \cdot \gamma_{D-2\beta}(z-D+1) = (-1)^D \cdot \gamma_0(z+1).$$

Thus the coefficient of $e_D = X^D Y^D$ in $\tilde{T}_1(z)\tilde{T}_2(z-D+1)(X^{2D}+Y^{2D})$ is

$$2^{-D+1} \cdot i^D \cdot \binom{2D}{D} \cdot \gamma_0(z+1) \cdot \gamma_0(z).$$

But $(X+Y)^{2D} + (-1)^D \cdot (X-Y)^{2D}$ has $2\binom{2D}{D}$ as coefficient of $X^D Y^D$. Thus it is clear that $X^D Y^D$ has the same coefficient on both sides of the formula (9.17), so that the theorem is proved in this case.

9.5.7.7 Computation with the − sign

Similarly we get $\tilde{T}_1(z)\tilde{T}_2(z-D+1)(X^{2D}-Y^{2D})$ is

$$2^{-D+1} \cdot \sum_{c=0}^{2D} \sum_{\beta=0}^{D-1} (-i)^{2D-2\beta-1-c} \cdot \binom{2D}{c} \cdot C^{2D}_{c,2\beta+1} \cdot \gamma_{D-2\beta-1}(z-D+1) \cdot \gamma_{D-c}(z) \cdot e_c.$$

The formula

$$\begin{aligned}(X+Y)^{D-1} \cdot (X-Y)^{D+1} &= (X^2+Y^2-2XY) \cdot (X^2-Y^2)^{D-1} \\ &= (X^2+Y^2-2XY) \cdot \sum_{\beta=0}^{D-1} (-1)^{D-1-\beta} \binom{D-1}{\beta} X^{2\beta} \cdot Y^{2D-2-2\beta}\end{aligned}$$

implies $C^{2D}_{D-1,2\beta+1} = 2^{-D+1} \cdot (-1)^{D-\beta} \cdot \binom{D-1}{\beta}$. Thus, the coefficient of $X^{D-1}Y^{D+1}$ in $\tilde{T}_1(z)\tilde{T}_2(z-D+1)(X^{2D}-Y^{2D})$ is

$$\begin{aligned}&2^{-D+1} \cdot \sum_{\beta=0}^{D-1} (-i)^{D-2\beta} \cdot \binom{2D}{D-1} \cdot C^{2D}_{D-1,2\beta+1} \cdot \gamma_{D-2\beta-1}(z-D+1) \cdot \gamma_1(z) \\ =\ & 2^{-2D+2} \cdot \gamma_1(z) \cdot i^D \cdot \binom{2D}{D-1} \cdot \sum_{\beta=0}^{D-1} \binom{D-1}{\beta} \cdot \gamma_{D-2\beta-1}(z-D+1) \\ =\ & 2^{-D+1} \cdot \gamma_0(z+1) \cdot i^D \cdot \binom{2D}{D-1} \cdot \gamma_0(z),\end{aligned}$$

where we used the relation (9.5) once and the relation (9.3) plus the Pascal triangle repeatedly. Now $(X+Y)^{2D} - (-1)^D \cdot (X-Y)^{2D}$ has $2\binom{2D}{D-1}$ as coefficient of $X^{D-1}Y^{D+1}$, and again we can deduce the claim of the theorem, which is now proved for $n=3$ and $J=\{1\}$.

9.5.7.8 The case $n = 3,\ J = \{2\}$

Let χ be a J-admissible character, described by $(s_1, s_2, s_3) \in \mathbb{C}^3$ and $\underline{\epsilon} = (\epsilon_1, \epsilon_2, \epsilon_3) \in (\mathbb{Z}/2\mathbb{Z})^3$. Then we have $s_2 - s_3 = D - 1 \geq 0$ for some integer $D \geq 1$ and $\epsilon_2 + \epsilon_3 \cong D \mod 2$. Thus $s_2 = s_3 + D - 1$. We put $z = s_1 - s_3$ and thus get $s_1 - s_2 = z - D + 1$.

The only nontrivial J-admissible permutation is the cycle $(1, 3, 2) \in S_3$. It is represented by

$$w' = \begin{pmatrix} 0 & -1 & 0 \\ 0 & 0 & -1 \\ 1 & 0 & 0 \end{pmatrix} = w_2 \cdot w_1 = R_2\left(\frac{\pi}{2}\right) \cdot R_1\left(\frac{\pi}{2}\right) = \pi(\tilde{A}'),$$

where

$$\tilde{A}' = \frac{1+i}{2} \cdot \begin{pmatrix} 1 & 1 \\ i & -i \end{pmatrix} = \begin{pmatrix} \cos(\frac{\pi}{4}) & i\sin(\frac{\pi}{4}) \\ i\sin(\frac{\pi}{4}) & \cos(\frac{\pi}{4}) \end{pmatrix} \cdot \begin{pmatrix} e^{i\frac{\pi}{4}} & 0 \\ 0 & e^{-i\frac{\pi}{4}} \end{pmatrix}.$$

Then we have $S^{>}_{J,w} = \{(\{1\}, \{2, 3\})\}$. Thus

$$\begin{aligned} L(\chi, J, w, s) &= L(\rho_{s_1}^{(-1)^{\epsilon_1}} \otimes \rho^{\vee}_{(s_2,s_3)}, s) \\ &= L(\rho_{(s_1-s_2, s_1-s_3)}, s) = L(\rho_{(z-D+1,z)}, s). \end{aligned}$$

Consequently, we get $L(\chi, J, w, s) = \Gamma_{\mathbb{C}}(z + s)$ again, and (9.16) is still valid in this case.

If $D \equiv \epsilon_2 + \epsilon_3 \mod 2$, then $v = 2^{-D} \cdot \left((X+Y)^{2D} + \epsilon \cdot (X-Y)^{2D}\right)$ with $\epsilon = (-1)^{\epsilon_1+\epsilon_3}$ is an averaged highest weight vector for the torus $K_2 = \{R_2(\alpha)\}$ in $V^{\underline{\epsilon}}$. We have $v = A \cdot v_0$ with $v_0 = X^{2D} + \epsilon \cdot Y^{2D}$.

Now an easy calculation shows $A \cdot \tilde{A} = \tilde{A}' \cdot A$. For the rest of this section we denote the images of $A, \tilde{A}, \tilde{A}' \in \mathrm{SU}_2$ in SO_3 by the same symbol. Furthermore, we have $A \cdot R_1(\alpha) \cdot A^{-1} = R_2(\alpha)$ and vice versa, since $A^2 = 1$. Thus the matrix A transforms the eigenspaces of K_1 into the eigenspaces of K_2. We get for the elementary intertwining operators:

$$\tilde{T}_2(z) = A \cdot \tilde{T}_1(z) \cdot A^{-1} \qquad \text{and} \qquad \tilde{T}_1(z) = A \cdot \tilde{T}_2(z) \cdot A^{-1}.$$

Consequently, if we put $Q = \frac{L(\chi,J,w,0)}{L(\chi,J,w,1)}$ and use the formula in the case $J = \{1\}$, then we get

$$\begin{aligned} \tilde{T}_{w'} v &= \tilde{T}_2(z) \circ \tilde{T}_1(z-D+1) v = (A \cdot \tilde{T}_1(z) \cdot A^{-1}) \cdot (A \cdot \tilde{T}_2(z-D+1) \cdot A^{-1}) \cdot A v_0 \\ &= A \cdot \tilde{T}_1(z) \cdot \tilde{T}_2(z-D+1)\, v_0 = Q \cdot A \cdot \tilde{A} v_0 = Q \cdot \tilde{A}' \cdot A v_0 = Q \cdot \tilde{A}' v. \end{aligned}$$

But this is the claim of the theorem, whose proof is now complete. □

9.6 APPLICATIONS TO COHOMOLOGY

9.6.1 Peter–Weyl and the Lie algebra cohomology

Let V be a finite-dimensional algebraic G-module and ρ an admissible $(\mathfrak{g}, K)$-module (Harish-Chandra module). The $(\mathfrak{g}, K)$-cohomology $H^p(\mathfrak{g}, K, V \otimes \rho)$ is the cohomology of the following complex:

$$\mathrm{Hom}_K\left(\Lambda^m(\mathfrak{g}/\mathfrak{k}), V \otimes \rho\right) \quad \cong \quad \mathrm{Hom}_K\left(\Lambda^m(\mathfrak{g}/\mathfrak{k}) \otimes V^*, \rho\right).$$

The differential is defined by

$$\begin{aligned}(\partial f)(g_0 \wedge \ldots \wedge g_m) &= \sum_{j=0}^{m}(-1)^j \cdot g_j \cdot f(g_0 \wedge \ldots \hat{g_j} \ldots \wedge g_m)\\ &+ \sum_{0\le j<k\le m} (-1)^{j+k} \cdot f([g_j, g_k] \wedge g_0 \wedge \ldots \hat{g_j} \wedge \ldots \hat{g_k} \wedge \ldots \wedge g_m).\end{aligned}$$

Let $\Lambda^m(\mathfrak{g}/\mathfrak{k}) \otimes V^* \cong \bigoplus_{j=1}^R V_j$ be a decomposition into irreducible K-modules. Then we have

$$\mathrm{Hom}_K\left(\Lambda^m(\mathfrak{g}/\mathfrak{k}), V \otimes \rho\right) \quad = \quad \bigoplus_{j=1}^{R} \mathrm{Hom}_K\left(V_j, \rho\right).$$

For $\chi \in X$ let now $\rho = I_\chi$ be the Harish-Chandra module attached to the unitarily induced representation $Ind_B^G \chi$. Recall the Peter–Weyl isomorphism of K-modules:

$$i_{PW} : I_\chi \quad \xleftarrow{\sim} \quad \bigoplus_{\vartheta \in \hat{K}} (V_\vartheta^*)^\epsilon \otimes V_\vartheta.$$

By Schur's Lemma i_{PW} induces a canonical isomorphism: $\mathrm{Hom}_K\left(V_j, I_\chi\right) \cong (V_j^*)^\epsilon$ and thus

$$\mathrm{Hom}_K\left(\Lambda^m(\mathfrak{g}/\mathfrak{k}), V \otimes \rho\right) \quad = \quad \left((\Lambda^m(\mathfrak{g}/\mathfrak{k}) \otimes V^*)^*\right)^\epsilon = \left(\Lambda^m(\mathfrak{g}/\mathfrak{k})^* \otimes V\right)^\epsilon. \tag{9.18}$$

If we have $\rho = \Delta_J(I_\chi)$ in the sense of 9.4, we get similarly

$$\mathrm{Hom}_K\left(\Lambda^m(\mathfrak{g}/\mathfrak{k}), V \otimes \Delta_J(I_\chi)\right) \quad = \quad \Delta_J\left(\Lambda^m(\mathfrak{g}/\mathfrak{k})^* \otimes V\right)^\epsilon.$$

Since we may replace $\mathrm{O}_n(\mathbb{R})$ by $\mathrm{SO}_n(\mathbb{R})$ in the Iwasawa decomposition, this isomorphism is still valid for $K = \mathrm{SO}_n(\mathbb{R})$ in the sense that the ϵ-eigenspace is built with respect to the smaller group Σ^0.

9.6.2 Frobenius reciprocity for parabolics

These isomorphisms are compatible with the isomorphisms coming from Frobenius reciprocity: Let $P \subset G$ be a parabolic with Levi subgroup M containing T, and let σ be an admissible $(\mathfrak{m}, K^M)$-module, where $\mathfrak{m} = \mathrm{Lie}(M)$ and $K^M = K \cap M(\mathbb{R}) = K \cap P(\mathbb{R})$. Extend σ to an admissible $(\mathfrak{p}, K^M)$-module by letting the nilpotent radical $\mathfrak{n}$ of $\mathfrak{p} = \mathrm{Lie}(P)$ act trivially on σ. Consider the arithmetically induced module ${}^{\mathrm{a}}\mathrm{Ind}_P^G\sigma$. Then we have an isomorphism of complexes:

$$r_P: \quad \mathrm{Hom}_K\left(\Lambda^m(\mathfrak{g}/\mathfrak{k}), V \otimes {}^{\mathrm{a}}\mathrm{Ind}_P^G\sigma\right) \quad \tilde{\to} \quad \mathrm{Hom}_{K^M}\left(\Lambda^m(\mathfrak{g}/\mathfrak{k}), V \otimes \sigma\right), \tag{9.19}$$

which may be described in the following way: We may shift V to the first variable of the *Hom*. Introducing the notation $W^m = \Lambda^m(\mathfrak{g}/\mathfrak{k}) \otimes V^*$ we then have to construct an isomorphism:

$$\tilde{r}_P: \quad \mathrm{Hom}_K(W^m, {}^{\mathrm{a}}\mathrm{Ind}_P^G\sigma) \quad \tilde{\to} \quad \mathrm{Hom}_{K^M}(W^m, \sigma).$$

The map $\tilde{r}_P$ from the left to the right-hand side consists in evaluating elements in ${}^{\mathrm{a}}\mathrm{Ind}_P^G\sigma$ on the identity. Conversely if $\Phi \in \mathrm{Hom}_{K^M}(W^m, \sigma)$ is given, then the corresponding $\tilde{\Phi}$ on the left-hand side has the property, that it maps $w \in W^m$ to the map $\tilde{\Phi}(w)$ on G, which maps pk to $p\,\Phi(kw)$ for $p \in P(\mathbb{R})$ and $k \in K$. By the Iwasawa decomposition this describes $\tilde{\Phi}(w)$ completely. It is easy to see that $\tilde{\Phi}$ is well defined and that the map $\Phi \mapsto \tilde{\Phi}$ is the inverse of the restriction map $\tilde{r}_P$.

If one replaces $\Lambda^m(\mathfrak{g}/\mathfrak{k})$ on the right-hand side of (9.19) by $\Lambda^m(\mathfrak{p}/\mathfrak{k}^M)$, (i.e. represents elements in $\mathfrak{g}/\mathfrak{k}$ by elements of $\mathfrak{p}$), then the right-hand side turns into a complex and r_P is an isomorphism of complexes.

Lemma 9.16. *In the case that $P = B$ is a Borel subgroup and $\sigma = \chi \otimes \delta$ we have ${}^{\mathrm{a}}\mathrm{Ind}_P^G\sigma = I_\chi$ and the isomorphism r_B from (9.19) coincides with the former isomorphism (9.18), which is induced from the Peter–Weyl isomorphism.*

Proof. At first we remark, that the right-hand sides of (9.18) and (9.19) are canonically isomorphic: In the case $P = B$ we have $M = T$ and $K^M = \Sigma$. Furthermore ϵ is the restriction of χ to Σ. Then the K^M equivariant elements in $Hom\,(\Lambda^m(\mathfrak{g}/\mathfrak{k}), V \otimes \sigma)$ are those elements in $Hom\,(\Lambda^m(\mathfrak{g}/\mathfrak{k}), V \otimes \sigma) = \Lambda^m(\mathfrak{g}/\mathfrak{k})^* \otimes V$, which transform under Σ with ϵ. The claim of the lemma then follows easily from the definitions. □

9.6.3 Specification of maximal parabolics in G

From now on we call the large group $G = \mathrm{GL}_N$ and assume that $N = n + n'$ is a sum of positive integers n, n'. Consider the following parabolic subgroups of $G = \mathrm{GL}_N$:

$$\begin{aligned} P &= \left\{ \begin{pmatrix} A & B \\ 0 & D \end{pmatrix} \middle| \; A \in \mathrm{GL}_n,\ B \in Mat_{n \times n'},\ D \in \mathrm{GL}_{n'} \right\} \\ P^- &= \left\{ \begin{pmatrix} A & 0 \\ C & D \end{pmatrix} \middle| \; A \in \mathrm{GL}_n,\ C \in Mat_{n' \times n},\ D \in \mathrm{GL}_{n'} \right\} \\ Q &= \left\{ \begin{pmatrix} A & B \\ 0 & D \end{pmatrix} \middle| \; A \in \mathrm{GL}_{n'},\ B \in Mat_{n' \times n},\ D \in \mathrm{GL}_n \right\}. \end{aligned}$$

We consider the Levi subgroup $M = P \cap P^- \simeq \mathrm{GL}_n \times GL_{n'}$. Let $\mathfrak{m}$ be the Lie algebra of M. We introduce the following elements of $N_c = \mathrm{O}_N \cap N_T(\mathbb{R})$:

$$\begin{aligned} w_G: &\quad e_j \mapsto e_{N+1-j} \\ w_M: &\quad e_j \mapsto e_{n+1-j} \quad \text{for } j \le n, \quad e_j \mapsto e_{n+N+1-j} \quad \text{for } n+1 \le j \le N \\ w_0: &\quad e_j \mapsto e_{n'+j} \quad \text{for } j \le n, \quad e_j \mapsto e_{j-n} \quad \text{for } n+1 \le j \le N. \end{aligned}$$

Then we have $w_G w_M = w_0$ and $Q = w_0 \cdot P^- \cdot w_0^{-1}$. Furthermore, $\det(w_0) = (-1)^{nn'}$. Let N (resp., N_-) be the unipotent radical of P (resp., P^-), and $\mathfrak{n}$ (resp., $\mathfrak{n}_-$) its Lie algebra. Thus we have

$$\mathfrak{n} = \left\{ \begin{pmatrix} 0 & B \\ 0 & 0 \end{pmatrix} \middle| \; B \in Mat_{n \times n'} \right\}, \qquad \mathfrak{n}_- = \left\{ \begin{pmatrix} 0 & 0 \\ C & 0 \end{pmatrix} \middle| \; C \in Mat_{n' \times n} \right\}.$$

9.6.4 The isomorphisms ι_- and ι_{w_0}

Let $\tilde{\sigma}$ be a $(\mathfrak{m}, K^M)$-module, where the underlying space and the action of K^M are the same as for σ. With W^m as earlier we get isomorphisms of vector spaces $r_{P^-}^{-1} \circ r_P$:

$$\begin{aligned} \mathrm{Hom}_K(W^m, {}^{\mathrm{a}}\mathrm{Ind}_P^G \sigma) \xrightarrow{\sim} \mathrm{Hom}_{K^M}(W^m, \sigma) &= \mathrm{Hom}_{K^M}(W^m, \tilde{\sigma}) \\ &\xrightarrow{\sim} \mathrm{Hom}_K(W^m, {}^{\mathrm{a}}\mathrm{Ind}_{P^-}^G \tilde{\sigma}). \end{aligned}$$

If we shift the V back to the second argument of the Hom-functor, we get

$$\iota_- : \quad \mathrm{Hom}_K(\Lambda^m(\mathfrak{g}/\mathfrak{k}), V \otimes {}^{\mathrm{a}}\mathrm{Ind}_P^G \sigma) \xrightarrow{\sim} \mathrm{Hom}_K(\Lambda^m(\mathfrak{g}/\mathfrak{k}), V \otimes {}^{\mathrm{a}}\mathrm{Ind}_{P^-}^G \tilde{\sigma}). \tag{9.20}$$

Recall $Q = w_0 \cdot P^- \cdot w_0^{-1}$. Then $M_Q = w_0 \cdot M \cdot w_0^{-1}$ is a Levi subgroup of

Q. Let $\tilde{\sigma}'$ be the $(\mathfrak{m}_Q, K^{M_Q})$-module, whose space is the same as the space of $\tilde{\sigma}$, and where $w_0 m w_0^{-1}$ acts in the same way as m of the representation $\tilde{\sigma}$ for $m \in \mathfrak{m}$ and for $m \in K^M$. We get an isomorphism:

$$l_{w_0} : \quad {}^{\mathrm{a}}\mathrm{Ind}_Q^G \tilde{\sigma}' \quad \tilde{\rightarrow} \quad {}^{\mathrm{a}}\mathrm{Ind}_{P^-}^G \tilde{\sigma}, \qquad f \mapsto f(w_0 \cdot\ ).$$

The composition of $l_{w_0}^{-1} \circ \iota_-$ gives the isomorphism:

$$\iota_{w_0} : \quad \mathrm{Hom}_K(\Lambda^m(\mathfrak{g}/\mathfrak{k}), V \otimes {}^{\mathrm{a}}\mathrm{Ind}_P^G \sigma) \tilde{\rightarrow} \mathrm{Hom}_K(\Lambda^m(\mathfrak{g}/\mathfrak{k}), V \otimes {}^{\mathrm{a}}\mathrm{Ind}_Q^G \tilde{\sigma}'). \tag{9.21}$$

We remark that this identification does not give an isomorphism of complexes and therefore does not induce isomorphisms on cohomology in general. The first reason is that we made no further assumptions between the compatibility of the $\mathfrak{m}$-actions on σ and on $\tilde{\sigma}$. Another and more serious reason is that we have to represent elements of $\mathfrak{g}/\mathfrak{k}$ by elements in $\mathfrak{p}$ for the isomorphism r_P and by elements in $\mathfrak{p}^-$ for the isomorphism r_{P^-}. Therefore the differentials in the two complexes are not defined in a compatible way.

Now let us assume that $\tilde{\sigma}$ is the twist of σ with the character δ_P^{-2}, which is the absolute value of the character ν, with which $M(\mathbb{R})$ acts on $\Lambda^D(\mathfrak{n}^*)$. Thus we have $\sigma = \sigma_0 \otimes \delta_P$ and $\tilde{\sigma} = \sigma_0 \otimes \delta_P^{-1}$. Then we may write ${}^{\mathrm{a}}\mathrm{Ind}_P^G \sigma = Ind_P^G \sigma_0$ and ${}^{\mathrm{a}}\mathrm{Ind}_{P^-}^G \tilde{\sigma} = Ind_{P^-}^G \sigma_0$. Then we have the intertwining operator $T_{w_0} : Ind_P^G \sigma_0 \rightarrow Ind_Q^G \sigma_0'$, which is the composition of an intertwining operator from $Ind_P^G \sigma_0$ to $Ind_{P^-}^G \sigma_0$ given by integration over $N_-(\mathbb{R})$ and of the operator $l_{w_0}^{-1}$. Here $\tilde{\sigma}' = \sigma_0' \otimes \delta_Q$. This intertwining operator induces a map of complexes

$$T_{w_0} : \mathrm{Hom}_K(\Lambda^m(\mathfrak{g}/\mathfrak{k}), V \otimes {}^{\mathrm{a}}\mathrm{Ind}_P^G \sigma) \rightarrow \mathrm{Hom}_K(\Lambda^m(\mathfrak{g}/\mathfrak{k}), V \otimes {}^{\mathrm{a}}\mathrm{Ind}_Q^G \tilde{\sigma}'). \tag{9.22}$$

9.6.5 Preparation of the second theorem

Now we assume that σ_0 is a submodule of a Borel induced representation $Ind_{B_M}^M \chi$, where $B_M = B \cap M$ is a Borel subgroup of M. To be more precise we assume $\sigma_0 = \Delta_J(Ind_{B_M}^M \chi)$, where J is a set of mutually orthogonal simple roots for M (and therefore also for G). By induction via stages we can write ${}^{\mathrm{a}}\mathrm{Ind}_P^G \sigma = Ind_P^G \sigma_0 = \Delta_J\left(Ind_B^G \chi\right)$. As we have already remarked we get an identification:

$$i_P : \quad \mathrm{Hom}_K\left(\Lambda^m(\mathfrak{g}/\mathfrak{k}), V \otimes {}^{\mathrm{a}}\mathrm{Ind}_P^G \sigma\right) \quad \tilde{\rightarrow} \quad \Delta_J\left(\mathrm{Hom}(\Lambda^m(\mathfrak{g}/\mathfrak{k}), V)\right)^{\epsilon}.$$

Now we can formulate the second version of our main theorem.

Theorem 9.17. *Let $N = n + n'$ as above, $J \subset \{1, 2, \ldots, N-1\}$ be a maxi-*

mal orthogonal subset with $n \notin J$, χ *be a* J*-admissible character and let* $w_0 \in N_T(\mathbb{R}) \cap K$ *be as above. Here* K *is either* $\mathrm{O}_N(\mathbb{R})$ *or in the case of even* nn' *it may be* $\mathrm{SO}_N(\mathbb{R})$. *Assume that the irreducible* K*-submodule* $V_\vartheta \subset \mathrm{Hom}(\Lambda^m(\mathfrak{g}/\mathfrak{k}), V)$ *is a* (J,χ)*-minimal* K*-type and* $v \in \Delta_J(V_\vartheta)^\epsilon$ *is an averaged highest weight vector. Let* $\tilde{v} \in \mathrm{Hom}_K\left(\Lambda^m(\mathfrak{g}/\mathfrak{k}), V \otimes {}^{\mathrm{a}}\mathrm{Ind}_P^G \sigma\right)$ *correspond to* v *via the isomorphism* i_P. *Then we have*

$$T_{w_0}\tilde{v} \quad = \quad \frac{L(\chi, J, w_0, 0)}{L(\chi, J, w_0, 1)} \cdot \iota_{w_0}(\tilde{v}).$$

Proof. Of course we reduce to Thm. 9.13. At first we note, that w_0 is J-admissible, since $n \notin J$ and since α_n is the only simple root that is transformed into a negative root by w_0. Now we consider variants of the isomorphism i_P: the first one, where ${}^{\mathrm{a}}\mathrm{Ind}_P^G\sigma$ on the left-hand side is replaced by ${}^{\mathrm{a}}\mathrm{Ind}_{P^-}^G\tilde{\sigma}$ and the right-hand side is unchanged, will be denoted i_{P^-}. The second one reads

$$i_Q: \quad \mathrm{Hom}_K\left(\Lambda^m(\mathfrak{g}/\mathfrak{k}), V \otimes {}^{\mathrm{a}}\mathrm{Ind}_Q^G\tilde{\sigma}'\right) \quad \tilde{\to} \quad \Delta_{w_0(J)}\left(\mathrm{Hom}(\Lambda^m(\mathfrak{g}/\mathfrak{k}), V)\right)^{w_0(\epsilon)}.$$

Then it is clear from (9.12), that $\tilde{T}_{w_0}v \in V_\vartheta$ corresponds to $T_{w_0}\tilde{v}$ via the isomorphism i_Q. To prove that $\iota_{w_0}(\tilde{v})$ corresponds to $w_0(v)$ via i_Q, we first observe that we have $i_{P^-} \circ \iota_- = i_P$. Secondly we remark that $l_{w_0}^{-1}$ acts by left translations with w_0^{-1} on the induced representations. Identifying the induced representations with spaces of functions on K it still acts by left translations. Under the Peter–Weyl isomorphism this transforms to the action of w_0 on V_ϑ^* tensored with the identity on V_ϑ. Thus we get

$$i_Q(\iota_{w_0}(\tilde{v})) \;=\; i_Q(l_{w_0}^{-1}(\iota_-(\tilde{v}))) \;=\; w_0(i_{P^-}(\iota_-(\tilde{v}))) = w_0(i_P(\tilde{v})) = w_0(v).$$

This concludes the proof of the theorem. □

Now we can compare this with what is needed for global applications.

9.6.6 Delorme's method

We continue using the notations of 9.6.3. The decomposition $\mathfrak{g}/\mathfrak{k} \cong \mathfrak{p}/\mathfrak{k}^M \cong \mathfrak{m}/\mathfrak{k}^M \oplus \mathfrak{n}$ allows a further refinement (see [9, Chap. III]):

$$\mathrm{Hom}_K\left(\Lambda^m(\mathfrak{g}/\mathfrak{k}), V \otimes {}^{\mathrm{a}}\mathrm{Ind}_P^G\sigma\right)$$
$$\cong \bigoplus_{p+q=m} \mathrm{Hom}_{K^M}\left(\Lambda^p(\mathfrak{m}/\mathfrak{k}^M), \mathrm{Hom}(\Lambda^q(\mathfrak{n}), V) \otimes \sigma\right).$$

Similarly, we get an isomorphism

$$\mathrm{Hom}_K\left(\Lambda^m(\mathfrak{g}/\mathfrak{k}), V\otimes {}^{\mathrm{a}}\mathrm{Ind}_{P^-}^G\tilde{\sigma}\right)$$
$$\cong \bigoplus_{p+q=m} \mathrm{Hom}_{K^M}\left(\Lambda^p(\mathfrak{m}/\mathfrak{k}^M), \mathrm{Hom}(\Lambda^q(\mathfrak{n}_-), V)\otimes\tilde{\sigma}\right).$$

If we identify the left-hand sides using the isomorphism (9.20), then the isomorphism of the right-hand sides is induced from an isomorphism $\iota_t : \mathfrak{n}\tilde{\to}\mathfrak{n}_-$, which comes from the fact, that under the embeddings $\mathfrak{n}\hookrightarrow\mathfrak{g}/\mathfrak{k}\hookleftarrow\mathfrak{n}_-$ we get the same subspace in $\mathfrak{g}/\mathfrak{k}$. The map ι_t is induced from the transposition map $A\mapsto {}^tA$, since $\mathfrak{k}$ is built up by skew symmetric matrices $A-{}^tA$. But the transposition map is equivariant for the group O_N and not for the full GL_N. Similarly the isomorphism ι_t is equivariant with respect to $M(\mathbb{R})\cap\mathrm{O}_N$, but not with respect to the full group $M(\mathbb{R})$. Therefore the induced isomorphism ι_t for the vector spaces on the right-hand side is not an isomorphism of complexes.

In fact ι_t is not the isomorphism which is used in the rest of the monograph. So we have to compare it with the construction in the preceding sections.

9.6.7 M-equivariant isomorphisms

The trace form $\beta:\mathfrak{g}\times\mathfrak{g}\to\mathbb{C}, (A,B)\mapsto tr(A\cdot B)$ is a perfect pairing and invariant under the adjoint action of G. It induces a perfect pairing $\beta:\mathfrak{n}\times\mathfrak{n}_-\to\mathbb{C}$, which is invariant under the adjoint action of M. Therefore we may identify the dual $\mathfrak{n}^*$ of $\mathfrak{n}$ as an M-module with $\mathfrak{n}_-$.

Let $D=\dim(\mathfrak{n})=n\cdot n'$. Then we have for each $0\le d\le D$ a canonical M-equivariant isomorphism $\Lambda^d(\mathfrak{n})\cong\mathrm{Hom}(\Lambda^{D-d}(\mathfrak{n}),\Lambda^D(\mathfrak{n}))$, which is given by the map $a\mapsto(b\mapsto a\wedge b)$. From this we deduce the M-equivariant isomorphism

$$\begin{aligned}\iota_M:\ \Lambda^d(\mathfrak{n}) &\cong \mathrm{Hom}(\Lambda^{D-d}(\mathfrak{n}),\Lambda^D(\mathfrak{n})) \cong \Lambda^{D-d}(\mathfrak{n})^*\otimes\Lambda^D(\mathfrak{n}) \\ &\cong \Lambda^{D-d}(\mathfrak{n}^*)\otimes\Lambda^D(\mathfrak{n}) \cong \Lambda^{D-d}(\mathfrak{n}_-)\otimes\Lambda^D(\mathfrak{n}).\end{aligned}\tag{9.23}$$

To be more explicit, we may find $\iota_M\left(\Lambda_{j=1}^d e_j\right)$ for linearly independent $e_1,\dots,e_d\in\mathfrak{n}$ in the following way: At first complete to a basis $\{e_j\}_{1\le j\le D}$ of $\mathfrak{n}$. Next determine the dual basis f_k of $\mathfrak{n}_-$; i.e. $\beta(e_j,f_k)=\delta_{jk}$. Then we have

$$\iota_M\left(\Lambda_{j=1}^d e_j\right) = \Lambda_{j=d+1}^D f_j\otimes\Lambda_{j=1}^D e_j.$$

If a nontrivial $\omega\in\Lambda^D(\mathfrak{n})^*=\Lambda^D(\mathfrak{n}_-)$ is fixed, we can use it to define a contraction isomorphism $c_\omega:\Lambda^{D-d}(\mathfrak{n}_-)\otimes\Lambda^D(\mathfrak{n})\simeq\Lambda^{D-d}(\mathfrak{n}_-)$. Composition with ι_M gives

$$\iota_{M,\omega}=c_\omega\circ\iota_M:\quad \Lambda^d(\mathfrak{n})\tilde{\to}\Lambda^{D-d}(\mathfrak{n}_-).$$

This isomorphism is equivariant with respect to the group

$$M' = \{(A,B) \in \mathrm{GL}_n \times \mathrm{GL}_{n'} \mid \det(A)^{n'} = 1 = \det(B)^n\},$$

which includes the derived group $M_{der} = \mathrm{SL}_n \times \mathrm{SL}_{n'}$. Dualizing (9.23) we get an isomorphism

$$\iota_M^* : \Lambda^d(\mathfrak{n})^* \otimes \Lambda^D(\mathfrak{n}) \cong \Lambda^{D-d}(\mathfrak{n}_-)^*. \tag{9.24}$$

We remark that even in the case $d = \frac{D}{2}$ the isomorphism $\iota_{M,\omega}$ is completely different from the isomorphism ι_t. But we will see, that these morphisms may agree for suitable choices of ω on vectors, which are related to averaged highest weight vectors.

9.6.8 The irreducible M-modules

Let V be an irreducible finite-dimensional representation of $G = \mathrm{GL}_N$ of highest weight λ. By Kostant's theorem [46] the cohomology $H^\bullet(\mathfrak{n}, V)$ is the direct sum over irreducible M-modules of highest weight $w(\lambda+\delta)-\delta$, where w runs over W^P. In fact, the weight $w(\lambda+\delta)-\delta$ appears in the complex $\mathrm{Hom}(\Lambda^q(\mathfrak{n}), V)$ only in degree $q = l(w)$ and here with multiplicity one and thus generates an irreducible M-submodule $V_{w(\lambda+\delta)-\delta}$ of $\mathrm{Hom}(\Lambda^{l(w)}(\mathfrak{n}), V)$, which projects injectively into cohomology, and all the cohomology is obtained in this way. By the isomorphism (9.24) we get an injection

$$V_{w(\lambda+\delta)-\delta} \otimes \Lambda^D(\mathfrak{n}) \subset \mathrm{Hom}(\Lambda^{l(w)}(\mathfrak{n}), V) \otimes \Lambda^D(\mathfrak{n}) \cong \mathrm{Hom}(\Lambda^{D-l(w)}(\mathfrak{n}_-), V),$$

and this projects injectively into the cohomology.

Observe that M acts on $\Lambda^D(\mathfrak{n})$ with the character $\chi : M = \mathrm{GL}_n \times \mathrm{GL}_{n'} \ni (A,B) \mapsto \det(A)^{n'} \cdot \det(B)^{-n} \in \mathbb{G}_m$. Now let $V_{w(\lambda+\delta)-\delta,\mathrm{Tate}}$ be the representation of M which has the same underlying space as $V_{w(\lambda+\delta)-\delta}$, but where the M-action is twisted by the character χ. So $V_{w(\lambda+\delta)-\delta,\mathrm{Tate}}$ is a concrete realization of the highest weight module $V_{w(\lambda+\delta)-\delta+\chi}$. If some $\omega \in \Lambda^D(\mathfrak{n})$ is fixed, we thus get an M-equivariant isomorphism

$$V_{w(\lambda+\delta)-\delta,\mathrm{Tate}} \xrightarrow{\sim} V_{w(\lambda+\delta)-\delta} \otimes \Lambda^D(\mathfrak{n}), \quad v \mapsto v \otimes \omega,$$

and consequently we get an embedding

$$V_{w(\lambda+\delta)-\delta,\mathrm{Tate}} \hookrightarrow H^{D-l(w)}(\mathfrak{n}_-, V). \tag{9.25}$$

9.6.9 Considerations with ι_t

Let us assume that we are in the situation of Thm. 9.17. So we have an averaged highest weight vector $v \in \Delta_J(V_\vartheta)^\epsilon$, where $V_\vartheta \subset \mathrm{Hom}(\Lambda^m(\mathfrak{g}/\mathfrak{k}), V) =: W$ is irreducible. Then v is a linear combination of eigenvectors $v = \sum_{\lambda \in X} v_\lambda$ for the compact torus $T_c \subset K^M \subset K$. Here X is a certain Σ-orbit of characters of T_c. We assume that each character λ appears with multiplicity 1 in the K-module W. We can identify

$$W \cong \bigoplus_{p+q=m} \Lambda^p(\mathfrak{m}/\mathfrak{k}^M)^* \otimes \Lambda^q(\mathfrak{n})^* \otimes V \tag{9.26}$$

as modules for the group $M(\mathbb{R}) \cap K$, which has $T_c^\Sigma = \Sigma \cdot T_c$ as a subgroup. Since each character $\lambda \in X$ of T_c appears with multiplicity 1 in the sum of tensor products, there has to be a unique pair (p, q) such that each v_λ has to be a pure tensor product $v_\lambda = v_{\lambda,M} \otimes v_{\lambda,N} \otimes v_{\lambda,V}$, where $v_{\lambda,M} \in \Lambda^p(\mathfrak{m}/\mathfrak{k}^M)^*$, $v_{\lambda,N} \in \Lambda^q(\mathfrak{n})^*$ and $v_{\lambda,V} \in V$ all generate one-dimensional T_c-eigenspaces. Furthermore, the different $v_{\lambda,N}$ for $\lambda \in X$ belong to eigenspaces of one Σ-orbit of characters of T_c. This means that there is an irreducible representation $V_q \subset \Lambda^q(\mathfrak{n})^*$ of the group $T_c^\Sigma = \Sigma \cdot T_c$, which contains all the $v_{\lambda,N}$ for different $\lambda \in X$. Also, the representation V_q appears with multiplicity one in $\Lambda^q(\mathfrak{n})^*$.

Since T_c^Σ is a subgroup of $M(\mathbb{R}) \cap K$, the transposition map $\iota_t : \mathfrak{n} \to \mathfrak{n}_-$ is T_c^Σ-equivariant. Therefore it maps V_q to an irreducible T_c^Σ-submodule of $\Lambda^q(\mathfrak{n}_-)$. So if we write $v = \sum_{\lambda \in X} v_{\lambda,M} \otimes v_{\lambda,N} \otimes v_{\lambda,V}$ with respect to (9.26), then

$$v = \sum_{\lambda \in X} v_{\lambda,M} \otimes \iota_t(v_{\lambda,N}) \otimes v_{\lambda,V}$$

with respect to the decomposition

$$W \cong \bigoplus_{p+q=m} \Lambda^p(\mathfrak{m}/\mathfrak{k}^M)^* \otimes \Lambda^q(\mathfrak{n}_-)^* \otimes V.$$

9.6.10 Intersection with SO_{2n_1}

Now let $T_c^{\Sigma,0} = T_c^\Sigma \cap M'$. Until otherwise mentioned, assume that $n = 2n_1$ is even and $n' = 2n_2 + 1$ is odd. Then we have $T_c^{\Sigma,0} = T_c \cdot \Sigma_0$, where

$$\Sigma_0 = \{(\epsilon_1, \ldots, \epsilon_N) \in \Sigma \mid \prod_{i=1}^{n} \epsilon_i = 1\}.$$

Then Σ_0 has index 2 in Σ and $T_c^{\Sigma,0}$ has index 2 in T_c^Σ.

Restricted to $T_c^{\Sigma,0}$ the representation V_q remains either irreducible or decomposes into the sum of two inequivalent irreducible $T_c^{\Sigma,0}$-representations: $V_q = V_q^+ \oplus V_q^-$. Let $V_q^?$ denote either V_q (if it remains irreducible) or one of these submodules if V_q decomposes. In all cases $V_q^?$ is an irreducible $T_c^{\Sigma,0}$-submodule, which appears with multiplicity one in $\Lambda^q(\mathfrak{n})^*$. Again, $\iota_t(V_q^?)$ is an irreducible subrepresentation of $\Lambda^q(\mathfrak{n}_-)^*$ that appears with multiplicity one.

But now $T_c^{\Sigma,0}$ is a subgroup of $M'(\mathbb{R})$ and thus $\iota_{M,\omega}$ for a nontrivial $\omega \in \Lambda^D(\mathfrak{n})^*$ is another $T_c^{\Sigma,0}$-equivariant isomorphism between these irreducible representations. By Schur's lemma it coincides with ι_t up to a constant factor. Therefore there exists a unique $\omega^?$, such that $\iota_{M,\omega^?}$ and ι_t agree on $V_q^?$.

Remark 9.18. In the case $V_q = V_q^+ \oplus V_q^-$ we have $\omega^- = -\omega^+$.

Proof: Let $\omega = \omega^+$ be chosen as above. We have to prove $\iota_{M,-\omega}(v^-) = \iota_t(v^-)$ for a given $v^- \in V^-$. Fix some $\sigma \in \Sigma - \Sigma_0$. Thus $\sigma(-\omega) = \omega$. Then we have $V^- = \sigma(V^+)$. Thus $v^- = \sigma(v)$ for some $v \in V^+$. Now we use the fact that ι_t and ι_M are T_c^{Σ}-equivariant since T_c^{Σ} is included in K and in $M(\mathbb{R})$. Thus

$$\iota_{M,-\omega}(v^-) = c_{-\omega}(\iota_M(\sigma v)) = c_{-\omega}(\sigma(\iota_M v)) = \sigma(c_{\sigma(-\omega)}(\iota_M v)).$$

But now $\sigma(-\omega) = \omega$ implies

$$\iota_{M,-\omega}(v^-) = \sigma(c_\omega(\iota_M v)) = \sigma(\iota_{M,\omega}(v)) = \sigma(\iota_t(v)) = \iota_t(\sigma(v)) = \iota_t(v^-). \quad \square$$

Lemma 9.19. *With the preceding notations we have*

$$\omega = i^{n_1} \bigwedge_{1 \leq j \leq n, n+1 \leq k \leq N} E_{jk}$$

for a suitable ordering of the E_{jk} factors in the wedge product.

Proof. We sketch the proof: It suffices to check $\iota_{M,\omega}(v) = \iota_t(v)$ for some T_c-eigenvector $v \in \Lambda^d(\mathfrak{n})$. Over $\mathbb{C}$ we may conjugate T_c into the standard split Torus $T(\mathbb{C})$, and we may assume that v then becomes an eigenvector with respect to the restriction of the character $w\delta - \delta$, where $w \in W^P$ satisfies the condition of the combinatorial lemma (see Lem. 7.14). In particular, we have $l(w) = \frac{nn'}{2} = d = \frac{D}{2}$. As in the proof of Lem. 7.16 we may partition the index set $S = \{(j,k),\ 1 \leq j \leq n, n+1 \leq k \leq N\}$ for the standard basis E_{jk} of $\mathfrak{n}$ into subsets of order 4 and 2, which are denoted Q_{jk} there. They are of the form $Q_{jk} = \{(j,k),(j,l),(m,k),(m,l)\}$ resp. $Q_{jk} = \{(j,k),(m,k)\}$. The vector v may then be obtained as $\wedge$-product of elements of the form

$$(E_{jk} + iE_{jl}) \wedge (E_{mk} + iE_{ml}) \qquad \text{(first subcase of the case } \#(Q_{jk}) = 4\text{)},$$

$$(E_{jk} + iE_{mk}) \wedge (E_{jl} + iE_{ml}) \qquad \text{(second subcase of the case } \#(Q_{jk}) = 4\text{)},$$

$$E_{jk} + iE_{mk} \qquad \text{(in the case } \#(Q_{jk}) = 2\text{)}.$$

The effect of $\iota_{M,\omega}(v)$ for $\omega = i^{n_1} \cdot \bigwedge_{1 \leq j \leq r, r+1 \leq k \leq n} E_{jk}$ may be calculated for each individual factor of v of this $\wedge$ product with respect to a corresponding factor of ω and compared with the effect of ι_t. We get that these effects agree, if we take as a corresponding factor $E_{jk} \wedge E_{jl} \wedge E_{mk} \wedge E_{ml}$ in the first subcase and $E_{jk} \wedge E_{mk} \wedge E_{jl} \wedge E_{ml}$ in the second subcase of the case $\#(Q_{jk}) = 4$ and $i \cdot E_{jk} \wedge E_{mk}$ in the case $\#(Q_{jk}) = 2$. Since there are exactly n_1 sets of order 2, we get an ω as in the claim. The order in which we wedge the contributions of the different Q_{jk} does not matter, since they lie in the even part of the exterior algebra. □

9.6.11 Final Considerations

From (7.41) we have an isomorphism

$$T_{\mathrm{loc}}(\sigma)(\epsilon') : \quad H^b(\mathfrak{g}, K, {}^{\mathrm{a}}\mathrm{Ind}_P^G \sigma \otimes V_\lambda)(\epsilon') \xrightarrow{\sim} H^b(\mathfrak{g}, K, {}^{\mathrm{a}}\mathrm{Ind}_Q^G \tilde{\sigma}' \otimes V_\lambda)(\epsilon'), \tag{9.27}$$

which should be compared with the isomorphism induced from our map ι_{w_0} in (9.21). Here $K = K_N = \mathrm{SO}_N \cdot \mathbb{R}^*$. The ϵ' denotes a character of the group of connected components $\pi_0(\mathrm{GL}_N(\mathbb{R}))$ and (ϵ') denotes the eigenspace under this group. In order to apply Delorme's method one has to fix embeddings of the M-modules $V_\mu = V_{w(\lambda+\delta)-\delta}$ into $H^d(\mathfrak{n}, V_\lambda)$ and of the twisted version $V_{w(\lambda+\delta)-\delta,Tate}$ into $H^d(\mathfrak{n}_Q, V_\lambda) = H^d(\mathfrak{n}_-, V_\lambda)$. (Here we switch freely between the parabolics P_- and Q by conjugation with w_0.) As already explained in Sect. 9.6.8 this isomorphism is fixed once $\omega \in \Lambda^D(\mathfrak{n})$ is fixed, since the second embedding is fixed by the first one and by ω. Furthermore, the identification of the right-hand sides of (7.39) and (7.40) involves the map $T^{\epsilon'}(\mu)$, which is defined in (5.8).

The right-hand sides of (7.39) and (7.40) involve, via the Künneth theorem, a factor $H^{b_n}(\mathfrak{g}_n, K_n, \sigma_n \otimes V_{\mu,n},)(\epsilon')$, respectively its Tate twist with $\det^{n'}$. Recall that we have assumed $n = 2n_1$ is even and n' is odd. The map $T^{\epsilon'}(\mu)$ is an isomorphism from $H^{b_n}(\mathfrak{g}_n, K_n, \sigma_n \otimes V_{\mu,n})(\epsilon')$ to the space $H^{b_n}(\mathfrak{g}_n, K_n, \sigma_n \otimes V_{\mu,n})(-\epsilon')$, and the odd Tate twist (which is the identity map on the underlying space) brings this back to the ϵ' eigenspace of the Tate-twisted cohomology.

The isomorphism $T^{\epsilon'}(\mu)$ in (5.8) may be interpreted in the following way. The representation σ_n reproduces under twisting with the character sign(det):

$$\tau : \quad \sigma_n \quad \xrightarrow{\sim} \quad \sigma_n \otimes \mathrm{sign}(\det).$$

So if σ_n and $\sigma_n \otimes \mathrm{sign}(\det)$ have the same underlying space, the map τ will be a nonscalar linear automorphism of this space. As a representation of the group $G^0 = \mathrm{GL}_n(\mathbb{R})^0 = \{g \in \mathrm{GL}_n(\mathbb{R}) | \det(g) > 0\}$ the space decomposes $\sigma_n = \sigma_n^+ \oplus \sigma_n^-$. Then we get a τ as the linear map, which is multiplication with a scalar $c \in \mathbb{C}^*$ on σ_n^+ and with the scalar $-c$ on σ_n^-. We denote this map by τ_c.

For $c = i^{n_1}$ this map induces the map $T^{\epsilon'}(\mu)$ on cohomology as defined in (5.8). And this is a good choice for another reason: If σ_n is defined over $\mathbb{Q}$, it follows from the work of Januszewski ([41], [42, Theorem B]), that the modules σ_n^+ and σ_n^- are defined over $\mathbb{Q}$ if and only if n_1 is even (and thus n is divisible by 4). But then our τ_c is defined over $\mathbb{Q}$ in all cases. In the case $n = 2$ this is already stated in (8.34) and the following lines, and it is an exercise to generalize it to arbitrary even n using the induction procedure therein. Using this τ_c and the "standard" $\omega = \pm\bigwedge_{1\le j\le n, n+1\le k\le N} E_{jk}$ (for a suitable choice of sign), one can show that the map $T_{\mathrm{loc}}(\sigma)(\epsilon')$ coincides with the map induced by our map ι_{w_0} in (9.21). Of course the standard ω and its negative are the only choices, which are defined over $\mathbb{Z}$ and have a pairing 1 with a $\mathbb{Z}$-generator of the dual space. So they are canonical up to sign.

To be more precise, $T_{\mathrm{loc}}(\sigma)(\epsilon')$ coincides with our map ι_{w_0} if $\tilde{\sigma}'$ is the (Tate)-twist with χ of σ (after conjugation with w_0). But now the characters χ and its absolute value $|\chi|$ differ by the character $\mathrm{GL}_n(\mathbb{R}) \times \mathrm{GL}_{n'}(\mathbb{R}) \ni (A, B) \mapsto \mathrm{sign}(\det(A))$, (recall that n is even and n' is odd). Since the isomorphism class of σ_n is invariant under the self twist with sign(det) via τ_c we get that we may replace in the right-hand side of (9.27) the algebraic Tate twist $\tilde{\sigma}'$ by the twist with the square of the modulus character. Thus we are in the situation, where our analytically defined intertwining operator T_{w_0}, which is called $T_{st}(-\frac{N}{2}, \sigma)^*(\epsilon')$ in 7.3.2.2 (up to a minor but obvious little change in notation that has been implicit in this chapter), induces a map on cohomology and then the main result reads as follows.

9.6.12 The final formula

With the notations introduced earlier we have the following identity:

$$T_{w_0} \quad = \quad \frac{L(\chi, J, w_0, 0)}{L(\chi, J, w_0, 1)} \cdot T_{\mathrm{loc}}(\sigma)(\epsilon')$$

between maps from $H^b(\mathfrak{g}, K, {}^{\mathrm{a}}\mathrm{Ind}_P^G \sigma \otimes V_\lambda)(\epsilon')$ to $H^b(\mathfrak{g}, K, {}^{\mathrm{a}}\mathrm{Ind}_Q^G \tilde{\sigma}' \otimes V_\lambda)(\epsilon')$.

Remarks on the case where n and n' are both even. If $n = 2n_1$ and $n' = 2n_2$ are both even, the final formula is still valid. Let us briefly mention the necessary modifications in this case: We have $T_c^\Sigma \subset M'$ and thus $T_c^{\Sigma,0} = T_c^\Sigma$. So V_q is irreducible as an $T_c^{\Sigma,0}$-module. Furthermore the subsets Q_{jk} in the proof of Lem. 9.19 are all of order 4, so that the statement of this lemma is still valid with the modification that no power of i appears in front of the wedge product of the E_{jk}. In fact one can summarize both cases to the formula

$$\omega = \pm i^{nn'/2} \cdot \bigwedge_{1\le j\le n, n+1\le k\le N} E_{jk}.$$

The final considerations are still valid for even n' with the simplifying modification, that the twisting character χ takes positive values on $M(\mathbb{R})$ and thus coincides with its absolute value $|\chi|$. So there is no need to switch between different ϵ'-eigenspaces and to consider self-twists with the sign character.

Bibliography

[1] W. Ballmann, *Lectures on Kähler manifolds.* ESI Lectures in Mathematics and Physics, European Mathematical Society (EMS), Zürich, 2006.

[2] C. Bhagwat, *On Deligne's periods for tensor product motives.* C. R. Math. Acad. Sci. Paris 353 (2015), no. 3, 191–195.

[3] C. Bhagwat and A. Raghuram, *Ratios of periods for tensor product motives.* Math. Res. Lett., 20 (2013), no. 4, 615–628.

[4] C. Bhagwat and A. Raghuram, *Special values of L-functions for orthogonal groups.* Comptes Rendus Math. Acad. Sci. Paris., 355 (2017), no. 3, 263–267.

[5] C. Bhagwat and A. Raghuram, *Endoscopy and the cohomology of GL(n).* Bull. Iranian Math. Soc., 43 (2017), no. 4, 317–335.

[6] A. Borel, *Stable real cohomology of arithmetic groups II.* Manifolds and Lie groups (Notre Dame, Ind., 1980), pp. 21–55, Progress in Mathematics, 14, Birkhäuser, Boston, 1981.

[7] A. Borel and W. Casselman, *L^2-cohomology of locally symmetric manifolds of finite volume.* Duke Math. J., 50 (1983), no. 3, 625–647.

[8] A. Borel and H. Garland, *Laplacian and the discrete spectrum of an arithmetic group.* Amer. J. Math. 105 (1983), no. 2, 309–335.

[9] A. Borel and N. Wallach, *Continuous cohomology, discrete subgroups, and representations of reductive groups.* Second edition. Mathematical Surveys and Monographs, 67, American Mathematical Society, Providence, RI, 2000.

[10] A. Borel and J.-P. Serre, *Corners and arithmetic groups.* Commen. Math. Helvetici 48 (1973), 436–491.

[11] H. Bruinier, G. van der Geer, G. Harder, D. Zagier, *The 1-2-3 of Modular Forms.* Lectures from the Summer School on Modular Forms and their

Applications held in Nordfjordeid, June 2004. Edited by Kristian Ranestad. Universitext. Springer-Verlag, Berlin, 2008.

[12] W. Casselman and F. Shahidi, *On irreducibility of standard modules for generic representations.* Ann. Sci. École Norm. Sup., (4) 31 (1998), no. 4, 561–589.

[13] J. Cogdell, *Lectures on L-functions, Converse Theorems, and Functoriality for* GL(n). Lectures on automorphic L-functions. Fields Institute Monographs, 20, American Mathematical Society, Providence, RI, 2004.

[14] L. Clozel, *Motifs et formes automorphes: applications du principe de fonctorialité.* Automorphic forms, Shimura varieties, and L-functions, Vol. I (Ann Arbor, 1988), 77–159, Perspectives in Mathematics, 10, Academic Press, Boston, 1990.

[15] P. Deligne, *Valeurs de fonctions L et périodes d'intégrales (French),* With an appendix by N. Koblitz and A. Ogus. Proceedings of Symposia in Pure Mathematics, XXXIII, Automorphic forms, representations and L-functions (Oregon State, Corvallis, 1977), Part 2, pp. 313–346, American Mathematical Society, Providence, RI, 1979.

[16] P. Deligne and A. Raghuram, *Functoriality, Motives and Periods,* Preprint in preparation.

[17] M. Demazure and A. Grothendieck, *Structure des schémas en groupes réductifs.* SGA 3, TOME III, Lecture Notes in Mathematics, 153, Springer-Verlag, Berlin 1970.

[18] G. de Rham, *Differentiable manifolds. Forms, currents, harmonic forms.* Grundlehren der Mathematischen Wissenschaften [Fundamental Principles of Mathematical Sciences], 266, Springer-Verlag, Berlin, 1984.

[19] J. Franke and J. Schwermer, *A decomposition of spaces of automorphic forms, and the Eisenstein cohomology of arithmetic groups,* Math. Ann., 311 (1998), 765–790.

[20] W.T. Gan and A. Raghuram, *Arithmeticity for periods of automorphic forms,* In "Automorphic Representations and L-functions," 187–229, Tata Institute of Fundamental Research Studies of Mathematics, 22, Tata Institute Fundamental Resesearch, Mumbai, 2013.

[21] L. Grenié, *Critical values of automorphic L-functions for* $\mathrm{GL}(r) \times \mathrm{GL}(r)$, Manuscripta Math., 110 (2003), no. 3, 283–311.

[22] H. Grobner, *A cohomological injectivity result for the residual automorphic spectrum of* GL(n). Pacific J. Math., 268 (2014), no. 1, 33–46.

[23] H. Grobner and A. Raghuram, *On the arithmetic of Shalika models and the critical values of L-functions for* GL($2n$). *With an appendix by Wee Teck Gan.* Amer. J. Math. 136 (2014), no. 3, 675–728.

[24] G. Harder, *Eisenstein cohomology of arithmetic groups. The case* GL_2. Invent. Math. 89, no. 1, 37–118 (1987).

[25] G. Harder, *Some results on the Eisenstein cohomology of arithmetic subgroups of* GL_n, in Cohomology of arithmetic groups and automorphic forms (ed. J.-P. Labesse and J. Schwermer), Springer Lecture Notes in Mathematics, 1447, 85–153, 1990.

[26] G. Harder, *Lectures on Algebraic Geometry I. Sheaves, cohomology of sheaves, and applications to Riemann surfaces.* Second revised edition. Aspects of Mathematics, E35. Vieweg + Teubner Verlag, Wiesbaden, 2011.

[27] G. Harder, *Interpolating coefficient systems and p-ordinary cohomology of arithmetic groups.* Groups Geom. Dyn., 5 (2011), no. 2, 393–444.

[28] G. Harder *A congruence between a Siegel and an elliptic modular form.* The 1-2-3 of modular forms, 247–262, Universitext, Springer-Verlag, Berlin, 2008.

[29] G. Harder, *Arithmetic aspects of Rank one Eisenstein Cohomology,* in "Cycles, Motives and Shimura Varieties," 131–190, Tata Institute Fundamental Research Studies of Mathematics, 21, TIFR, Mumbai, 2010.

[30] G. Harder, *Cohomology of arithmetic groups,* Book in preparation. Preliminary version available at:
`http://www.math.uni-bonn.de/people/harder/Manuscripts/buch/`

[31] G. Harder, *Eisenstein cohomology for* $\mathrm{SL}_2(\mathbb{Z}[i])$ *and special values of L-functions,* Cohomology of arithmetic groups, 51–82, Springer Proceedings in Mathematics and Statistics, 245, Springer, Cham, 2018.

[32] G. Harder, *Secondary operations in the cohomology of Harish-Chandra modules,* Preprint available at *http://www.math.uni-bonn.de/people/harder/Manuscripts/Eisenstein/SecOPs.pdf*

[33] G. Harder and A. Raghuram, *Eisenstein cohomology and ratios of critical values of Rankin–Selberg L-functions.* C. R. Math. Acad. Sci. Paris 349 (2011), no. 13–14, 719–724.

[34] M. Harris, *Beilinson-Bernstein localization over* $\mathbb{Q}$ *and periods of automorphic forms.* Int. Math. Res. Not., IMRN 2013, no. 9, 2000–2053.

[35] Harish-Chandra, *Automorphic Forms on Semissimple Lie Groups,* Lecture Notes in Mathematics, 62, Springer, Berlin, 1968

[36] H. Jacquet, *Principal L-functions of the linear group.* Proceedings of Symposia in Pure Mathematics, XXXIII, Automorphic forms, representations and L-functions (Oregon State, Corvallis, 1977), Part 2, pp. 63–86, American Mathematical Society, Providence, RI, 1979.

[37] H. Jacquet, *On the residual spectrum of* $\mathrm{GL}(n)$. Lie group representations, II (College Park, Md., 1982/1983), Lecture Notes in Math., 1041, Springer, Berlin, 1984.

[38] H. Jacquet and J.A. Shalika, *On Euler products and the classification of automorphic representations. I.* Amer. J. Math. 103 (1981), no. 3, 499–558.

[39] H. Jacquet and J.A. Shalika, *On Euler products and the classification of automorphic forms. II.* Amer. J. Math. 103 (1981), no. 4, 777–815.

[40] H. Jacquet, I. Piatetski-Shapiro and J. Shalika. *Rankin–Selberg convolutions.* Amer. J. Math. 105 (1983), no. 2, 367–464.

[41] F. Januszewski, *On rational structures of automorphic representations,* Math. Ann., 370, (2018) 1805–1861.

[42] F. Januszewski, *On Period Relations for Automorphic L-functions I,* Trans. Amer. Math. Soc., 371 (2019), no. 9, 6547–6580.

[43] H. Kasten and C.-G. Schmidt *The critical values of Rankin-Selberg convolutions.* Int. J. Number Theory 9 (2013), no. 1, 205–256.

[44] D. Kazhdan; B. Mazur and C.-G. Schmidt *Relative modular symbols and Rankin-Selberg convolutions.* J. Reine Angew. Math., 519 (2000), 97–141.

[45] H. Kim, *Automorphic L-functions.* Lectures on automorphic L-functions, 97–201, Fields Institute Monographs, 20, American Mathematical Society, Providence, RI, 2004.

[46] B. Kostant, *Lie algebra cohomology and the generalized Borel-Weil theorem* Annals of Math., 74, no. 2 (1961) 329–387.

[47] A. Knapp, *Local Langlands correspondence: the Archimedean case,* Motives (Seattle, WA, 1991), 393–410, Proceedings of Symposia in Pure Mathematics, 55, Part 2, American Mathematical Society, Providence, RI, 1994.

[48] M. Krishnamurthy and A. Raghuram, *Eisenstein cohomology for unitary groups and the arithmetic of Asai L-functions,* In preparation.

[49] R. Langlands, *Euler products,* Yale University Press, New Haven, CT, 1971.

[50] R. Langlands, *On the functional equations satisfied by Eisenstein series.* Lecture Notes in Mathematics, 544., Springer-Verlag, Berlin, 1976.

[51] C. Mœglin, *Representations of* $\mathrm{GL}(n)$ *over the real field.* Representation theory and automorphic forms (Edinburgh, 1996), 157–166, Proceedings of Symposia in Pure Mathematics, 61, American Mathematical Society, Providence, RI, 1994.

[52] C. Mœglin and J.-L. Waldspurger, *Le spectre résiduel de* $\mathrm{GL}(n)$. *(French) [The residual spectrum of* $\mathrm{GL}(n)$*]* Ann. Sci. École Norm. Sup. (4) (1989), 22, no. 4, 605–674.

[53] A. Raghuram, *On the special values of certain Rankin-Selberg L-functions and applications to odd symmetric power L-functions of modular forms,* Int. Math. Res. Not., (2010) 334–372, doi:10.1093/imrn/rnp127.

[54] A. Raghuram, *Comparison results for certain periods of cusp forms on* $\mathrm{GL}(2n)$ *over a totally real number field.* The legacy of Srinivasa Ramanujan, 323–334, Ramanujan Mathematical Society, Lecture Notes Series, 20, Mysore, 2013.

[55] A. Raghuram, *Critical values of Rankin–Selberg L-functions for* $\mathrm{GL}_n \times \mathrm{GL}_{n-1}$ *and the symmetric cube L-functions for* GL_2. Forum Math., 28, no. 3, 2016; 457–489.

[56] A. Raghuram and F. Shahidi, *On certain period relations for cusp forms on* GL_n, Int. Math. Res. Not., (2008), doi:10.1093/imrn/rnn077.

[57] A. Raghuram and N. Tanabe, *Notes on the arithmetic of Hilbert modular forms.* J. Ramanujan Math. Soc., 26, no. 3 (2011), 261–319.

[58] A. Raghuram, *Eisenstein Cohomology for* GL_N *and the special values of Rankin–Selberg L-functions–II.* Preprint (2019).

[59] J. Schwermer, *Kohomologie arithmetisch definierter Gruppen und Eisensteinreihen.* (German) [Cohomology of arithmetically defined groups and Eisenstein series] Lecture Notes in Mathematics, 988. Springer-Verlag, Berlin, 1983.

[60] J. Schwermer, *Cohomology of arithmetic groups, automorphic forms and L-functions.* Cohomology of arithmetic groups and automorphic forms

(Luminy-Marseille, 1989), 1–29, Lecture Notes in Mathematics, 1447, Springer-Verlag, Berlin, 1990.

[61] J.-P. Serre, *Facteurs locaux des fonctions zeta des variétés algébriques (définitions et conjectures)*, in Séminaire Delange-Pisot-Poitou, 1969/70, Collected Papers II, 19, 581–592.

[62] F. Shahidi, *Whittaker models for real groups.* Duke Math. J. 47 (1980), no. 1, 99–125.

[63] F. Shahidi, *On certain L-functions.* Amer. J. Math.,103 (1981), no. 2, 297–355.

[64] F. Shahidi, *Local coefficients as Artin factors for real groups.* Duke Math. J. 52 (1985), no. 4, 973–1007.

[65] F. Shahidi, *Eisenstein series and automorphic L-functions.* American Mathematical Society Colloquium Publications, 58. American Mathematical Society, Providence, RI, 2010.

[66] G. Shimura, *The special values of the zeta functions associated with Hilbert modular forms.* Duke Math. J. 45 (1978), no. 3, 637–679.

[67] B. Speh, *Some results on principal series for* $\mathrm{GL}(n, \mathbb{R})$ Ph. D. dissertation, M.I.T.

[68] B. Sun, *The nonvanishing hypothesis at infinity for Rankin-Selberg convolutions.* J. Amer. Math. Soc., 30, no. 1 (2017), 1–25.

[69] D. Vogan and G. Zuckerman, *Unitary representations with nonzero cohomology.* Comp. Math. 53, no. 1 (1984), 51–90.

[70] J.-L. Waldspurger; *La formule de Plancherel pour les groupes p-adiques (d'après Harish-Chandra).* J. Inst. Math. Jussieu 2, no. 2 (2003), 235–333.

[71] N. Wallach, *Real Reductive Groups. Vol. I.* Pure and Applied Mathematics, 132, Academic Press, Boston, 1988.

[72] U. Weselmann, *Siegel Modular Varieties and the Eisenstein Cohomology of* PGL_{2g+1}, Manuscripta Math. 145, no. 1-2 (2014), 175–220.

[73] H. Yoshida, *Motives and Siegel modular forms.* Amer. J. Math., 123, no. 6 (2001), 1171–1197.

[74] D. Zagier, *Appendix: On Harder's* $\mathrm{SL}(2, \mathbb{R}) - \mathrm{SL}(3, \mathbb{R})$*-identity* in Cycles, Motives and Shimura Varieties TIFR 2010, 191–195 (2010).

Index

Key Words/Phrases

Notations

GPSR Authorized Representative: Easy Access System Europe - Mustamäe tee 50, 10621 Tallinn, Estonia, gpsr.requests@easproject.com

www.ingramcontent.com/pod-product-compliance
Lightning Source LLC
Chambersburg PA
CBHW060407220326
41598CB00023B/3053